Volume 6

APPLIED PHYSICS AND ENGINEERING

An International Series

Dynamic Stability of Bodies Containing Fluid

Dynamic Stability of Bodies Containing Fluid

N. N. Moiseyev and V. V. Rumyantsev

Edited by N. H. Abramson
SOUTHWEST RESEARCH INSTITUTE
SAN ANTONIO, TEXAS

Translated by Scripta Technica

SPRINGER-VERLAG NEW YORK INC. · 1968

Originally published under the title, Dinamika
Tela s Polostyami, Soderzhashchimi Zhidkost', by Nauka Press, Moscow, 1965

ISBN-13: 978-3-642-86454-4 e-ISBN-13: 978-3-642-86452-0
DOI: 10.1007/978-3-642-86452-0

Softcover reprint of the hardcover 1st edition 1968
Library of Congress Card Number 68-22688

Title No. 3896

Translation Editor's Preface

The dynamics of bodies containing fluids is a subject of long-standing importance in many technical applications. The stability of motion of such bodies, in particular, has been the subject of study by Soviet engineers and applied mathematicians who have brought their full powers of analysis to bear on the problem, and have succeeded in developing a very well-founded body of theory. It is difficult to find a more striking example anywhere of the application of the classical methods of analytical mechanics, together with more modern concepts of stability analysis, in such a comprehensive and elegent form as that presented by Profs. Moiseyev and Rumyantsev. Therefore, it is highly significant that this recent monograph has been translated and made available to the English-speaking community.

H. NORMAN ABRAMSON

San Antonio
July, 1967

Foreword

During the last 15–20 years, problems of dynamics of rigid bodies with fluid-filled cavities have increasingly attracted the attention of scientists. These problems, which have long been of theoretical interest, have also turned out to be very important from the practical viewpoint. The stimulus for intensifying studies in this field was probably the development of missiles and rockets. The large amount of fluid contained in a missile creates a number of problems that are pertinent to this theory; the development of spaceships taking off from orbits of artificial satellites (orbiting space-stations) will even further enlarge the range of practical problems. However, it would be incorrect to regard missiles and rockets as the only source for technical problems of dynamics of fluid-filled bodies. Considerable work in this field is devoted to problems of earthquake resistance for various types of fluid-storage reservoirs. Similar problems occur in the theory of ships and submarines; they are pertinent to the theory of aircraft wing flutter, etc. Hence during the last few years, hundreds of publications devoted to analysis of various aspects of the theory of motion of a fluid-filled body have appeared. In the opinion of the authors the extent of the results obtained and the interest expressed in the theory make timely the writing of a monograph that would review these studies. However, the authors have not attempted such an extensive task, but have concerned themselves only with some problems that are close to their particular fields of interest.

This book consists of two independent parts. The first, written by V. V. Rumyantsev, presents the general problems of dynamics and the theory of stability of a fluid-filled body, which involve the systematical application of the methods of analytic mechanics, particularly the second method of Lyapunov. Here the body and the fluid contained in its cavity are considered a single mechanical system and nonlinear equations of motion are considered. The second part was written by N. N. Moiseyev and is devoted to the theory of small vibrations of fluids and of a fluid-filled body. Only linear problems are considered. The core of this part is the traditional problem of small vibrations, i.e., the structure of the spectrum. No effort was made to give a unified presentation; each part of the book is written in the style of the respective author.

In the first chapter the equations of motion of the system are derived from the principle of least action in the Hamilton-Ostrogradskiy form. Variational formulations of problems of dynamics have a certain theoretical merit (for example, from the viewpoint of substantiating the necessity and sufficiency of the derived equations and boundary conditions) and also are of great importance in practice since direct methods are most efficient for numerical calculations. The mechanical meaning of the equations thus obtained is clarified, and formulas are derived for the forces through which the ideal fluid in the cavity acts on the body. The equations of motion are represented in several different forms, particularly in the form of Lagrangian equations. The conditions under which the equations of motion have one or another solution are pointed out. The chapter closes with a discussion of the equations of motion when the liquid is viscous.

The second chapter is concerned with the motion of a rigid body with cavities entirely filled with an ideal fluid. It is assumed that the motion of the fluid in the cavity is vortex-free, or that it is a homogeneous vortex motion. These cases are elementary in the sense that the motion of the system is described by ordinary differential equations and by the Laplace equation. After the latter is solved, the motion of the fluid is completely defined by a finite number of variables. First the fundamental results of N. Ye. Zhukovskiy are presented [15]. These were obtained in a study of vortex-free motion of a fluid in a cavity and in this case, the motion of the rigid body. Then the homogeneous vortex motion of the fluid contained in an ellipsoidal cavity and to the corresponding motion of the body is considered; the statement of the problem of stability in the sense of Lyapunov is presented, and theorems on stability and instability are given. For the elementary cases considered in this chapter, it is natural to present the problem of stability as that of Lyapunov for systems with a finite number of degrees of freedom. The chapter gives a solution to a number of problems, stated in this manner, of stability of motion of a fluid-containing rigid body. By constructing Lyapunov's functions, sufficient stability of conditions are obtained; in many cases, these are also identical to the necessary conditions.

The third chapter considers the stability of motion of a rigid body with a cavity partially or completely filled by an ideal or viscous fluid, making no assumptions on the type of motion of the fluid, except for that of continuity. Here the state of the system is described by an infinite number of variables, and the problem of stability becomes quite complex. However, in this case it is also possible to state the stability problem with respect to a finite number of variables by introducing certain quantities that integrally describe the motion of the fluid. Two theorems on stability with respect to a part of the variables, which can be regarded as modifications of the Lyapunov stability theorem, are proven. Sufficient stability conditions for a number of problems are obtained by constructing functions satisfying these theorems.

The fourth chapter presents a second method for studying stability of steady motion of a fluid-containing rigid body. This method involves ideas developed by Lyapunov in the theory of equilibrium patterns of a rotating fluid. The problem of stability is stated with respect to parameters defining the motion of the rigid body and to the mode of the equilibrium of the fluid. The chapter contains the proof of a number of theorems that reduce the problem of stability to finding the minimum of a certain expression, for which a solution is given. The chapter ends with the presentation of the solutions of several problems of stability of steady motion of fluid-filled bodies.

The fifth chapter is concerned with the statement of the problem of oscillations of an ideal fluid subjected to gravity and clarifies certain features of the motion. It also analyzes elementary examples and cases that demonstrate physical context of the theory most clearly.

The sixth chapter gives a rigorous presentation of the theory of small vibrations. It is shown that, from the mathematical viewpoint, the theory of small vibrations of a rigid or elastic body that satisfies the hypothesis of plane sections (beam) containing a fluid and situated in a field of conservative forces is quite elementary and is fully within the scope of the spectral theory of linear, fully continuous, self-adjoint operators. One significant property of these systems is clarified in detail. As was shown by N. Ye. Zhukovskiy, if the fluid completely fills the cavity of the body, then this system is equivalent to some other rigid body. If the fluid has a free surface, then this theorem does not apply. Nevertheless, it is possible to introduce, an equivalent rigid body in a certain sense; i.e., it turns out that for stability of the equilibrium position it is necessary and sufficient for some other rigid body to be stable. Here the term "stability" is used to denote that all the principal vibrations are bounded. The concept of stability thus defined differs from the one used in the initial chapters. In general these definitions are not equivalent, although in some cases (e.g., equilibrium position of a pendulum) the same condition imposed on the geometric characteristics of the system is sufficient for stability of the position of the system equilibrium in either sense. Knowing the values of the natural frequencies of the oscillating fluid and having a set of eigenfunctions describing the vibrations of the fluid in a stationary cavity, the problem of motion of a fluid-filled body can be reduced to a system of integro-differential equations in terms of the generalized coordinates of the solid body. This system of equations can be studied by well-known methods (operational methods, reduction to infinite systems, etc.). Hence the basic problem that arises here is the development of numerical methods for calculating the spectrum of natural frequencies and eigenfunctions. These problems cannot be examined in detail within the scope of this book. We have restricted ourselves to remarks on the expediency of using the Ritz method for this purpose (this fact is substantiated by the computing practice of Soviet, American, and Japanese specialists) and we also give a substantiation of it. It is precisely the role of the Ritz theory that determines the style of the presentation; it is based on Hamilton's principle.

In the fifth and sixth chapters a special place is given to various problems in the theory of perturbations. Experience shows that skillful utilization of the methods of this theory frequently makes it possible in applied problems to avoid time-consuming numerical calculations. Among these are all problems where the energy of wave motions of the body is small in comparison with the energy of the system in which the free surface is replaced by a solid cover.

During the last few years, in connection with the launching of artificial earth satellites, the problem of oscillation of a fluid placed in weak gravitational fields, previously regarded as exotic, acquired practical significance. Here surface tension forces begin to play an important role. The seventh chapter is devoted to these problems. The theory of motion of a body containing a weightless fluid which partially fills its cavities is still very little developed. Nonlinear problems are of greatest interest here. Even moderate accelerations can result in breakups in continuity. However, just these problems are of the greatest interest to engineers, but have not as yet been studied. This chapter considers the simplest problems, such as the theory of small oscillations of a fluid subjected to the action of surface tension forces, some problems of the theory of equilibrium patterns, and the asymptotic behavoir at large Bond numbers.

The eighth and final chapter is devoted to a study of the effect of viscosity on the oscillations of a fluid and of a fluid-filled body. Problems of prime importance in this theory involve the study of limiting cases at large and small Reynolds numbers. These problems are presented by using examples of elementary motions.

As can be seen from this outline the authors have concentrated their efforts on certain selected problems; an attempt has been made to present the subject systematically. This selection of problems makes it possible to consider this monograph an introduction to the dynamics of bodies with fluid-filled cavities.

The authors are sincerely grateful to V. A. Pal'mov, who carefully reviewed the manuscript and made a number of useful remarks.

N. N. MOISEYEV AND V. V. RUMYANTSEV

Introduction

The problem of the motion of a body with cavities completely filled with a fluid first began to attract the attention of scientists over one hundred years ago.

Stokes was apparently the first (1842–1847) who turned his attention to this interesting problem of mechanics; it occupied Helmholtz (1860), Lubeck (1873) and Lamb (1873), who considered a number of particular cases, as well as Neumann (1883), who considered this problem in connection with the study of motion of solid bodies in a fluid.

The first detailed general study of the dynamics of a rigid body with cavities completely filled with homogeneous incompressible fluid was performed by N. Ye. Zhukovskiy (1885). He showed that the potential motion of the fluid in such a cavity is determined by the motion of the body, while the body proper moves as though the fluid were replaced by an equivalent solid body.

At the same time that the motion of solid bodies with fluid-filled cavities was being studied, scientists became interested in the stability of motion of such bodies. In experiments with a fluid-filled thin-walled spheroidal top, Kelvin (1877) discovered that the top is stable if it spins rapidly and if the shape of the cavity is sufficiently reduced. However, if the top is slightly extended in shape, it becomes highly unstable, no matter what the angular speed with which it rotates. The mathematical explanation of this phenomenon also occupied Greenhill (1880), F. A. Sludskiy (1895), Gaf (1895), Poincaré (1910) and Basset (1911), who considered homogeneous vortex motion of a fluid in an ellipsoidal cavity. In particular, Gaf studied the small oscillations in the vicinity of the state of uniform rotation of a rigid body, and the fluid filling its cavity, considering them as the oscillation of one rigid body about the principal axis of intertia, obtained the necessary stability conditions and analyzed them for the case of a shell with a negligible small mass. Poincaré also considered this problem, taking into account the elasticity of the shell and inhomogeneity of the fluid.

Great interest in the problem of motion of rigid bodies with fluid-filled cavities has arisen again in our time, particularly during the last 15–20 years;

here, together with the study of the motions of bodies with completely fluid-filled cavities, there has arisen the new problem of the motions of a body with cavities incompletely filled.

The development of this theory was stimulated by the appearance of a variety of applied problems. These include problems of the dynamics of fluid-filled missiles and rockets, as well as problems involving considerations of the strength of reservoirs and other containments subjected to earthquake loads. A large number of similar problems has been published in connection with hydraulic engineering and with the theory of ships. Because of these circumstances, a large number of studies published in various countries has been devoted to different aspects of the dynamics of fluid-filled bodies. In the Soviet Union, many results of this theory were obtained by S. L. Sobolev, N. G. Chetayev, L. N. Sretenskiy, D. Ye. Okhotsimskiy, G. S. Nariman, A. Yu. Ishlinskiy, M. Ye. Temchenko, S. V. Malashenko, B. I. Rabinovich, I. M. Rapoport and many other authors. A bibliography of publications in the field contains so many titles that it is impossible to enumerate all the authors.

In this tremendous number of publications it is possible, however, to discern three basic trends: 1) study of the linearized equations of motion, utilizing the methods of the theory of small oscillations and the spectral theory of operators; 2) study of the complete nonlinear equations of motion, utilizing the methods of analytic mechanics; 3) experimental studies.

The greatest number of publications are devoted to various linear aspects of the theory. Foremost are problems of small oscillations about the equilibrium position of a rigid body containing an ideal, incompressible, heavy fluid. This is the simplest part of the theory, and one with extensive practical applications. Here the solution of the corresponding mathematical problems have been exhaustively clarified; *i.e.*, the structure of the spectrum has been determined, the existence of the solution of Cauchy's problem has been proved, *etc.* A large number of mechanical features of vibrations of a fluid-filled body have been clarified. It was established, for example, that for stability of vibrations of such a system with an infinite number of degrees of freedom, it is necessary and sufficient that a certain rigid body be stable. The computational aspect of this theory has also been developed. The solution of many problems of the theory of vibrations has been reduced to the level of standard programs.

The case when the fluid is acted upon by surface tension forces is more complicated. Purely mathematical problems of the theory of small oscillations have also not posed difficulties, since the problem can be reduced to an operational equation with completely continuous self-adjoint operators, and the entire mathematical context is easily reducible to elementary facts of the spectral theory of linear operators. However, the computational aspect remains almost undeveloped.

The theory of small oscillations of a rigid body containing a viscous fluid is even less developed. Only individual results have been obtained in this field; for example, asymptotic methods that make it possible to study some dynamic problems for large Reynolds numbers have been developed. In addition, some success was attained in developing methods for the study of motions of a rigid body having cavities which contain a highly viscous fluid.

Much less has been published concerning the second trend of the studies. Until recently no attempts were made to develop efficient numerical methods for the study of nonlinear equations of motion; this is understandable in view of the state of numerical methods for solving the Navier-Stokes equations. Attention was concentrated principally on the study of general properties of motions (consideration of various forms of nonlinear equations describing a rigid body containing a fluid, clarification of their equivalence, conditions of existence of first integrals of the equations of motion, *etc.*), primarily stability of the motion. Here the successes attained are almost exclusively related to the application and development of Lyapunov's methods.

The basic problem in stating problems of stability of motion of a fluid-filled rigid body that forms a system with an infinite number of degrees of freedom is that of defining the concept of stability. Three main approaches have been proposed for the study of stability of motion of fluid-filled rigid bodies when the problem is stated in nonlinear terms. The simplest is the case when the motion of the fluid in the cavity can be described by a finite number of variables. Then the stability problem naturally reduces to that of Lyapunov for systems with a finite number of degrees of freedom. In the general case the state of a system is described by an infinite number of variables. However, it turns out to be possible to state the problem of stability with respect to a finite number of variables by introducing certain quantities that describe the motion of the fluid in an integral manner. It is also possible to use effectively the method of Lyapunov's functions. Finally, the third approach involves Lyapunov's ideas in the theory of equilibrium patterns of a rotating fluid and reduces to a generalization of the theorems of Lagrange and Routh. Then the problem of stability of equilibrium or of steady motion is reduced to the problem of a minimum of some expression that generally is a functional.

This book examines certain basic problems pertaining to the first two trends and is not concerned with the third.

Contents

Part Two. Vibrations of a Fluid and of the Body Containing the Fluid

Part One

Dynamics and Stability of Rigid Bodies Containing a Fluid

I

Equations of Motion of a Rigid Body with Fluid-Containing Cavities

In this chapter are derived the equations of motion of fluid-containing rigid bodies, regarded as a single mechanical system. The dynamics are based on the principle of least action in the Hamilton-Ostrogradskiy form. The variational formulation of dynamics problems has definite advantages, for example, from the point of view of establishing the necessary and sufficient nature of the derived equations and boundary conditions, while the study of the body and the fluid as a single system permits the achievement of a well-recognized compactness. The mechanical meaning of the equations, which are represented in several different forms, particularly in the form of Lagrangian equations, is explained. The problem of solutions of the equations of motion is examined together with the conditions under which they are valid. The chapter concludes with an examination of the equations of motion of a rigid body with an internal cavity containing a viscous fluid.

[1-1] THE HAMILTON-OSTROGRADSKIY PRINCIPLE

As is well known, the variational principle of least action in the form due to Hamilton and Ostrogradskiy can be used as the basis of the mechanics of holonomous systems. (According to this principle, for the absolute motion of any mechanical system with ideal geometric constraints:

$$\int_{t_0}^{t_1} \left(\delta T + \sum_{\nu} \boldsymbol{F}_{\nu} \cdot \delta \boldsymbol{r}'_{\nu} \right) dt = 0 .) \tag{1-1}$$

Here, and throughout the book, t designates the time, t_0 and t_1 denote the limits of integration, T designates the kinetic energy of the system, $\boldsymbol{F}_\nu$ is the vector of the given applied force acting on the νth point with the summation taken over all of the points of the system, $\boldsymbol{r}'_\nu = \sum_{i=1}^{3} x'_{\nu i} \boldsymbol{i}'_i$ is the radius vector of a point in the system relative to the origin O' of a given stationary (inertial), orthogonal coordinate system $O'x'_1x'_2x'_3$. The unit (base) vectors defining the directions of the x'_s axes are denoted by $\boldsymbol{i}'_s$ $(s=1, 2, 3)$. The symbol δ denotes the variation or change in the corresponding quantity in front of which it stands, corresponding to any possible displacement of the system. Any actual motion between two fixed positions of the system takes place simultaneously ($\delta t=0$) with any motion being compared to it (imagined) that also takes place between the same two fixed points and is permitted by the restraints imposed on the system. Hence the variations of the radius vectors of all the points of the system should become zero in these end positions, *i.e.*, when $t=t_0$ and $t=t_1$, $\delta\boldsymbol{r}'=0$.

The principle given by Eq. (1–1) can be obtained by integration from the d'Alembert-Lagrange principle, which is valid for any material system with ideal constraints. By virtue of this, the Hamilton-Ostrogradskiy principle is also valid for any mechanical system with holonomous constraints, including a continuous medium. It can also be used to derive the equations of motion of rigid bodies with fluid-containing cavities.

We shall study the motion of a perfectly rigid body, *i.e.*, a body in which the distances between any two points remain unchanged in time. Let the body have one or more cavities, partially or completely filled with homogeneous, incompressible ideal fluids. In our further discussion we shall restrict ourselves, for simplicity, to considering the case of only one cavity, since in the case of several cavities it will only be necessary to sum, in the equations of motion, the quantities pertaining to the different fluids over all the cavities, which does not introduce any difficulties of principle. The rigid body taken together with the fluid will be regarded as a single mechanical system.

Let ϱ denote the constant density of the fluid, τ the volume of the region occupied by fluid at the given time instant, S' the boundary of region τ, and σ the surface area of the walls of the cavity (Fig. 1).

If the fluid fills the cavity completely, then obviously S' is identical with σ.

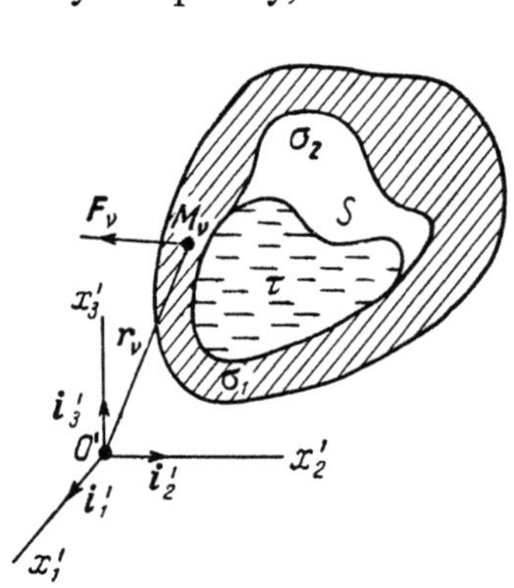

Fig. 1.

If, however, the fluid fills the cavity only partially, then the surface S' consists of the free fluid surface S and the part σ_1 of the surface σ with which the fluid comes into contact at the given time instant; *i.e.*,

$$S' = S + \sigma_1 , \quad \sigma = \sigma_1 + \sigma_2 ,$$

where σ_2 denotes the part of the surface σ that is not in contact with the fluid at the given instant.

In this case the remaining part of the cavity is either filled with air, which is bounded by the surface $S+\sigma_2$ and the pressure p_0 of which is regarded as constant while its mass is disregarded, or it is considered to be a vacuum with pressure $p_0=0$. We denote the closed line of intersection of the surfaces S and σ by s. We further assume that the surface S' is smooth, or that it consists of a finite number of smooth-surfaced segments.

Since the fluid is assumed to be incompressible, the variations of the radius vectors at each point in it at any given time, must satisfy the condition of incompressibility

$$\operatorname{div} \delta \boldsymbol{r}' = \frac{\partial \delta x_1'}{\partial x_1'} + \frac{\partial \delta x_2'}{\partial x_2'} + \frac{\partial \delta x_3'}{\partial x_3'} = 0 .$$

Multiplying this equation by the indeterminate Lagrange multiplier[1] $p(\boldsymbol{r}', t)$, integrated over the volume τ of the fluid and adding the result to Eq. (1–1), we will write the Hamilton-Ostrogradskiy equation for the mechanical system under study in the form

$$\int_{t_0}^{t_1} (\delta T + \sum_{\nu} \boldsymbol{F}_\nu \cdot \delta \boldsymbol{r}_\nu' + \int_{\tau} p \operatorname{div} \delta \boldsymbol{r}' d\tau) \, dt = 0 . \tag{1–2}$$

The symbol $\int_\tau (\ldots) \, d\tau$ denotes a triple integral taken over the volume τ of the region occupied by the fluid. From here on similar notation will be used for double and line integrals, taken over a given surface S, or line s.

We know from analytical mechanics that the Lagrangian multiplier is proportional to the reaction at the constraint. The constraint in the given case is the equation of incompressibility, and hence the Lagrangian multiplier $p(\boldsymbol{r}', t)$ becomes identical with the hydrodynamic pressure p, which can be regarded as the reaction of the constraints, *i.e.*, the incompressibility conditions[1,2].

[1-2] SOME FORMULAS OF KINEMATICS AND VECTOR ANALYSIS

In addition to the stationary coordinate system, we now introduce the moving coordinate system $Ox_1x_2x_3$ with its origin located at some point O

of the rigid body, rigidly fixed to the body, and directed similarly (*i.e.*, either right- or left-handed) to the stationary coordinate system. The unit vectors of the moving coordinate system are denoted by $\boldsymbol{i}_s$ ($s=1, 2, 3$). The radius vector of any point of the system relative to point O is

$$\boldsymbol{r} = \sum_{s=1}^{3} x_s \boldsymbol{i}_s ,$$

while

$$\boldsymbol{r}'_\nu = \boldsymbol{r}'_0 + \boldsymbol{r}_\nu , \qquad (1\text{–}3)$$

where $\boldsymbol{r}'_0$ is the radius vector of the point O relative to the stationary point O' (Fig. 2).

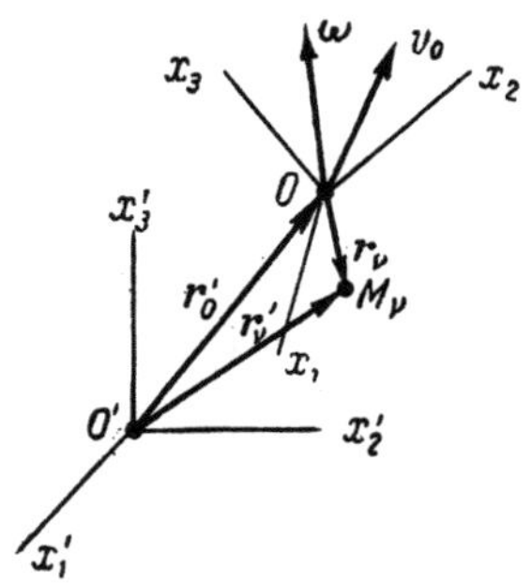

Fig. 2.

The position of the $Ox_1x_2x_3$ coordinate system, together with the rigid body in space $O'x'_1x'_2x'_3$ can be represented by the radius vector $\boldsymbol{r}'_0 = \sum_{s=1}^{3} x'_{0s} \boldsymbol{i}'_s$ and the cosines $\gamma_{sr} = \boldsymbol{i}' \cdot \boldsymbol{i}_r$ ($s, r = 1, 2, 3$) of the angles between the stationary and moving axes. The nine cosines γ_{sr} are interrelated by the expressions

$$\sum_{r=1}^{3} \gamma_{sr}^2 = 1 , \quad \sum_{j=1}^{3} \gamma_{sj}\gamma_{rj} = 0 \quad (s \neq r = 1, 2, 3) ,$$

so that only any three of cosines γ_{sr} ($s, r = 1, 2, 3$) will be independent.

From now on the motion of our mechanical system is relative to the stationary coordinate system will be called *absolute*, and that relative to the moving coordinate system will be called *relative*. The equations of motion will be referred to the moving coordinate system.

By the velocity addition theorem, the absolute velocity vector $\boldsymbol{v}$ of any point of the system can be represented in the form

$$\boldsymbol{v} = \boldsymbol{v}_0 + \boldsymbol{\omega} \times \boldsymbol{r} + \boldsymbol{u} , \qquad (1\text{–}4)$$

where $\boldsymbol{v}_0$ is the velocity vector of the point O, $\boldsymbol{\omega}$ is the instantaneous angular velocity vector of the body, and $\boldsymbol{u}$ is the relative velocity vector. By definition,

$\boldsymbol{u}=d\boldsymbol{r}/dt$, with the time derivative taken with respect to the moving coordinate system, $\boldsymbol{i}_s$ being fixed; it is obvious that for any point of a rigid body, $\boldsymbol{u}=0$. We recall that the time derivative $(d\boldsymbol{a}/dt)'$ of some vector $\boldsymbol{a}$ in the stationary coordinate system is related to the local derivative $d\boldsymbol{a}/dt$ (derivative in the moving coordinate system) by the relationship

$$\left(\frac{d\boldsymbol{a}}{dt}\right)' = \frac{d\boldsymbol{a}}{dt} + \boldsymbol{\omega} \times \boldsymbol{a}\,.$$

From this, in particular, we obtain, Poisson's differential equations for the unit vectors 0 of the stationary axes $\boldsymbol{i}'_s$

$$\frac{d\boldsymbol{i}'_s}{dt} + \boldsymbol{\omega} \times \boldsymbol{i}'_s = 0 \quad (s = 1, 2, 3)\,. \tag{1–5}$$

The projections of any vector $\boldsymbol{a}$ on the moving coordinate axes will be denoted by a_i $(i=1, 2, 3)$.

Transformation of our mechanical system undergoing actual motion from its instantaneous position to its simultaneous position in the motion being compared, can be carried out by imparting to the entire system, regarded as a single rigid body, an infinitesimal translational motion $\delta\boldsymbol{l}$ and a rotational motion $\delta\boldsymbol{\theta}$. In addition, we must impart to incompressible fluid particles the relative displacements $\delta_1\boldsymbol{r}(\delta x_1, \delta x_2, \delta x_3)$ which are independent of $\delta\boldsymbol{l}$ and $\delta\boldsymbol{\theta}$. The symbol δ_1 denotes a variation with $\boldsymbol{i}_s$ fixed. Obviously the incompressibility equation written in the stationary coordinate system in Sec. 1 is equivalent to

$$\operatorname{div} \delta_1 \boldsymbol{r} = \sum_{s=1}^{3} \frac{\partial \delta x_s}{\partial x_s} = 0$$

by virtue of Eq. (1–3).

We now denote the vectors of the instantaneous translational and angular velocities of the rigid body in the motion being compared by $\boldsymbol{v}_0+\delta\boldsymbol{v}_0$ and $\boldsymbol{\omega}+\delta\boldsymbol{\omega}$ and shall find expressions for $\delta\boldsymbol{v}_0$ and $\delta\boldsymbol{\omega}$ in terms of the vectors $\delta\boldsymbol{l}$ and $\delta\boldsymbol{\theta}$. Let $\boldsymbol{i}^*_s$ denote the unit vectors of the moving coordinate system in the motion being compared, where $\boldsymbol{i}^*_s=\boldsymbol{i}_s+\delta\boldsymbol{i}_s$, $\delta\boldsymbol{i}_s=\delta\boldsymbol{\theta}\times\boldsymbol{i}_s\,(s=1, 2, 3)$ [Fig. 1-3]. Then, for example, the projection of the vector $\boldsymbol{\omega}+\delta\boldsymbol{\omega}$ on the x_1 axis in the position being compared is

$$\omega_1 + \delta\omega_1 = (\boldsymbol{\omega} + \delta\boldsymbol{\omega})\cdot\boldsymbol{i}^*_1 = {}_3 + \omega_3\omega_2\delta\theta - \omega_3\delta\theta_2 + \frac{d\delta\theta_1}{dt},$$

since, to within second-order infinitesimals,

$$\boldsymbol{i}_1\cdot\boldsymbol{i}^*_1 = 1\,, \quad \boldsymbol{i}_2\cdot\boldsymbol{i}^*_1 = \delta\theta_3\,, \quad \boldsymbol{i}_3\cdot\boldsymbol{i}^*_1 = -\,\delta\theta_2\,, \quad \delta\boldsymbol{\theta}\cdot\boldsymbol{i}^*_1 = \delta\theta_1\,.$$

In a similar manner we obtain the two other equalities from which we find the expression for the vector $\delta\boldsymbol{\omega}$:

$$\delta\boldsymbol{\omega} = \frac{d\delta\boldsymbol{\theta}}{dt} + \boldsymbol{\omega} \times \delta\boldsymbol{\theta}\,. \tag{1–6}$$

Let $\boldsymbol{r}_0' + \delta\boldsymbol{r}_0'$ be the radius vector of the point O of the body in the motion being compared. Then, for example,

$$v_{01} + \delta v_{01} = \left(\frac{d\boldsymbol{r}_0'}{dt} + \frac{d\delta\boldsymbol{r}_0'}{dt}\right)' \cdot \boldsymbol{i}_1^* =$$
$$= v_{01} + v_{02}\delta\theta_3 - v_{03}\delta\theta_2 + \omega_2\delta l_3 - \omega_3\delta l_2 + \frac{d\delta l_1}{dt}\,,$$

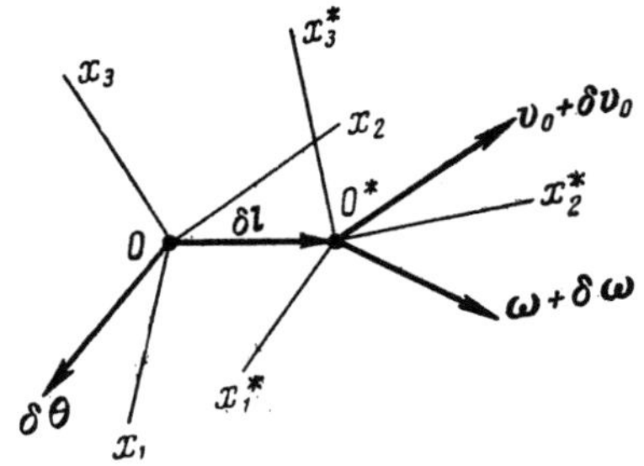

Fig. 3.

since, to within second-order infinitesimals,

$$\left(\frac{d\boldsymbol{r}_0'}{dt}\right)' \cdot \boldsymbol{i}_1^* = v_{01} + v_{02}\delta\theta_3 - v_{03}\delta\theta_2\,, \quad \delta\boldsymbol{r}_0' \cdot \boldsymbol{i}_1^* = \delta l_1\,,$$
$$\left(\frac{d\delta\boldsymbol{r}_0'}{dt}\right)' \cdot \boldsymbol{i}_1^* = \frac{d\delta l_1}{dt} + \omega_2\delta l_3 - \omega_3\delta l_2\,.$$

In a similar manner we obtain the two other equalities from which we find the expression for the vector $\delta\boldsymbol{v}_0$:

$$\delta\boldsymbol{v}_0 = \frac{d\delta\boldsymbol{l}}{dt} + \boldsymbol{\omega} \times \delta\boldsymbol{l} + \boldsymbol{v}_0 \times \delta\boldsymbol{\theta}\,. \tag{1–7}$$

We note that Eqs. (1–6) and (1–7) can be obtained in a different manner, for example, by equating the derivatives of the variational quasi-coordinates δl_i and $\delta\boldsymbol{\theta}_i$ to variations of the quasi-velocities $\delta\boldsymbol{v}_{0i}$ and $\delta\omega_i$[3]. We find from (1–3) that $\delta\boldsymbol{r}' = \delta\boldsymbol{r}_0' + \delta\boldsymbol{r}$. But $\delta\boldsymbol{r}_0' = \delta\boldsymbol{l}$,

$$\delta\boldsymbol{r} = \sum_{s=1}^{3} (x_s\delta\boldsymbol{i}_s + \delta x_s\boldsymbol{i}_s) = \delta\boldsymbol{\theta} \times \boldsymbol{r} + \delta_1\boldsymbol{r}\,,$$

so that

$$\delta\boldsymbol{r}' = \delta\boldsymbol{l} + \delta\boldsymbol{\theta} \times \boldsymbol{r} + \delta_1\boldsymbol{r}\,. \tag{1–8}$$

Further on we shall present, for reference purposes, certain formulas from vector analysis that we will use frequently[4].

Let S' be some closed surface bounding the region τ. We shall assume that S' is a smooth surface or that it consists of a finite number of smooth surface sectors, each of which has both a continuous normal and curvature. In region τ let there be specified a single-valued, finite and continuous vector function $\boldsymbol{a}(\boldsymbol{r})$ and a scalar function $\varphi(\boldsymbol{r})$, together with their first-order partial derivatives which are also single-valued, finite and continuous. Then the Gauss-Ostrogradskiy theorem will apply:

The vector flux through the closed surface is equal to the volume integral of the vector divergence

$$\int_{S'} a_n dS = \int_{\tau} \operatorname{div} \boldsymbol{a} d\tau, \tag{1–9}$$

where $a_n = \boldsymbol{a} \cdot \boldsymbol{n}$ is the projection of the vector $\boldsymbol{a}$ on the external normal to the surface S', which is defined by the unit vector $\boldsymbol{n}$. In particular, we also have

$$\int_{S'} \varphi n_i dS = \int_{\tau} \frac{\partial \varphi}{\partial x_i} d\tau, \quad \int_{S'} n_i dS = 0 \quad (i = 1, 2, 3).$$

Further let there be specified in region τ the continuous scalar functions $\varphi(\boldsymbol{r})$ and $\psi(\boldsymbol{r})$ with derivatives that are continuous up to the second order. Green's theorem applies under these conditions. The first form of this theorem is

$$\int_{\tau} (\nabla\varphi \cdot \nabla\psi) d\tau + \int_{\tau} \varphi \Delta\psi d\tau = \int_{S'} \varphi \frac{d\psi}{dn} dS. \tag{1–10}$$

It can also be written in the form

$$\int_{\tau} (\varphi\Delta\psi - \psi\Delta\varphi) d\tau = \int_{S'} \left(\varphi \frac{d\psi}{dn} - \psi \frac{d\varphi}{dn}\right) dS, \tag{1–11}$$

where

$$\nabla = \sum_{s=1}^{3} \frac{\partial}{\partial x_s} \boldsymbol{i}_s$$

is Hamilton's operator,

$$\nabla\varphi = \operatorname{grad} \varphi, \quad \Delta = \sum_{s=1}^{3} \frac{\partial^2}{\partial x_s^2}$$

is the Laplacian operator, and

$$\frac{d\varphi}{dn} = \sum_{s=1}^{3} \frac{\partial \varphi}{\partial x_s} n_s$$

is the derivative with respect to $\boldsymbol{n}$ of the function φ.

Let us now examine some surface S that is bounded by a closed curve (contour) s and to which is assigned a direction of traverse. For each element dS of this surface we shall construct a normal $\boldsymbol{n}$ oriented so that it forms with the direction of traverse of contour s a system that is oriented in the same manner as the coordinate system. Then Stokes' theorem becomes applicable

$$\int_S \operatorname{curl} \boldsymbol{a} dS = \int_s \boldsymbol{a} \cdot d\boldsymbol{s}, \tag{1–12}$$

where the curl of the vector $\boldsymbol{a}$ is defined as

$$\operatorname{curl} \boldsymbol{a} = \nabla \times \boldsymbol{a} = \left(\frac{\partial a_3}{\partial x_2} - \frac{\partial a_2}{\partial x_3}\right)\boldsymbol{i}_1 + \left(\frac{\partial a_1}{\partial x_3} - \frac{\partial a_3}{\partial x_1}\right)\boldsymbol{i}_2 + \left(\frac{\partial a_2}{\partial x_1} - \frac{\partial a_1}{\partial x_2}\right)\boldsymbol{i}_3 .$$

The integral in the right-hand side of Eq. (1–12) is called the circulation of the vector $\boldsymbol{a}$.

Stokes' theorem has the following corollary: If $\boldsymbol{a} = \operatorname{grad} \varphi$, then $\int_s \boldsymbol{a} \cdot d\boldsymbol{s} = 0$, and $\operatorname{curl}_n \boldsymbol{a} = 0$. Since this is valid for any direction,

$$\operatorname{curl} \operatorname{grad} \varphi = 0 .$$

Conversely, if $\operatorname{curl} \boldsymbol{a} = 0$, then the circulation of $\boldsymbol{a}$ over any contour that can be collapsed into a point is zero, *i.e.*, $\boldsymbol{a} = \operatorname{grad} \varphi$. Thus the necessary and sufficient condition for $\boldsymbol{a} = \operatorname{grad} \varphi$ consists in satisfaction of the equality $\operatorname{curl} \boldsymbol{a} = 0$.

Finally, for time derivatives of integrals taken over the fluid volume τ, or over the open fluid surface S in the case of an incompressible fluid, we have the formulas

$$\frac{d}{dt}\int_\tau \boldsymbol{a} d\tau = \int_\tau \frac{d\boldsymbol{a}}{dt} d\tau ,$$

$$\frac{d}{dt}\int_S a_n dS = \int_S \left[\frac{\partial a_n}{\partial t} + v_n \operatorname{div} \boldsymbol{a} + \operatorname{curl}_n (\boldsymbol{a} \times \boldsymbol{v})\right] dS .$$

[1-3] BASIC DYNAMIC PARAMETERS

The main dynamic parameters, i.e., the kinetic energy $T = \frac{1}{2}\sum m_\nu \boldsymbol{v}_\nu^2$, the linear momentum $\boldsymbol{Q} = \sum_\nu m_\nu \boldsymbol{v}_\nu$, and the angular momentum $\boldsymbol{G} = \sum_\nu \boldsymbol{r}_\nu \times m_\nu \boldsymbol{v}_\nu$, of the system are composed of the corresponding parameters for the rigid body and the fluid. From here on, the parameters pertaining to the rigid body will be denoted by the subscript 1 and those pertaining to the fluid by the subscript 2; the mass of the νth point of the system is denoted by m_ν. Using Eq. (1–4), we obtain

$$T_1 = \tfrac{1}{2}M_1 v_0^2 + M_1 \boldsymbol{v}_0 \cdot (\boldsymbol{\omega} \times \boldsymbol{r}_1) + \tfrac{1}{2}\boldsymbol{\omega} \cdot \boldsymbol{\Theta}^{(1)} \cdot \boldsymbol{\omega},$$

$$T_2 = \varrho \int_\tau \mathscr{T} d\tau = \tfrac{1}{2}M_2 v_0^2 + M_2 \boldsymbol{v}_0 \cdot (\boldsymbol{\omega} \times \boldsymbol{r}_2) + \tag{1–13}$$

$$+ \tfrac{1}{2}\boldsymbol{\omega} \cdot \boldsymbol{\Theta}^{(2)} \cdot \boldsymbol{\omega} + \boldsymbol{v}_0 \cdot \varrho \int_\tau \boldsymbol{u} d\tau + \boldsymbol{\omega} \cdot \varrho \int_\tau \boldsymbol{r} \times \boldsymbol{u} d\tau + \tfrac{1}{2}\varrho \int_\tau u^2 d\tau,$$

where $\mathscr{T} = \frac{1}{2}v^2$ is the kinetic energy density of the fluid,

$$M_i = \textstyle\sum_{(i)} m_\nu$$

are the masses, and

$$\boldsymbol{r}_i = \frac{1}{M_i} \sum_{(i)\,\nu}^{\nu} m_\nu \boldsymbol{r}_\nu$$

are the radius vectors of the mass centers of the rigid body ($i=1$) and of the fluid ($i=2$) relative to point O; here the symbol $\sum_{(i)\,\nu}$ denotes summation only over points of the rigid body ($i=1$) or of the fluid ($i=2$);

$$\boldsymbol{\Theta}^{(i)} = \sum_{r,s=1}^{3} \Theta_{rs}^{(i)} \boldsymbol{i}_r \boldsymbol{i}_s \quad (i = 1, 2)$$

are symmetrical inertia tensors of the body and of the fluid for the point O. The diagonal components of the inertia tensor

$$\Theta_{11}^{(i)} = A_i, \quad \Theta_{22}^{(i)} = B_i, \quad \Theta_{33}^{(i)} = C_i$$

represent moments of inertia, while the nondiagonal components, taken with the minus sign,

$$\Theta_{23}^{(i)} = \Theta_{32}^{(i)} = -D_i, \quad \Theta_{31}^{(i)} = \Theta_{13}^{(i)} = -E_i, \quad \Theta_{12}^{(i)} = \Theta_{21}^{(i)} = -F_i$$

are centrifugal moments of inertia, or the inertia products, relative to the x_1, x_2, x_3 axes[3]. Thus the kinetic energy of the system is

$$T = \tfrac{1}{2}Mv_0^2 + M\boldsymbol{v}_0 \cdot (\boldsymbol{\omega} \times \boldsymbol{r}_c) + \tfrac{1}{2}\boldsymbol{\omega} \cdot \boldsymbol{\Theta} \cdot \boldsymbol{\omega} + \\ + \boldsymbol{v}_0 \cdot \varrho \int_\tau \boldsymbol{u} d\tau + \boldsymbol{\omega} \cdot \varrho \int_\tau \boldsymbol{r} \times \boldsymbol{u} d\tau + \tfrac{1}{2}\varrho \int_\tau u^2 d\tau. \tag{1–14}$$

Here

$$M = M_1 + M_2, \quad \boldsymbol{r}_c = \frac{1}{M} \sum_{i=1}^{2} M_i \boldsymbol{r}_i$$

is the radius vector of the center of mass of the system about the point O, and $\boldsymbol{\Theta}=\boldsymbol{\Theta}^{(1)}+\boldsymbol{\Theta}^{(2)}$ is the inertia tensor of the system for the point O.

In certain cases Eqs. (1–13) and (1–14) can be simplified. For example, if we take the principal central axes of inertia of the rigid body as the moving coordinate axes, then $\boldsymbol{r}_1=0$, $\boldsymbol{\Theta}^1_{sr}=0$ $(s\neq r)$. In the case of an unrestrained rigid body with a cavity completely filled with a fluid it is expedient to take as the moving axes the principal central axes of inertia of the system; then

$$T = \tfrac{1}{2}Mv_0^2 + \tfrac{1}{2}\boldsymbol{\omega}\cdot\boldsymbol{\Theta}\cdot\boldsymbol{\omega} + \boldsymbol{v}_0\cdot\varrho\int_\tau \boldsymbol{u}d\tau + \boldsymbol{\omega}\cdot\varrho\int_\tau \boldsymbol{r}\times\boldsymbol{u}d\tau + \tfrac{1}{2}\varrho\int_\tau u^2d\tau,$$

where the inertia tensor of the system is diagonal in the form

$$\boldsymbol{\Theta} = \sum_{s=1}^{3} \Theta_{ss}\boldsymbol{i}_s\boldsymbol{i}_s,$$

where

$$\Theta_{11} = A, \quad \Theta_{22} = B, \quad \Theta_{33} = C$$

are the moments of inertia of the system relative to the x_1, x_2, x_3 axes, and are equal to the sum of the corresponding moments of inertia of the rigid body and the fluid.

The momentum vector of the system $\boldsymbol{Q}=\boldsymbol{Q}_1+\boldsymbol{Q}_2$ can be represented as

$$\boldsymbol{Q} = M(\boldsymbol{v}_0 + \boldsymbol{\omega}\times\boldsymbol{r}_c) + \varrho\int_\tau \boldsymbol{u}d\tau = M\boldsymbol{v}_c, \qquad (1\text{–}15)$$

where $\boldsymbol{v}_c$ is the velocity vector of the center of mass of the system.

Let us now compare Eqs. (1–14) and (1–15). Taking partial derivatives of the kinetic energy of the system with respect to the projection of the velocity vector of point O, we find[5]

$$Q_i = \frac{\partial T}{\partial v_{0i}} \quad (i = 1, 2, 3).$$

Thus the momentum vector of the system is the gradient of the scalar function T with respect to the projection of the velocity vector $\boldsymbol{v}_0$ of the point O:

$$\boldsymbol{Q} = \operatorname{grad}_{v_0} T. \qquad (1\text{–}16)$$

The angular momentum vector of the system $\boldsymbol{G}=\boldsymbol{G}_1+\boldsymbol{G}_2$ relative to the moving point O can be written

$$\boldsymbol{G} = \boldsymbol{r}_c\times M\boldsymbol{v}_0 + \boldsymbol{\Theta}\cdot\boldsymbol{\omega} + \varrho\int_\tau \boldsymbol{r}\times\boldsymbol{u}d\tau. \qquad (1\text{–}17)$$

Equating this expression with (1–14), we find that

$$G_i = \frac{\partial T}{\partial \omega_i} \quad (i = 1, 2, 3);$$

i.e., the angular momentum vector relative to the point O is equal to the gradient of the function T with respect to the projection of the instantaneous angular velocity $\boldsymbol{\omega}$:

$$G = \text{grad}_\omega T. \tag{1–18}$$

[1-4] DERIVATION OF THE EQUATIONS OF MOTION OF AN UNCONSTRAINED RIGID BODY CONTAINING A FLUID

Let us now turn to the derivation of the equations of motion of a system referred to the $Ox_1x_2x_3$ moving coordinate system. For this purpose we must represent the integrand in Eq. (1–2) in terms of quantities referred to the moving coordinate system.

Making use of Eqs. (1–16) and (1–18), we now write the variation of the kinetic energy (1–14) in the form

$$\delta T = \boldsymbol{Q}\cdot\delta\boldsymbol{v}_0 + \boldsymbol{G}\cdot\delta\boldsymbol{\omega} + \varrho\int\limits_\tau [\boldsymbol{v}\cdot\delta\boldsymbol{u} + (\boldsymbol{v}\times\boldsymbol{\omega})\cdot\delta_1\boldsymbol{r}]\,d\tau$$

and we replace the vectors $\delta\boldsymbol{v}_0$ and $\delta\boldsymbol{\omega}$ by their expressions (1–7) and (1–6):

$$\begin{aligned}\delta T = \boldsymbol{Q}\cdot\left(\frac{d\delta\boldsymbol{l}}{dt} + \boldsymbol{\omega}\times\delta\boldsymbol{l} + \boldsymbol{v}_0\times\delta\boldsymbol{\theta}\right) + \boldsymbol{G}\cdot\left(\frac{d\delta\boldsymbol{\theta}}{dt} + \boldsymbol{\omega}\times\delta\boldsymbol{\theta}\right) + \\ + \varrho\int\limits_\tau [\boldsymbol{v}\cdot\delta\boldsymbol{u} + (\boldsymbol{v}\times\boldsymbol{\omega})\cdot\delta_1\boldsymbol{r}]\,d\tau.\end{aligned} \tag{1–19}$$

Furthermore, we examine the sum of the elementary work applied to a system of effective forces $\boldsymbol{F}_\nu$ corresponding to the possible displacements $\Sigma\boldsymbol{F}_\nu\cdot\delta\boldsymbol{r}'_\nu$. Among the forces acting on the fluid we shall, as usual[6] differentiate between mass forces acting on each element $\varrho d\tau$ of the fluid mass and surface forces acting on elements dS of the fluid's surface. The mass force vector referred to unit mass is denoted by $\boldsymbol{F}$. Among the surface forces, we shall be primarily concerned with forces of the normal pressure $p_0\boldsymbol{n}$, where $\boldsymbol{n}$ denotes the unit vector of the external normal to the surface S, and in certain cases we shall also be concerned with the surface tension forces.

It is known from molecular theory[2] that the action of fluid particles on one another is determined not only by the pressure p as a result of the constancy of volume, but also by the binding forces acting between the particles. For internal elements of the fluid volume these forces cancel one another as a result of their average uniform directional distribution, and only for volume elements near the surface S' do they yield a nonzero resultant. Under "ordinary" conditions, when the dimensions of the cavity are sufficiently large and the surface tension forces are small in comparison with the mass forces, they are usually disregarded; however, in some cases, particularly under conditions of weightlessness, when mass forces are small, or when fluids are moving in capillaries, these forces must be taken into account.

In considering the surface forces we shall not dwell on the fine points of molecular theory but simply follow Gauss's concepts[7,8]. Namely, we shall assume that, as a result of contact between two different media r and s along some surface, tension forces arise with a potential that is equal to the product of the area of the surface of contact by the surface tension coefficient α_{rs} and is a function of the nature of both media; it is obvious that $\alpha_{rs}=\alpha_{sr}$. In the case under study, a fluid-containing rigid body will generally have three such media: the rigid body, the fluid and the air, which will be assigned subscripts 1, 2, and 3, respectively. For brevity we set $\alpha_{12}=\alpha_1$, $\alpha_{13}=\alpha_2$, $\alpha_{23}=\alpha$.

Referring to the above and making use of Eq. (1–8), we obtain the following for the sum of the elementary work performed by the effective forces corresponding to the possible displacements of the system:

$$\sum_\nu \boldsymbol{F}_\nu \cdot \delta \boldsymbol{r}'_\nu = \boldsymbol{K}\cdot\delta\boldsymbol{l} + \boldsymbol{L}\cdot\delta\boldsymbol{\theta} + \varrho\int_\tau \boldsymbol{F}\cdot\delta_1\boldsymbol{r}\,d\tau - \\ - p_0\int_S \boldsymbol{n}\cdot\delta_1\boldsymbol{r}\,dS - \alpha\delta S - \alpha_1\delta\sigma_1 - \alpha_2\delta\sigma_2\,, \tag{1–20}$$

where $\boldsymbol{K}$ and $\boldsymbol{L}$ denote the principal vector and the principal moment relative to the point O of all the effective forces applied to the system:

$$\boldsymbol{K} = \sum_\nu \boldsymbol{F}_\nu = \sum_\nu{}_{(1)} \boldsymbol{F}_\nu + \varrho\int_\tau \boldsymbol{F}\,d\tau\,, \\ \boldsymbol{L} = \sum_\nu \boldsymbol{r}_\nu \times \boldsymbol{F}_\nu = \sum_\nu{}_{(1)} \boldsymbol{r}_\nu \times \boldsymbol{F}_\nu + \varrho\int_\tau \boldsymbol{r}\times\boldsymbol{F}\,d\tau\,. \tag{1–21}$$

It is easy to see[8] that the changes corresponding to the possible displacements $\delta_1\boldsymbol{r}$ of the areas of the surfaces S and σ_i are defined as*

$$\delta S = \int_S \left(\frac{1}{R_1}+\frac{1}{R_2}\right)\delta\zeta\,dS + \int_s \delta\zeta_1\,ds\,, \\ \delta\sigma_1 = -\,\delta\sigma_2 = \int_s \delta\zeta_2\,ds\,. \tag{1–22}$$

* The derivation of Eqs. (1.22) is given on pp. 305–306.

Here R_1 denote the principal radii of curvature of the surface S at the given point, regarded as positive when the center of curvature lies on the same side of the surface as the fluid and negative in the opposite case;

$$\delta\zeta = \boldsymbol{n}\cdot\delta_1\boldsymbol{r}, \quad \delta\zeta_1 = \boldsymbol{n}_1\cdot\delta_1\boldsymbol{r}, \quad \delta\zeta_2 = \boldsymbol{n}_2\cdot\delta_1\boldsymbol{r},$$

$\boldsymbol{n}_1$ and $\boldsymbol{n}_2$ are the unit vectors of the external normals to the contour s of the surfaces S and σ_1, situated, respectively, in the tangent planes to these surfaces (Fig. 1-4). It is obvious that at the surface σ_1, as a result of the

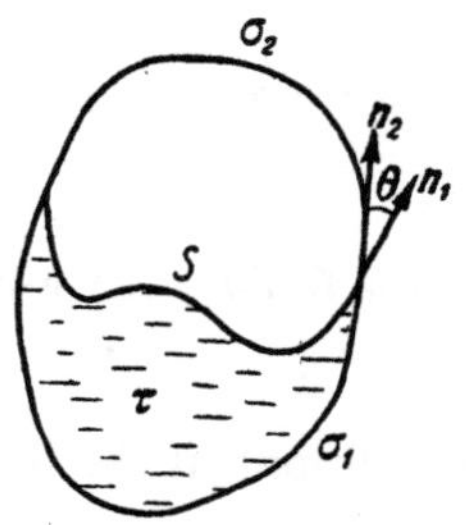

Fig. 4.

impermeability of the solid walls, $\delta\zeta = 0$. The angle between the normals $\boldsymbol{n}_1$ and $\boldsymbol{n}_2$ is denoted by θ. Assuming that the surface of the walls of the cavity σ is sufficiently smooth in the vicinity of line s, and that it has no sharp ribs, we find that

$$\delta\zeta_1 = \delta\zeta_2 \cos\theta . \tag{1–23}$$

Let us finally turn to the integral

$$\int_\tau p \operatorname{div} \delta\boldsymbol{r}' d\tau = \int_\tau p \operatorname{div} \delta_1\boldsymbol{r} d\tau .$$

Since[4]

$$p \operatorname{div} \delta_1\boldsymbol{r} = \operatorname{div}(p\delta_1\boldsymbol{r}) - \operatorname{grad} p\cdot\delta_1\boldsymbol{r},$$

then, using Eq. (1–9), we obtain

$$\int_\tau p \operatorname{div} \delta_1\boldsymbol{r} d\tau = \int_S p\boldsymbol{n}\cdot\delta_1\boldsymbol{r} dS - \int_\tau \operatorname{grad} p\cdot\delta_1\boldsymbol{r} d\tau . \tag{1–24}$$

We substitute Eqs. (1–19), (1–20), and (1–24) into Eq. (1–2), making use in the process of Eqs. (1–22) and (1–23), and we integrate by parts the terms

$\boldsymbol{Q}\cdot d\delta\boldsymbol{l}/dt$, $\boldsymbol{G}\cdot d\delta\boldsymbol{\theta}/dt$, $\boldsymbol{v}\cdot\delta\boldsymbol{u}$; as a result we obtain

$$\int_{t_0}^{t_1}\left\{\left(-\frac{d\boldsymbol{Q}}{dt}+\boldsymbol{Q}\times\boldsymbol{\omega}+\boldsymbol{K}\right)\cdot\delta\boldsymbol{l}+\left(-\frac{d\boldsymbol{G}}{dt}+\boldsymbol{G}\times\boldsymbol{\omega}+\boldsymbol{Q}\times\boldsymbol{v}_0+\boldsymbol{L}\right)\cdot\right.$$
$$\cdot\delta\boldsymbol{\theta}+\varrho\int_\tau\left(-\frac{d\boldsymbol{v}}{dt}+\boldsymbol{v}\times\boldsymbol{\omega}+\boldsymbol{F}-\frac{1}{\varrho}\operatorname{grad}p\right)\cdot$$
$$\cdot\delta_1\boldsymbol{r}d\tau+\int_S\left[p-p_0-\alpha\left(\frac{1}{R_1}+\frac{1}{R_2}\right)\right]\delta\zeta dS- \qquad (1\text{–}25)$$
$$\left.-\int_s(\alpha\cos\theta+\alpha_1-\alpha_2)\,\delta\zeta_2 ds\right\}dt+$$
$$+\left(\boldsymbol{Q}\cdot\delta\boldsymbol{l}+\boldsymbol{G}\cdot\delta\boldsymbol{\theta}+\varrho\int_\tau\boldsymbol{v}\cdot\delta_1\boldsymbol{r}d\tau\right)_{t_0}^{t_1}=0\,.$$

By virtue of Eq. (1–8), the conditions at the limits of integration will take on the form

$$\text{when}\quad t=t_0,\ t=t_1\quad \delta\boldsymbol{l}=\delta\boldsymbol{\theta}=\delta_1\boldsymbol{r}=0\,,$$

as a result of which the terms outside the time integral in Eq. (1–25) cancel out.

For an unconstrained rigid body the vectors $\delta\boldsymbol{l}$ and $\delta\boldsymbol{\theta}$ are independent and arbitrary. Even the vector $\delta_1\boldsymbol{r}$ is independent of them, and in addition, by introducing the multiplier p, it can be selected arbitrarily. By virtue of the arbitrariness of these vectors, Eq. (1–25) yields the following equations of motion of the system:

$$\frac{d\boldsymbol{Q}}{dt}+\boldsymbol{\omega}\times\boldsymbol{Q}=\boldsymbol{K}; \qquad (1\text{–}26)$$

$$\frac{d\boldsymbol{G}}{dt}+\boldsymbol{\omega}\times\boldsymbol{G}+\boldsymbol{v}_0\times\boldsymbol{Q}=\boldsymbol{L}; \qquad (1\text{–}27)$$

$$\frac{d\boldsymbol{v}}{dt}+\boldsymbol{\omega}\times\boldsymbol{v}=\boldsymbol{F}-\frac{1}{\varrho}\operatorname{grad}p; \qquad (1\text{–}28)$$

which should be supplemented by the incompressibility equation (equation of constraint), which for the actual motion of a system has the form

$$\operatorname{div}\boldsymbol{v}=\frac{\partial v_1}{\partial x_1}+\frac{\partial v_2}{\partial x_2}+\frac{\partial v_3}{\partial x_3}=0\,. \qquad (1\text{–}29)$$

From Eq. (1–25) we also obtain the boundary conditions for the hydrodynamic pressure on the free surface S (the Laplace formula [2]):

$$p = p_0 + \alpha\left(\frac{1}{R_1} + \frac{1}{R_2}\right) \tag{1–30}$$

and for the contact angle θ at the contour s:

$$\cos\theta = \frac{\alpha_2 - \alpha_1}{\alpha}; \tag{1–31}$$

which must be satisfied at any given instant of time. It is seen from the last equation, by the way, that in the case of motion (and equilibrium) the free fluid surfaces form a constant contact angle θ, with the solid wall σ, if, of course, the values of α_{rs} are regarded as constant (Fig. 1-5).

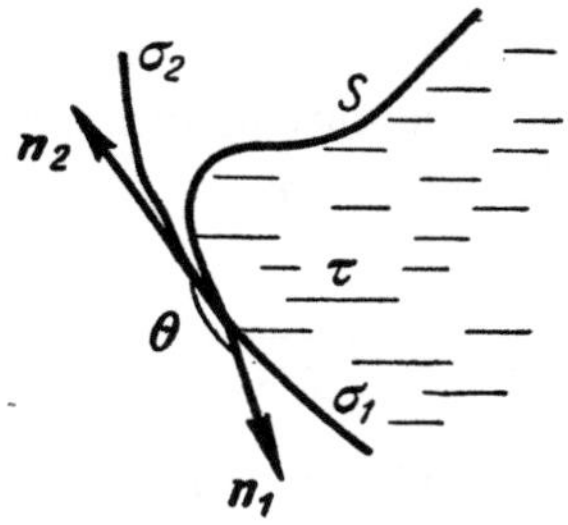

Fig. 5.

The conditions (1–30) and (1–31) should be supplemented by the kinematic conditions [6]

$$\text{on} \quad \sigma_1 \qquad u_n = \boldsymbol{u}\cdot\boldsymbol{n} = 0, \tag{1–32}$$

$$\text{on} \quad S \qquad \frac{\partial f}{\partial t} + \boldsymbol{u}\cdot\operatorname{grad} f = 0, \tag{1–33}$$

where $f(x_1, x_2, x_3, t) = 0$ is the equation of the free surface of the fluid.

Condition (1–32) expresses the impermeability of the solid walls for the actual motion of the fluid, while condition (1–33) means that the velocity of displacement of any point on the free surface and the velocity of a fluid particle situated on the surface have identical projections on the normal to the surface.

In the case when surface tension can be disregarded, all of the surface tension coefficients α_{rs} will be assumed to be equal to zero. Then condition (1–31) disappears, while condition (1–30) for the hydrodynamic pressure on the free surface takes the form

$$p = p_0 . \tag{1–34}$$

If the fluid fills the cavity completely we obviously have only the one boundary condition (1–32).

Let us clarify the mechanical meaning of Eqs. (1–26)–(1–28). The left-hand side of Eq. (1–26) is the absolute derivative with respect to time of the system's momentum vector, while the right-hand side is the principal vector of all of the effective forces applied to the system. Thus the vector equation (1–26) expresses the theorem of the system's momentum[9]; it is equivalent to the three scalar equations

$$\frac{d}{dt}\left(\frac{\partial T}{\partial v_{01}}\right) + \omega_2 \frac{\partial T}{\partial v_{03}} - \omega_3 \frac{\partial T}{\partial v_{02}} = K_1 \quad (123). \tag{1–35}$$

Here and below, the symbols (123) mean that the two other equations may be obtained from the written one by cyclical transposition of the numbers in the parentheses.

It is easy to see that Eq. (1–27) expresses the theorem of the angular momentum of the system for the moving center O. This vector equation is equivalent to the three scalar equations

$$\frac{d}{dt}\left(\frac{\partial T}{\partial \omega_1}\right) + \omega_2 \frac{\partial T}{\partial \omega_3} - \omega_3 \frac{\partial T}{\partial \omega_2} + v_{02} \frac{\partial T}{\partial v_{03}} - v_{03} \frac{\partial T}{\partial v_{02}} = L_1 \quad (123). \tag{1–36}$$

Finally, Eq. (1–28), which is equivalent to the three scalar equations

$$\frac{dv_1}{dt} + \omega_2 v_3 - \omega_3 v_2 = F_1 - \frac{1}{\varrho} \frac{\partial p}{\partial x_1} \quad (123), \tag{1–37}$$

is Euler's hydrodynamic equation referred to the moving coordinate axes. This equation, as well as Eq. (1–29), can be rewritten, making use of Eq. (1–4), in the form of equations of relative motion of the fluid:

$$\frac{d\boldsymbol{u}}{dt} + 2\boldsymbol{\omega} \times \boldsymbol{u} = \boldsymbol{F} - \frac{1}{\varrho} \operatorname{grad} p - \frac{d\boldsymbol{\omega}}{dt} \times \boldsymbol{r} - \frac{d\boldsymbol{v}_0}{dt} - \tag{1–38}$$
$$- \boldsymbol{\omega} \times \boldsymbol{v}_0 - \boldsymbol{\omega} \times (\boldsymbol{\omega} \times \boldsymbol{r}), \quad \operatorname{div} \boldsymbol{u} = 0.$$

We note that in the case of vortex motion of the fluid in the body's cavities, Eqs. (1–28) and (1–29) can be transformed into equations defining the change in the vortices with time. If the mass forces have a potential, then the equations for vortices have the form of Helmholtz's equations[6]:

$$\frac{d\Omega}{dt} + \boldsymbol{\omega} \times \boldsymbol{\Omega} = (\boldsymbol{\Omega} \cdot \nabla)\, \boldsymbol{v}, \quad \boldsymbol{\Omega} = \operatorname{curl} \boldsymbol{v}. \tag{1–39}$$

We recall that in general for a fluid, $\boldsymbol{v} = \boldsymbol{v}(\boldsymbol{r}, t)$, $\boldsymbol{u} = \boldsymbol{u}(\boldsymbol{r}, t)$, and $\boldsymbol{\Omega} = \boldsymbol{\Omega}(\boldsymbol{r}, t)$ and hence the derivatives $d\boldsymbol{v}/dt$, $d\boldsymbol{u}/dt$, and $d\boldsymbol{\Omega}/dt$ in Eqs. (1–28), (1–38), and

(1–39) are total or substantial time derivatives, for example,

$$\frac{d\boldsymbol{v}}{dt} = \frac{\partial \boldsymbol{v}}{\partial t} + (\boldsymbol{u} \cdot \nabla)\, \boldsymbol{v}\,.$$

We also note that in the case when the fluid completely fills the cavity and the center of inertia of the system is taken as the origin of the moving coordinate axes, Eqs. (1–26) and (1–27) take on the simpler form

$$\begin{aligned} M\left(\frac{d\boldsymbol{v}_0}{dt} + \boldsymbol{\omega} \times \boldsymbol{v}_0\right) &= \boldsymbol{K}\,, \\ \frac{d\boldsymbol{G}}{dt} + \boldsymbol{\omega} \times \boldsymbol{G} &= \boldsymbol{L}\,. \end{aligned} \tag{1–40}$$

Two basic problems in the study of the motion of a rigid body with fluid-containing cavities can be stated. The first is: Given a motion of the body and the forces acting on the fluid, it is required to investigate the motion of the fluid in the cavities and to determine the forces that must be applied to the solid body in order for it to perform the specified motion. This problem is basically hydrodynamic and reduces to the study of Euler's equations (1–28) and (1–29) or the equivalent equations (1–38) or (1–39), with the applicable boundary conditions, and subsequent determination from Eqs. (1–26) and (1–27) of the principal vector and principal moment of forces applied to the rigid body.

The second problem is: Given the forces acting on the rigid body and on the fluid, it is required to investigate the motion of the system under the action of the given forces. This problem is much more complicated than the first, and it also includes the problem of hydrodynamics; it reduces to the simultaneous study of the system of equations (1–26)–(1–29). This is a system of simultaneous ordinary and partial differential equations. Its solutions will contain arbitrary functions and arbitrary constants that must be subjected to boundary and initial conditions to obtain unique solutions to specific problems. The boundary conditions (1–30) to (1–34) must be satisfied at the fluid boundary at every instant of time. The initial conditions must be satisfied at the time $t = t_0$ at points of the space that are occupied by the rigid body and the fluid. The initial conditions usually consist of the requirement that the positions $\boldsymbol{v}_0$, and velocities $\boldsymbol{\omega}$ of the rigid body, the shape of the surface S, and the velocity field of the fluid $\boldsymbol{v}(\boldsymbol{r})$ are specified when $t = t_0$.

The effective forces $\boldsymbol{F}_\nu$ acting on the system can in general depend not only on the positions and velocities of the points of the system and on the time, but also on certain parameters. In this case the equations of motion of the system should also be supplemented by the equations of the parameters, and it is necessary to consider simultaneously the complete system of equations.

In many cases of interest in practice the forces acting on the system depend on the cosines γ_{rs} of the angles between the axes of the stationary and moving coordinate systems. In these cases the equations of motion of the system should be supplemented by Eqs. (1–5), which in projections on the moving axes have the form

$$\frac{d\gamma_{r1}}{dt} + \omega_2\gamma_{r3} - \omega_3\gamma_{r2} = 0 \quad (123) \quad (r = 1, 2, 3). \tag{1–41}$$

In a sufficiently extensive class of cases, the effective forces permit the force function $U = U(\gamma_1, \gamma_2, \gamma_3)$, where γ_i denote projections of the unit vector $\boldsymbol{i}_3'$ of the stationary x_3' axis on the moving axes, $\gamma_i = \gamma_{3i}$ $(i = 1, 2, 3)$. From now on the above function U will be assumed to be continuous and to possess continuous first-order partial derivatives with respect to γ_i. Then the moment of the effective forces relative to the point O is represented in the form[10] $\boldsymbol{L} = -\boldsymbol{i}_3' \times \operatorname{grad} U$.

For example, for a homogeneous gravity field with acceleration g

$$U(\gamma_1, \gamma_2, \gamma_3) = -Mg(x_{c1}\gamma_1 + x_{c2}\gamma_2 + x_{c3}\gamma_3), \tag{1–42}$$

where the x_3' axis is directly vertically upward.

In the case of the motion of a missile with a very flat trajectory, the moment of the pitching couple depends on the angle between the velocity of the center of mass, which is assumed to be constant, and the axis of symmetry x_3 of the projectile[11]. Here

$$U(\gamma_3) = -a\gamma_3, \tag{1–43}$$

where the x_3' axis is directed along the velocity vector of the center of mass, and $a = \text{const}$. For the case of a central Newtonian force field

$$U(\gamma_1, \gamma_2, \gamma_3) = \int\limits_v \frac{\mu\varrho(x_1, x_2, x_3)\, dv}{[R^2 + 2R(x_1\gamma_1 + x_2\gamma_2 + x_3\gamma_3) + x_1^2 + x_2^2 + x_3^2]^{1/2}},$$

if the x_3' axis is directed away from the center of attraction O' along the straight line $O'O$. Here μ is the gravitational constant, $\varrho(x_1, x_2, x_3)$ and v are the density and volume of the body and of the fluid, and R is the distance $O'O$.

If the distance R is large in comparison with the dimensions of the body, it is possible to assume the following approximate expression for the force function[12]:

$$\begin{aligned} U = \mu\frac{M}{R} - \mu\frac{M}{R^2}(x_{c1}\gamma_1 + x_{c2}\gamma_2 + x_{c3}\gamma_3) - \\ -\frac{3}{2}\frac{\mu}{R^3}\left(A\gamma_1^2 + B\gamma_2^2 + C\gamma_3^2 - \frac{A+B+C}{3}\right). \end{aligned} \tag{1–44}$$

We note, in conclusion, that derivation of the equations of motion from the

Hamilton-Ostrogradskiy principle makes it possible to claim not only necessity but also sufficiency of the equations thus obtained for describing the motions of a solid body with cavities partially or completely filled with an ideal fluid.

It should be noted that during the last few years mechanics has been making increasing use of methods of group theory, since the motion of a mechanical system can be represented as a sequence of infinitesimal transformations forming a group. It is pertinent that the equations of motion (1–26)–(1–28) belong to the class of Poincaré equations[13] in terms of group variables[14].

[1-5] LAGRANGE'S EQUATIONS

The equations of motion (1–26)–(1–29) can also be used in a number of cases of motion of a constrained rigid body containing a fluid provided that we add in the right-hand sides of Eqs. (1–26) and (1–27) the constraint reaction forces and the moments of the reaction forces, respectively. For example, if the rigid body has a stationary point, it is natural to use O of the moving axes as the origin, when $\boldsymbol{v}_0=0$. In this case Eq. (1–26) will determine the reaction of the stationary point, and Eq. (1–27) will take on the form of the second of Eqs. (1–40), the right-hand side of which will again contain only the moment of the effective forces $\boldsymbol{L}$. This equation should be supplemented by Eqs. (1–28) and (1–29) with boundary conditions (1–30)–(1–34).

The equations of motion of a rigid body with a fluid, subject to holonomic constraints, can also be represented in the form of Lagrange's equations. We denote the Lagrangian coordinates of the body by $q_j (j=1, 2, \ldots, n \leqslant 6)$ and their generalized velocities by $\dot{q}_j$; the Cartesian coordinates x_i $(i=1, 2, 3)$ will serve as the coordinates of the fluid particles. The vectors $\boldsymbol{v}_0$ and $\boldsymbol{\omega}$ can be represented in the form of certain linear functions of $\dot{q}_j$ with coefficients that are functions of q_j. Substituting these expressions into Eqs. (1–13) for the kinetic energy of the rigid body and for the density of the kinetic energy of the fluid, we obtain quadratic functions of $\dot{q}_j$:

$$T_1 = T_1(q_j, \dot{q}_j, t), \quad \mathscr{T} = \mathscr{T}(q_j, \dot{q}_j, x_i, u_i, t).$$

Taking the variation of the expression for the kinetic energy of the system $T = T_1 + \varrho \int_\tau \mathscr{T} d\tau$, we obtain

$$\delta T = \sum_{j=1}^{n} \left(\frac{\partial T}{\partial q_j} \delta q_j + \frac{\partial T}{\partial \dot{q}_j} \delta \dot{q}_j \right) + \varrho \sum_{i=1}^{3} \int_\tau \left(\frac{\partial \mathscr{T}}{\partial x_i} \delta x_i + \frac{\partial \mathscr{T}}{\partial u_i} \delta u_i \right) d\tau .$$

Let us now calculate the sum of the elementary work applied to the system of effective forces corresponding to the possible displacement of the system. Since for points of the rigid body and for fluid particles $\boldsymbol{r}'=\boldsymbol{r}'(q_j, x_i, t)$, the sum of the elementary work of the effective forces is

$$\sum_{\nu} \boldsymbol{F}_{\nu}\cdot\delta\boldsymbol{r}'_{\nu} = \sum_{j=1}^{n} \Phi_j\delta q_j + \varrho\int_{\tau}\sum_{i=1}^{3} F_i\delta x_i d\tau -$$
$$- p_0\int_{S}\boldsymbol{n}\cdot\delta_1\boldsymbol{r}dS - \alpha\delta S - \alpha_1\delta\chi_1 - \alpha_2\delta\sigma_2,$$

where

$$\Phi_j = \sum_{\nu}\boldsymbol{F}_{\nu}\cdot\frac{\partial\boldsymbol{r}'_{\nu}}{\partial q_j} \quad (j=1,\dots,n)$$

are the generalized forces.

Let us substitute these expressions, as well as Eqs. (1–24) into Eq. (1–2) and integrate the terms

$$\frac{\partial T}{\partial\dot{q}_j}\delta\dot{q}_j \quad\text{and}\quad \frac{\partial\mathcal{T}}{\partial u_i}\delta u_i$$

by parts, remembering that the conditions at the limits of integration are equivalent to the following: when $t=t_0$ and $t=t_1$

$$\delta q_j = 0(j=1,\dots,n),\ \delta x_i = 0(i=1,2,3).$$

As a result we obtain equations of motion for the rigid body and the fluid contained in its cavity in the form of Lagrange's equations:

$$\frac{d}{dt}\frac{\partial T}{\partial\dot{q}_j} - \frac{\partial T}{\partial q_j} = \Phi_j \qquad (j=1,\dots,n), \tag{1–45}$$

$$\frac{d}{dt}\frac{\partial\mathcal{T}}{\partial u_i} - \frac{\partial\mathcal{T}}{\partial x_i} = F_i - \frac{1}{\varrho}\frac{\partial p}{\partial x_i} \quad (i=1,2,3), \tag{1–46}$$

as well as the boundary conditions (1–30), or (1–34) and (1–31), for the pressure on the free surface and at the contact angle. These equations should be supplemented by the incompressibility equation (1–29), as well as by the kinematic boundary conditions (1–32) and (1–33). Since

$$\frac{\partial\mathcal{T}}{\partial u_1} = v_1,\quad \frac{\partial\mathcal{T}}{\partial x_1} - \omega_3 v_2 - \omega_2 v_3 \quad (123),$$

Eqs. (1–46) are Euler's hydrodynamic equations (1–37).

As an example of the application of Eqs. (1–45) we now derive the equations of motion of a rigid body containing fluid and with a stationary point O. The principal inertia axes of the body for this point will serve as the moving coordinate axes. The position of a rigid body can be defined by the three Euler's angles ϑ, φ and ψ. We then have Euler's kinematic equations

$$\begin{aligned}\omega_1 &= \dot{\psi}\sin\vartheta\sin\varphi + \dot{\vartheta}\cos\varphi\,,\\ \omega_2 &= \dot{\psi}\sin\vartheta\cos\varphi - \dot{\vartheta}\sin\varphi\,,\\ \omega_3 &= \dot{\varphi} + \dot{\psi}\cos\vartheta\,.\end{aligned} \tag{1–47}$$

The kinetic energy of the system is

$$T = \tfrac{1}{2}(A_1\omega_1^2 + B_1\omega_2^2 + C_1\omega_3^2) + \tfrac{1}{2}\varrho\int_\tau \{(\omega_2 x_3 - \omega_3 x_2 + u_1)^2 + \\ + (\omega_3 x_1 - \omega_1 x_3 + u_2)^2 + (\omega_1 x_2 - \omega_2 x_1 + u_3)^2\}\,d\tau\,.$$

We now set up an equation in the form of (1–45) for the coordinate φ. We obtain

$$\frac{\partial T}{\partial \dot{\varphi}} = \frac{\partial T}{\partial \omega_3}\frac{\partial \omega_3}{\partial \dot{\varphi}} = G_3\,,$$

$$\frac{\partial T}{\partial \varphi} = \frac{\partial T}{\partial \omega_1}\frac{\partial \omega_1}{\partial \varphi} + \frac{\partial T}{\partial \omega_2}\frac{\partial \omega_2}{\partial \varphi} = G_1\omega_2 - G_2\omega_1\,,$$

so that the equation for φ becomes

$$\frac{dG_3}{dt} + \omega_1 G_2 - \omega_2 G_1 = \Phi_3\,.$$

It is easy to see that $\Phi_3\delta\varphi$ is the elementary work of forces applied to the system corresponding to an infinitesimal displacement $\delta\varphi$; *i.e.*, $\Phi_3 = L_3$ is the moment of effective forces about the x_3 axis.

The two other equations are obtained from the one just developed by cyclical transposition of the indices 1, 2, and 3; we thus obtain three scalar equations which are equivalent to the second of Eqs. (1–40). Obviously Eqs. (1–45)–(1–46) can also be used to describe the motion of a non-constrained rigid body containing a fluid.

[1-6] FORCES EXERTED BY THE FLUID ON THE RIGID BODY

We are considering the rigid body and the fluid contained in its cavity as a single mechanical system in which it is not necessary, in setting up the equations of motion, to calculate the forces of interaction between the body and the fluid. However, in many cases it may become necessary to calculate the forces that the fluid or the air as both exert on the rigid body[5]. It is easy to see that the principal vector and the principal moment of the pressure forces are represented by the same terms in Eqs. (1–26) and (1–27), respectively, which pertain to the fluid, since the corresponding equations of motion for one rigid body should have the form

$$\begin{aligned} &\frac{d\boldsymbol{Q}_1}{dt} + \boldsymbol{\omega} \times \boldsymbol{Q}_1 = \boldsymbol{K}_1 + \boldsymbol{P}, \\ &\frac{d\boldsymbol{G}_1}{dt} + \boldsymbol{\omega} \times \boldsymbol{G}_1 + \boldsymbol{v}_0 \times \boldsymbol{Q}_1 = \boldsymbol{L}_1 + \boldsymbol{N}, \end{aligned} \tag{1–48}$$

where $\boldsymbol{K}_1$ and $\boldsymbol{P}$, and $\boldsymbol{L}_1$ and $\boldsymbol{N}$ denote the principal vectors and the principal moments, respectively, relative to the point O, of the effective forces applied to points of the rigid body and of the forces produced by the pressure of the fluid and the air. This can also be checked by direct calculation of $\boldsymbol{P}$ and $\boldsymbol{N}$.

In fact, by definition,

$$\boldsymbol{P} = \int_\sigma p\boldsymbol{n}d\sigma, \quad \boldsymbol{N} = \int_\sigma \boldsymbol{r} \times p\boldsymbol{n}d\sigma.$$

If the fluid completely fills the cavity, the surface σ becomes identical with the fluid surface S'. However, as in the case of a partially fluid-filled cavity, these integrals can be extended over the fluid surface S'. Assuming that the air pressure in the cavity is constant, and applying the Gauss-Ostrogradskiy theorem, we transform the expression for $\boldsymbol{P}$ and $\boldsymbol{N}$ into

$$\boldsymbol{P} = \int_{S'} p\boldsymbol{n}dS' = \int_\tau \operatorname{grad} p d\tau,$$

$$\boldsymbol{N} = \int_{S'} \boldsymbol{r} \times p\boldsymbol{n}dS' = \int_\tau \boldsymbol{r} \times \operatorname{grad} p d\tau.$$

We now replace grad p in these expressions, using Eq. (1–28), so as to obtain

$$\boldsymbol{P} = \varrho \int_\tau \left(\boldsymbol{F} - \frac{d\boldsymbol{v}}{dt} - \boldsymbol{\omega} \times \boldsymbol{v}\right) d\tau,$$

$$\boldsymbol{N} = \varrho \int_\tau \boldsymbol{r} \times \left(\boldsymbol{F} - \frac{d\boldsymbol{v}}{dt} - \boldsymbol{\omega} \times \boldsymbol{v}\right) d\tau.$$

But

$$\varrho \int_\tau \boldsymbol{v}d\tau = \boldsymbol{Q}_2, \quad \varrho \int_\tau \boldsymbol{r} \times \boldsymbol{v}d\tau = \boldsymbol{G}_2,$$

$$\boldsymbol{u} \times \boldsymbol{v} - \boldsymbol{r} \times (\boldsymbol{\omega} \times \boldsymbol{v}) = \boldsymbol{v} \times \boldsymbol{v}_0 - \boldsymbol{\omega} \times (\boldsymbol{r} \times \boldsymbol{v}),$$

so that we have

$$P = \varrho \int_\tau F d\tau - \frac{dQ_2}{dt} - \omega \times Q_2 ,$$
$$N = \varrho \int_\tau r \times F d\tau - \frac{dG_2}{dt} - \omega \times G_2 - v_0 \times Q_2 . \tag{1–49}$$

We have thus established that the principal vector and the principal moment of the forces produced by the fluid (and air) pressure on the body are expressed by Eqs. (1–49) in terms of the linear and angular momenta of the fluid, their time derivatives, and the principal vector and the principal moment of the mass forces acting on the fluid.

We note also that Eqs. (1–48) can be considered a different form of the equations of motion of a fluid-containing rigid body. It is natural to use the mass center of the rigid body as the origin O of the moving coordinate system. Then

$$Q_1 = M_1 v_0 , \quad G_1 = \sum_{\nu}^{(1)} r_\nu \times m_\nu (\omega \times r_\nu) = \Theta^{(1)} \cdot \omega$$

and Eqs. (1–48) become

$$M_1 \left(\frac{dv_0}{dt} + \omega \times v_0 \right) = K_1 + P ,$$
$$\Theta^{(1)} \cdot \frac{d\omega}{dt} + \omega \times \Theta^{(1)} \cdot \omega = L_1 + N . \tag{1–50}$$

These equations should be considered simultaneously with Eqs. (1–28) and (1–29), which describe the motion of the fluid under the assumed boundary conditions, making use in the process of Eqs. (1–49) for the vectors P and N.

It is similarly possible to obtain from Eqs. (1–45) the equations of motion of a rigid body in the form of Lagrange's equations:

$$\frac{d}{dt} \frac{\partial T_1}{\partial \dot{q}_j} - \frac{\partial T_1}{\partial q_j} = \Phi_j^{(1)} + P_j , \tag{1–51}$$

where $\Phi_j^{(1)}$ and P_j ($j = 1, \ldots, n$) are the generalized forces originating in the effective forces applied to points of the rigid body and in the forces due to the pressures of the fluid and air on the wall of the cavity, where

$$P_j = - \frac{d}{dt} \frac{\partial T_2}{\partial \dot{q}_j} + \frac{\partial T_2}{\partial q_j} + \varrho \int_\tau F \cdot \frac{\partial r'}{\partial q_j} d\tau . \tag{1–52}$$

In fact, by definition,

$$P_j = \int_\sigma p n \cdot \frac{\partial r'}{\partial q_j} d\sigma .$$

Applying the Gauss-Ostrogradskiy formula, and making use of the fact that

$$\operatorname{div}\left(p\frac{\partial \boldsymbol{r}'}{\partial q_j}\right) = p\operatorname{div}\frac{\partial \boldsymbol{r}'}{\partial q_j} + \frac{\partial \boldsymbol{r}'}{\partial q_j}\cdot \operatorname{grad} p, \quad \text{where} \quad \operatorname{div}\frac{\partial \boldsymbol{r}'}{\partial q_j} = 0,$$

we obtain

$$P_j = \int_\tau \frac{\partial \boldsymbol{r}'}{\partial q_j}\cdot \operatorname{grad} p\,d\tau,$$

since for air $p = p_0$. Replacing grad p by its expression from Euler's equations, we have

$$P_j = \varrho\int_\tau \frac{\partial \boldsymbol{r}'}{\partial q_j}\cdot\left(\boldsymbol{F} - \frac{d\boldsymbol{v}}{dt}\right)d\tau.$$

We now consider the expression

$$\frac{\partial \boldsymbol{r}'}{\partial q_j}\cdot\frac{d\boldsymbol{v}}{dt} = \frac{d}{dt}\left(\boldsymbol{v}\cdot\frac{\partial \boldsymbol{r}'}{\partial q_j}\right) - \boldsymbol{v}\cdot\frac{d}{dt}\left(\frac{\partial \boldsymbol{r}'}{\partial q_j}\right).$$

It is easy to check that

$$\frac{\partial \boldsymbol{v}}{\partial \dot{q}_j} = \frac{\partial \boldsymbol{r}'}{\partial q_j}, \quad \frac{d}{dt}\left(\frac{\partial \boldsymbol{r}'}{\partial q_j}\right) = \frac{\partial \boldsymbol{v}}{\partial q_j}.$$

Therefore

$$\frac{\partial \boldsymbol{r}'}{\partial q_j}\cdot\frac{d\boldsymbol{v}}{dt} = \frac{d}{dt}\left(\boldsymbol{v}\cdot\frac{\partial \boldsymbol{v}}{\partial \dot{q}_j}\right) - \boldsymbol{v}\cdot\frac{\partial \boldsymbol{v}}{\partial q_j} = \frac{d}{dt}\frac{\partial \mathcal{T}}{\dot{q}\partial_j} - \frac{\partial \mathcal{T}}{\partial q_j}.$$

Substituting this expression, we obtain Eq. (1–52), which proves the validity of our assertion.

[1-7] EQUATIONS OF MOTION OF THE SYSTEM RELATIVE TO THE MASS CENTER

In the preceding sections we have considered the absolute motion of the system, *i.e.*, relative to a stationary coordinate system. However, it is often necessary to consider the motion of an unconstrained rigid body containing a fluid in a cavity relative to an orthogonal system of coordinates $C_1''e_2''e_3''e$, with its origin in the center of mass C of the system and its axes parallel to those of the stationary coordinate system (Fig. 6).

The velocity vector $\boldsymbol{v}_c$ of the mass center C of the system is determined by integrating Eq. (1–26), which, using Eq. (1–5), can be rewritten as

$$M\left(\frac{d\boldsymbol{v}_c}{dt} + \boldsymbol{\omega}\times\boldsymbol{v}_c\right) = K. \tag{1–53}$$

The radius vector of any point of the system relative to the mass center is denoted by $\boldsymbol{\varrho}_\nu$. The static moment of the system about the mass center is

$$\sum_\nu m_\nu \boldsymbol{\varrho}_\nu = 0 .$$

Differentiating this expression with respect to time, and denoting the velocity vector of a point about the mass center by $\boldsymbol{v}'_\nu = (d\boldsymbol{\varrho}_\nu/dt)'$, we find that

$$\boldsymbol{Q}' = \sum_\nu m_\nu \boldsymbol{v}'_\nu = 0 .$$

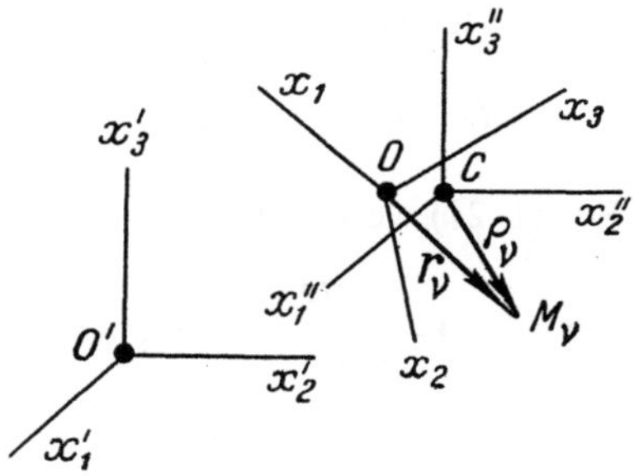

Fig. 6.

That is, the relative momentum of the system is equal to zero. As a result the relative angular momentum of the system will be the same for all points of the space[9].

The relative velocity of any point can be represented as

$$\boldsymbol{v}' = \boldsymbol{v}'_0 + \boldsymbol{\omega} \times \boldsymbol{r} + \boldsymbol{u} , \qquad (1\text{–}54)$$

where v'_0 denotes the relative velocity of the point O of the rigid body. In fact, using Eq. (1–4) and remembering that

$$\boldsymbol{v}_c = \boldsymbol{v}_0 + \boldsymbol{\omega} \times \boldsymbol{r}_c + \boldsymbol{u}_c ,$$

where $\boldsymbol{u}_c$ denotes the velocity of the mass center of the system relative to the rigid body, we find that

$$\boldsymbol{v}' = \boldsymbol{v} - \boldsymbol{v}_c = \boldsymbol{\omega} \times (\boldsymbol{r} - \boldsymbol{r}_c) + \boldsymbol{u} - \boldsymbol{u}_c .$$

But

$$\boldsymbol{r}_c = - \boldsymbol{\varrho}_0 , \quad \boldsymbol{v}'_0 = - \left(\frac{d\boldsymbol{r}_c}{dt}\right)' = \boldsymbol{r}_c \times \boldsymbol{\omega} - \boldsymbol{u}_c ,$$

from which we obtain Eq. (1–54).

The relative linear and angular momentum vectors are

$$\begin{aligned} \boldsymbol{Q}' &= M(\boldsymbol{v}_0' + \boldsymbol{\omega} \times \boldsymbol{r}_c) + \varrho \int_\tau \boldsymbol{u}\, d\tau = 0\,, \\ \boldsymbol{G}' &= M\boldsymbol{r}_c \times \boldsymbol{v}_0' + \boldsymbol{\Theta} \cdot \boldsymbol{\omega} + \varrho \int_\tau \boldsymbol{r} \times \boldsymbol{u}\, d\tau\,. \end{aligned} \tag{1–55}$$

Consequently the vectors $\boldsymbol{v}'$, $\boldsymbol{\omega}$, and $\boldsymbol{u}$ are not independent but are related by the first of these expressions.

The equations of relative motion of the system are obtained from Eqs. (1–26)–(1–28). The vector $\boldsymbol{G}$ is given by

$$\boldsymbol{G} = \boldsymbol{G}' + \boldsymbol{r}_c \times \boldsymbol{Q}\,. \tag{1–56}$$

Substituting this into Eq. (1–27) we obtain

$$\frac{d\boldsymbol{G}'}{dt} + \boldsymbol{\omega} \times \boldsymbol{G}' + \boldsymbol{u}_c \times \boldsymbol{Q} + \boldsymbol{r}_c \times \frac{d\boldsymbol{Q}}{dt} + \boldsymbol{\omega} \times (\boldsymbol{r}_c \times \boldsymbol{Q}) + \boldsymbol{v}_0 \times \boldsymbol{Q} = \boldsymbol{L}\,.$$

But

$$\boldsymbol{u}_c \times \boldsymbol{Q} + \boldsymbol{r}_c \times \frac{d\boldsymbol{Q}}{dt} + \boldsymbol{\omega} \times (\boldsymbol{r}_c \times \boldsymbol{Q}) + \boldsymbol{v}_0 \times \boldsymbol{Q} = \boldsymbol{r}_c \times \left(\frac{d\boldsymbol{Q}}{dt} + \boldsymbol{\omega} \times \boldsymbol{Q}\right),$$

$$\boldsymbol{r}_\nu = \boldsymbol{r}_c + \boldsymbol{\varrho}_\nu\,, \quad \boldsymbol{L} = \boldsymbol{r}_c \times \boldsymbol{K} + \boldsymbol{L}'\,, \quad \boldsymbol{L}' = \sum_\nu \boldsymbol{\varrho}_\nu \times \boldsymbol{F}_\nu\,,$$

so that our equations, making use of Eq. (1–26), take the form

$$\frac{d\boldsymbol{G}'}{dt} + \boldsymbol{\omega} \times \boldsymbol{G}' = \boldsymbol{L}'\,. \tag{1–57}$$

This equation expresses the theorem of the relative angular momentum of the system.

This equation should be supplemented by Eqs. (1–28) and (1–29) together with the applicable boundary conditions. Since $\boldsymbol{v} = \boldsymbol{v}_c + \boldsymbol{v}'$, Euler's equations for the moving coordinate system can be rewritten

$$\frac{d\boldsymbol{v}'}{dt} + \boldsymbol{\omega} \times \boldsymbol{v}' = \boldsymbol{F} - \frac{1}{\varrho}\operatorname{grad} p'\,, \quad \operatorname{div} \boldsymbol{v}' = 0\,, \tag{1–58}$$

where for brevity we have used the notation

$$p' = p + \varrho\left(\frac{d\boldsymbol{v}_c}{dt} + \boldsymbol{\omega} \times \boldsymbol{v}_c\right) \cdot \boldsymbol{r}\,.$$

Thus the study of the relative motion of a rigid body containing a fluid in a cavity reduces to the study of the simultaneous system of equations (1–57) and (1–58) with boundary conditions (1–30) – (1–34). Here Eqs. (1–55) should also be used.

[1-8] INTEGRALS OF THE EQUATIONS OF MOTION

The equations of motion of a rigid body containing a fluid in a cavity allow first integrals under certain conditions. They can be derived either from the equations of motion or from general theorems of dynamics, which are of course applicable to the mechanical system under study.

1. The Energy Integral

If the restraints imposed on the system are not explicitly time-dependent, while the specified effective forces $\boldsymbol{F}_\nu$ acting on points of the system permit the force function $U(\boldsymbol{r}'_\nu)$, in other words, $\boldsymbol{F}_\nu = \operatorname{grad}_{r'\nu} U$, then there exists the energy integral [9]

$$T - U = \text{const}. \tag{1–59}$$

Along with the force function $U(\boldsymbol{r})'_\nu$ we shall consider the potential energy of the system

$$\Pi(\boldsymbol{r}'_\nu) = - U(\boldsymbol{r}'_\nu) + \text{const}.$$

Then the energy integral (1–59) can be rewritten in the form

$$T + \Pi = \text{const}.$$

That is, the total mechanical energy of the system will be found to be constant during the entire motion.

The energy integral can easily be derived directly from the equations of motion in one form or another.

Let us first consider an unconstrained body. We take the scalar product of Eq. (1–26) and the vector $\boldsymbol{v}_0$, Eq. (1–27) and the vector $\boldsymbol{\omega}$, and Eq. (1–28) and the vector $\varrho \boldsymbol{u} d\tau$, integrate over the volume τ, and add up the equations obtained term by term. As a result we find

$$\frac{d}{dt}(\boldsymbol{Q}\cdot\boldsymbol{v}_0 + \boldsymbol{G}\cdot\boldsymbol{\omega}) - \boldsymbol{Q}\cdot\frac{d\boldsymbol{v}_0}{dt} - \boldsymbol{G}\cdot\frac{d\boldsymbol{\omega}}{dt} +$$
$$+ \varrho\int_\tau \boldsymbol{u}\cdot\left(\frac{d\boldsymbol{v}}{dt} + \boldsymbol{\omega}\times\boldsymbol{v}\right)d\tau = \boldsymbol{K}\cdot\boldsymbol{v}_0 + \boldsymbol{L}\cdot\boldsymbol{\omega} + \tag{1–60}$$
$$+ \varrho\int_\tau\left(\boldsymbol{F} - \frac{1}{\varrho}\operatorname{grad} p\right)\cdot\boldsymbol{u}\,d\tau.$$

It is easy to check that the expressions

$$Q\cdot v_0 + G\cdot \omega = 2T - \varrho \int_\tau u\cdot v d\tau ,$$
$$Q\cdot \frac{dv_0}{dt} + G\cdot \frac{d\omega}{dt} = \frac{dT}{dt} - \varrho \int_\tau v\cdot \left(\frac{du}{dt} + \omega \times u\right) d\tau \tag{1-61}$$

are applicable here. Further, using the second of Eqs. (1–38) and applying the Gauss-Ostrogradskiy formula, we find, using the boundary conditions (1–30) and (1–32), that

$$\int_\tau u\cdot \operatorname{grad} p d\tau = \iint_S \left[p_0 + \alpha\left(\frac{1}{R_1} + \frac{1}{R_2}\right)\right] u_n dS .$$

On the basis of formulas that can be obtained from Eqs. (1–22) and (1–23) by replacing δ by d/dt, and of Eq. (1–32), we have

$$\alpha \int_S \left(\frac{1}{R_1} + \frac{1}{R_2}\right) u_n dS = \frac{d\Pi_2^*}{dt} ,$$

where

$$\Pi_2^* = \alpha S + \alpha_1 \sigma_1 + \alpha_2 \sigma_2$$

denotes the potential energy of the surface tension forces. In addition, because of the conservation of the fluid volume τ, it is obvious that $\int_S u_n dS = 0$. Hence

$$\int_\tau u\cdot \operatorname{grad} p d\tau = \frac{d\Pi_2^*}{dt} . \tag{1-62}$$

Using the above expression and Eq. (1–61), we can write Eq. (1–60) in the form

$$\frac{dT}{dt} = K\cdot v_0 + L\cdot \omega + \varrho \int_\tau F\cdot u d\tau - \frac{d\Pi_2^*}{dt} . \tag{1-63}$$

The right-hand side of this equation is the power $\sum_\nu F_\nu \cdot v_\nu$ of all of the effective forces applied to the system. Thus Eq. (1–63) expresses the theorem of the system's kinetic energy.

Let us now assume that, in addition to the surface tension forces, the other effective forces applied to the system are also potential forces. In other words, there exists a scalar power function $U(r'_\nu)$ such that the sum of the elementary work of all the effective forces corresponding to actual displacements of the system is

$$\sum_\nu F_\nu \cdot dr'_\nu = dU .$$

Here, as can easily be checked by direct calculation,

$$\boldsymbol{K}\cdot\boldsymbol{v}_0 + \boldsymbol{L}\cdot\boldsymbol{\omega} + \varrho\int_\tau \boldsymbol{F}\cdot\boldsymbol{u}\,d\tau - \frac{d\Pi_2^*}{dt} = \frac{dU}{dt}$$

and from Eq. (1–63) we immediately obtain the energy integral (1–59).

In the case under study the force function $U(\boldsymbol{r}_\nu')$ can in general be represented as

$$U = U_1 + \varrho\int_\tau U_2 d\tau + U_2^*,$$

where U_1 denotes the force function of the effective forces applied to the rigid body, U_2 the force function of the mass forces acting on the fluid, and U_2^* the force function of the surface tension forces. Similarly, the potential energy of the system has the form

$$\Pi = \Pi_1 + \Pi_2 + \Pi_2^*$$

where the notation is obvious.

The theorem of kinetic energy, as is known, is also valid for the relative motion of the system about the mass center. In fact, according to König's theorem

$$T = \tfrac{1}{2}Mv_c^2 + T', \tag{1–64}$$

where

$$T' = \tfrac{1}{2}\sum_\nu m_\nu v_\nu'^2 = \tfrac{1}{2}\boldsymbol{\omega}\cdot\boldsymbol{\Theta}\cdot\boldsymbol{\omega} + \boldsymbol{\omega}\cdot\varrho\int_\tau \boldsymbol{r}\times\boldsymbol{u}\,d\tau - \tfrac{1}{2}M\boldsymbol{v}_0'^2 + \tfrac{1}{2}\varrho\int_\tau \boldsymbol{u}^2 d\tau$$

denotes the kinetic energy of the relative motion of the system. Substituting Eq. (1–64) into (1–63) and making use in the process of Eqs. (1–53) and (1–55), we obtain

$$\frac{dT'}{dt} = -\boldsymbol{K}\cdot\boldsymbol{u}_c + \boldsymbol{L}'\cdot\boldsymbol{\omega} + \varrho\int_\tau \boldsymbol{F}\cdot\boldsymbol{u}\,d\tau - \frac{d\Pi_2^*}{dt}. \tag{1–65}$$

It is easy to see that the right-hand side of this equation represents the sum of the elementary work of the effective forces applied to the system corresponding to the actual displacements of the system $\sum_\nu \boldsymbol{F}_\nu\cdot d\boldsymbol{\varrho}_\nu$, divided by dt. Thus Eq. (1–65) expresses the theorem of the kinetic energy of the system moving relative to the mass center.

It is clear that this equation can also be obtained directly from the equations of relative motion. In fact, let us take the scalar product of Eq. (1–57) and the vector $\boldsymbol{\omega}$:

$$\frac{d}{dt}(\boldsymbol{G}'\cdot\boldsymbol{\omega}) - \boldsymbol{G}'\cdot\frac{d\boldsymbol{\omega}}{dt} = \boldsymbol{L}'\cdot\boldsymbol{\omega}.$$

It can easily be shown that, by virtue of Eqs. (1–55), Eqs. (1–61) will assume the form

$$G' \cdot \omega + \varrho \int_\tau \boldsymbol{u} \cdot \boldsymbol{v}' d\tau = 2T',$$

$$\frac{dT'}{dt} = G' \cdot \frac{d\omega}{dt} + \varrho \int_\tau \boldsymbol{v}' \cdot \left(\frac{d\boldsymbol{u}}{dt} + \omega \times \boldsymbol{u} \right) d\tau .$$

Then, making use of Eq. (1–58), we again obtain Eq. (1–65).

If it is found that there is a scalar function $U'(\varrho_\nu)$ such that

$$\sum_\nu \boldsymbol{F}_\nu \cdot d\varrho_\nu = dU',$$

we obtain from Eq. (1–65) the energy integral for the relative motion of the system

$$T' - U' = \text{const}. \tag{1–66}$$

We note that if the points of the system are acted upon by forces that are proportional to the masses of the points and that reduce to a single resultant passing through the mass center of the system (for example, gravity forces), the sum of the elementary work of these forces corresponding to actual displacements of the system is zero, since $\sum_\nu m_\nu d\varrho_\nu = 0$, and they will not form a part of the force function U'.

Let us now turn to the equations of motion of a constrained rigid body containing a fluid. If these equations are used in a form similar to Eqs. (1–26) and (1–27), with their right-hand sides including the reactions of ideal constraints, the preceding derivation of the energy integral will again be valid since the sum of the elementary work of the reactions of the ideal steady-state constraints corresponding to the actual displacements of the system is zero. If, however, the equations of motion are used in the form of Lagrange's equations (1–45) and (1–46), we multiply (1–45) by $\dot{q}_j$ and add term by term, multiply (1–46) by $\varrho u_i d\tau$, add, integrate over the volume τ, and then add the two equations obtained, we obtain, making use of Eq. (1–62),

$$\sum_{j=1}^{n} \left\{ \frac{d}{dt} \left(\frac{\partial T}{\partial \dot{q}_j} \dot{q}_j \right) - \frac{\partial T}{\partial q_j} \dot{q}_j - \frac{\partial T}{\partial \dot{q}_j} \ddot{q}_j \right\} + \varrho \int_\tau \sum_{i=1}^{3} \left\{ \frac{d}{dt} \left(\frac{\partial \mathcal{T}}{\partial u_i} u_i \right) - \right.$$

$$\left. - \frac{\partial \mathcal{T}}{\partial u_i} \dot{u}_i - \frac{\partial \mathcal{T}}{\partial x_i} u_i \right\} d\tau = \sum_{j=1}^{n} \Phi_j \dot{q}_j + \varrho \int_\tau \boldsymbol{F} \cdot \boldsymbol{u} d\tau - \frac{d\Pi_2^*}{dt} .$$

It is easy to check that, in the case of steady-state constraints,

$$\frac{dT}{dt} = \sum_{j=1}^{n} \left(\frac{\partial T}{\partial q_j} \dot{q}_j + \frac{\partial T}{\partial \dot{q}_j} \ddot{q}_j \right) + \sum_{i=1}^{3} \varrho \int_{\tau} \left(\frac{\partial \mathcal{T}}{\partial u_i} \dot{u}_i + \frac{\partial \mathcal{T}}{\partial x_i} u_i \right) d\tau .$$

Here

$$\sum_{j=1}^{n} \frac{\partial T}{\partial \dot{q}_j} \dot{q}_j + \sum_{i=1}^{3} \varrho \int_{\tau} \frac{\partial \mathcal{T}}{\partial u_i} u_i d\tau = 2T ,$$

as a result of which our equation will assume the form

$$\frac{dT}{dt} = \sum_{j=1}^{n} \Phi_j \dot{q}_j + \varrho \int_{\tau} \boldsymbol{F} \cdot \boldsymbol{u} d\tau - \frac{d\Pi_2^*}{dt} . \tag{1–67}$$

If the effective forces permit the force function U, then

$$\sum_{j=1}^{n} \Phi_j \dot{q}_j + \varrho \int_{\tau} \boldsymbol{F} \cdot \boldsymbol{u} d\tau - \frac{d\Pi_2^*}{dt} = \frac{dU}{dt}$$

and from Eqs. (1–67) we again obtain the energy integral (1–59).

2. Linear and Angular Momentum Integrals

It follows from the theory of linear momentum that, if the constraints on the system permit translation of the rigid body along some stationary straight line, and if the sum of the projections of all the effective forces applied to the system on this straight line is zero, then the projection of the linear momentum of the system on this straight line remains constant.

In fact, let us use this straight line as our axis x_1' and let us take the scalar product of Eq. (1–26) and the unit vector $\boldsymbol{i}_1'$. We will have

$$\frac{d}{dt} (\boldsymbol{Q} \cdot \boldsymbol{i}_1') - \boldsymbol{Q} \cdot \left(\frac{d\boldsymbol{i}_1'}{dt} + \boldsymbol{\omega} \times \boldsymbol{i}_1' \right) = 0 ,$$

whence, making use of Eq. (1–5), we obtain the first integral

$$\boldsymbol{Q} \cdot \boldsymbol{i}_1' = \text{const} . \tag{1–68}$$

If the principal vector of the effective forces applied to the system is $\boldsymbol{K}=0$, then integrals such as (1–68) for a constrained rigid body containing a fluid apply for any stationary straight line; in other words, there exists a vectorial momentum integral

$$\boldsymbol{Q} = M \boldsymbol{v}_c = \textbf{const} . \tag{1–69}$$

It follows from the theory of angular momentum that, if the constraints on the system allow rotation of the system as a single rigid body about some stationary axis, and if the effective forces do not produce a moment about this axis, then the projection of the angular momentum of the system on this axis remains constant.

Let us consider an unconstrained rigid body containing a fluid, and let us assume that the effective forces do not produce a moment about axis x_3'. We take the vector product of Eq. (1–26) and the vector $\boldsymbol{r}_3'$ and add it to Eq. (1–27) to obtain an equation, expressing the theorem of angular momentum for a stationary point in the form

$$\frac{d\boldsymbol{G}_{o'}}{dt} + \boldsymbol{\omega} \times \boldsymbol{G}_{o'} = \boldsymbol{L} + \boldsymbol{r}_0' \times \boldsymbol{K},$$

where

$$\boldsymbol{G}_{o'} = \boldsymbol{G} + \boldsymbol{r}_0' \times \boldsymbol{Q}$$

is the angular momentum for the point O'. We now take the scalar product of this equation and the vector $\boldsymbol{i}_3'$, so that

$$\frac{d}{dt}(\boldsymbol{G}_{o'} \cdot \boldsymbol{i}_3') - \boldsymbol{G}_{o'} \cdot \left(\frac{d\boldsymbol{i}_3'}{dt} + \boldsymbol{\omega} \times \boldsymbol{i}_3'\right) = 0.$$

Since the vector $\boldsymbol{i}_3'$ satisfies Eq. (1–5), we immediately obtain the area integral

$$\boldsymbol{G}_{o'} \cdot \boldsymbol{i}_3' = \text{const}. \tag{1–70}$$

If the moment of the effective forces relative to the stationary point O' is zero, then we have the vectorial integral

$$\boldsymbol{G}_{o'} = \textbf{const} \tag{1–71}$$

or three scalar integrals such as (1–70).

Under certain circumstances the equation of relative motion of a system also yields the area integral. Assume that the moment about the Cx_3'' axis of the effective forces applied to the system is zero. Forming the scalar product of Eq. (1–57) and the unit vector of this axis and making use, in the process, of Eq. (1–5), we obtain the area integral in relative motion as

$$G'_{x''_3} = \text{const}. \tag{1–72}$$

If the moment of the effective forces about the mass center is $L' = 0$, then there exists a vectorial integral of the relative angular momentum

$$\boldsymbol{G}' = \textbf{const}. \tag{1–73}$$

In addition, if the principal vector of the effective forces is $\boldsymbol{K}=0$, we have the first integral

$$\boldsymbol{Q}\cdot\boldsymbol{G}' = \text{const}. \tag{1–74}$$

To satisfy ourselves that this is so, let us take the scalar product of Eq. (1–57) and the vector $\boldsymbol{Q}$. Then, using Eq. (1–26) we obtain for our conditions,

$$\frac{d}{dt}(\boldsymbol{Q}\cdot\boldsymbol{G}') = 0,$$

from which follows the existence of the integral (1–74).

If the central ellipsoid of inertia of the rigid body is an ellipsoid of revolution about the x_3 axis, the effective forces acting on the body have a moment relative to the dynamic axis of symmetry x_3 equal to zero, and the cavity has the shape of a body of revolution about the axis x_3, then the projection of the instantaneous angular velocity of the body on this axis is constant:

$$\omega_3 = \text{const}. \tag{1–75}$$

In fact, taking the scalar product of the second of equations (1–50) and the vector $\boldsymbol{i}_3$, we obtain, under the assumptions made:

$$C_1 \frac{d\omega_3}{dt} = \int_\sigma (\boldsymbol{r} \times p\boldsymbol{n})\cdot\boldsymbol{i}_3 d\sigma.$$

But $(\boldsymbol{r}\times\boldsymbol{n})\cdot\boldsymbol{i}_3 = x_1 n_2 - x_2 n_1 = 0$ for any point on the surface of revolution, and hence from this equation we immediately obtain the integral (1–75).

Integrals such as (1–68), (1–70), (1–72), and (1–75) can also be obtained from Eqs. (1–45) for cyclical coordinates q_α, for which, by definition,

$$\frac{\partial T}{\partial q_\alpha} = \Phi_\alpha = 0 \quad (\alpha = k+1, \ldots, n).$$

When these conditions are satisfied, Eqs. (1–45) yield the cyclical first integrals

$$\frac{\partial T}{\partial \dot{q}_\alpha} = \text{const} \quad (\alpha = k+1, \ldots, n), \tag{1–76}$$

which are analogous to the above.

We also recall that the direction cosines γ_{rs} satisfy the equations

$$\boldsymbol{i}_s'^2 = \sum_{r=1}^{3} \gamma_{sr}^2 = 1 \quad (s = 1, 2, 3), \tag{1–77}$$

which are the first integrals of Eqs. (1–5). In fact, forming the scalar product of (1–5) and the vector $\boldsymbol{i}_s'$, we immediately obtain Eq. (1–77).

3. Integrals of the Hydrodynamic Equations

Under some conditions the hydrodynamic equations (1–28) and (1–29) also permit integrals.

From here to the end of this section, we shall assume that the mass forces acting on the ideal incompressible fluid in the cavity of the body have a force function $U_2(\boldsymbol{r}')$. Under these conditions, as was pointed out in Sec. 4, it is possible to obtain from Eqs. (1–28) and (1–29) the Helmholtz equation (1–39) for $\boldsymbol{\Omega} = \operatorname{curl} \boldsymbol{v}$. The vector lines of the field $\boldsymbol{\Omega} = \boldsymbol{\Omega}(\boldsymbol{r})$ are called *vortex lines*. Taking a small closed contour in the fluid and passing a vortex line through each point in it, we obtain a vortex tube. The value of the velocity circulation over any closed contour embracing the given vortex tube is called *intensity* of the vortex tube.

Under the above assumptions the following theorems are valid.

Kelvin's Theorem

The velocity circulation over any closed circuit during the entire time of the motion of the fluid remains unchanged.

Lagrange's Theorem

If no vortices were present at some initial instant of time in some part of the fluid, then they were not present before and will not be present later in this part of the fluid.

Helmholtz's Theorem

Vortex lines and the intensity of vortex tubes possess the conservation property.

We note that Kelvin's theorem can be proven by using the formula

$$\frac{d}{dt}\int_s \boldsymbol{v}\cdot d\boldsymbol{s} = \int_s \frac{d\boldsymbol{v}}{dt}\cdot d\boldsymbol{s}$$

and Eq. (1–28). On the basis of this theorem and of Stokes' equation (1–12) it is also possible to prove Lagrange's and Helmholtz's theorems. For the purpose mentioned, Eq. (1–39) is the necessary and sufficient condition for the vortex lines and intensity of vortex tubes to be conservative.

Let us first consider irrotational motion of the fluid. Let there be no vortex at any point of the velocity field, *i.e.*,

$$\boldsymbol{\Omega} = \operatorname{curl} \boldsymbol{v} = 0\,.$$

As was noted in Sec. 2, the above expression is the necessary and sufficient condition for the velocity vector to be equal to the gradient of some function:

$$\boldsymbol{v} = \operatorname{grad} \Phi , \tag{1–78}$$

where the scalar function $\Phi(\boldsymbol{r}, t)$ is called the *velocity potential*. By virtue of Eq. (1–29), the function Φ should satisfy Laplace's equation

$$\operatorname{div} \boldsymbol{v} = \Delta \Phi = 0 .$$

Equation (1–28) then becomes

$$\operatorname{grad}\left[\frac{\partial \Phi}{\partial t} + \frac{1}{2} v^2 - \boldsymbol{v}\cdot(\boldsymbol{v}_0 + \boldsymbol{\omega} \times \boldsymbol{r}) - U_2 + \frac{p}{\varrho}\right] = 0 ,$$

whence follows Cauchy's integral

$$\frac{\partial \Phi}{\partial t} + \frac{1}{2} v^2 - \boldsymbol{v}\cdot(\boldsymbol{v}_0 + \boldsymbol{\omega} \times \boldsymbol{r}) - U_2 + \frac{p}{\varrho} = F(t) , \tag{1–79}$$

where $F(t)$ is an arbitrary function of time.

Consequently the problem of determining the irrotational motion of an incompressible fluid reduces to finding one function Φ satisfying Laplace's equation, the boundary conditions following from (1–30)–(1–34), and the initial conditions. After the function Φ is found the hydrodynamic pressure p will be determined from Eq. (1–79), except for the arbitrary function $F(t)$, for which no data has been presented in this problem. This function can be determined when the time dependence of the pressure at one point of the field is specified[6].

Let us now consider the vortex motion of the fluid, i.e., $\boldsymbol{\Omega} \neq 0$. Equation (1–28) can then be rewritten as[6]

$$\operatorname{grad}\left[\frac{1}{2} v^2 - \boldsymbol{v}\cdot(\boldsymbol{v}_0 + \boldsymbol{\omega} \times \boldsymbol{r}) - U_2 + \frac{p}{\varrho}\right] = - \frac{\partial \boldsymbol{v}}{\partial t} + \boldsymbol{u} \times \Omega .$$

Integrating this equation along some line s, we obtain[15]

$$\frac{p}{\varrho} - U_2 + \tfrac{1}{2} v^2 - \boldsymbol{v}\cdot(\boldsymbol{v}_0 + \boldsymbol{\omega} \times \boldsymbol{r}) = F(t) - \int_s \left(\frac{\partial \boldsymbol{v}}{\partial t} + \boldsymbol{u} \times \boldsymbol{\Omega}\right)\cdot d\boldsymbol{s} . \tag{1–80}$$

The integrand is a total differential, as is easily proved since the integrability condition reduces to Eq. (1–39).

When the motion of the fluid is steady, $\partial \boldsymbol{v}/\partial t = 0$; taking the streamline in relative motion of the fluid, for which $d\boldsymbol{s}/\boldsymbol{u} = \text{const}$, as the line s, we obtain Bernoulli's integral from (1–80) as

$$\frac{p}{\varrho} + \tfrac{1}{2} v^2 - \boldsymbol{v}\cdot(\boldsymbol{v}_0 + \boldsymbol{\omega} \times \boldsymbol{r}) - U_2 = \Gamma , \tag{1–81}$$

where Γ is a quantity that is constant at the given streamline.

[1-9] VISCOUS FLUIDS

In the preceding sections we have assumed that the fluid filling the cavity of the body is ideal. This means that the internal friction forces are disregarded, and the forces applied to surface elements are therefore normal pressure forces. However, all real fluids possess to some degree the property of internal friction, *i.e.*, viscosity.

For a viscous fluid, in general, the surface forces have tangential components in addition to normal ones. It can be shown[6] that for any isolated area in the fluid with an external normal $\boldsymbol{n}$, the surface force stress is

$$\boldsymbol{p}_n = \boldsymbol{p}_1 n_1 + \boldsymbol{p}_2 n_2 + \boldsymbol{p}_3 n_3 ,$$

where $\boldsymbol{p}_i$ denotes the stresses of the surface forces for areas with external normals which are collinear with the unit vectors $\boldsymbol{i}_i$, respectively.

The three vectors $\boldsymbol{p}_i$ $(i=1, 2, 3)$ or, equivalently, the nine quantities $p_{ij}=\boldsymbol{p}_i \cdot \boldsymbol{i}_j$ $(i,j=1, 2, 3)$, define the stress tensor $\boldsymbol{P}$; here we can show that $p_{ij}=p_{ji}$. It is assumed in the hydrodynamics of a Newtonian incompressible fluid that the stress tensor has the components[6]

$$p_{ij} = - p\delta_{ij} + \mu\left(\frac{\partial v_i}{\partial x_j} + \frac{\partial v_j}{\partial x_i}\right) \quad (i, j = 1, 2, 3), \tag{1–82}$$

where the Kronecker delta is defined as

$$\delta_{ij} = \begin{cases} 1, & i = j , \\ 0, & i \neq j , \end{cases}$$

and μ is the coefficient of internal friction, or the viscosity coefficient, which in general is a function of the composition, pressure, and temperature of the fluid. In the subsequent discussion, μ will be assumed to be constant for a given fluid.

If we isolate an arbitrary volume in the fluid and equate to zero the principal vector and principal moment of all of the forces including inertia forces applied to particles of this volume, it is easy to obtain from the D'Alembert-Lagrange principle the basic equation of the mechanics of a continuous medium:

$$\left(\frac{d\boldsymbol{v}}{dt}\right)' = \boldsymbol{F} + \frac{1}{\varrho}\,\mathrm{div}\,\boldsymbol{P} .$$

For the tensor $\boldsymbol{P}$ with components given by (1–82)

$$\mathrm{div}\,\boldsymbol{P} = - \,\mathrm{grad}\, p + \mu \Delta \boldsymbol{v} .$$

The equations of motion of a viscous incompressible fluid referred to the moving axes x_1, x_2, x_3, have the form

$$\frac{d\boldsymbol{v}}{dt} + \boldsymbol{\omega} \times \boldsymbol{v} = \boldsymbol{F} - \frac{1}{\varrho}\operatorname{grad} p + \nu \Delta \boldsymbol{v}, \quad \operatorname{div} \boldsymbol{v} = 0. \tag{1–83}$$

The first of these equations is the Navier-Stokes equation, where $\nu = \mu/\varrho$ is the kinematic viscosity coefficient.

As can be seen, Eq. (1–83) differs from Eq. (1–28) by the additional term $\nu\Delta\boldsymbol{v}$ in the right-hand side, which can be treated as the viscous force per unit mass.

The boundary conditions for a viscous fluid also differ from those for an ideal fluid. At the free surface of the fluid the stress vector for areas tangent to the free surface should be directed along a normal to these areas toward the fluid and be numerically equal to the pressure obtained from Eq. (1–30), *i.e.*,[16]

$$\text{on} \quad S \quad \boldsymbol{p}_n = -\left[p_0 + \alpha\left(\frac{1}{R_1} + \frac{1}{R_2}\right)\right]\boldsymbol{n}. \tag{1–84}$$

It is also assumed that the viscous fluid adheres to the solid walls, so that the condition (1–32) is replaced, in the case of a viscous fluid, by the kinematic condition

$$\text{on} \quad \sigma_1 \quad \boldsymbol{u} = 0. \tag{1–85}$$

Condition (1–33), which states that the component of the fluid velocity normal to the free surface coincides with the velocity of the surface displacement, is of course retained for a viscous fluid.

The initial conditions for the motion of a body with a cavity filled with a viscous fluid do not differ from their counterparts for an ideal fluid. As was pointed out in Sec. 4, Eqs. (1–26) and (1–27) express the general theorems of mechanics concerning the linear and angular momenta of the system. They are obviously applicable when the fluid in the cavity is viscous. Thus the study of the motion reduces to the study of the system of simultaneous equations (1–26), (1–27), and (1–83) with boundary conditions (1–84), (1–85), (1–31), and (1–33). When the forces applied to the system are functions of some additional variables this system should be supplemented by appropriate equations for the parameters, for example, Eqs. (1–5).

Because the Navier-Stokes equation contains the term $\nu\Delta\boldsymbol{v}$ the motion of a viscous fluid is irreversible, since the forces $\nu\Delta\boldsymbol{\varrho}$ produce energy dissipation. The theorem on kinetic energy for a viscous fluid is expressed by an equation which differs from (1–63) by additional terms, that depend on the fluid's viscosity. To find them we take the scalar product of $1/\varrho \operatorname{div}\boldsymbol{P}$ and the vector $\varrho\boldsymbol{u}d\tau$, integrate with respect to τ, and apply the Gauss-Ostrogradskiy theorem.

We then have

$$\int_\tau \boldsymbol{u}\cdot\operatorname{div}\boldsymbol{P}d\tau = \int_S \boldsymbol{p}_n\cdot\boldsymbol{u}dS - \int_\tau \left(\boldsymbol{p}_1\cdot\frac{\partial \boldsymbol{u}}{\partial x_1} + \boldsymbol{p}_2\cdot\frac{\partial \boldsymbol{u}}{\partial x_2} + \boldsymbol{p}_3\cdot\frac{\partial \boldsymbol{u}}{\partial x_3}\right) d\tau .$$

But by virtue of the incompressibility of the fluid and of conditions (1–84) and (1–85)

$$\int_S \boldsymbol{p}_n\cdot\boldsymbol{u}dS = -\int_S \alpha\left(\frac{1}{R_1}+\frac{1}{R_2}\right) u_n dS = -\frac{d\Pi_2^*}{dt}.$$

Making use of (1–82), we can represent the expression

$$E = \boldsymbol{p}_1\cdot\frac{\partial \boldsymbol{u}}{\partial x_1} + \boldsymbol{p}_2\cdot\frac{\partial \boldsymbol{u}}{\partial x_2} + \boldsymbol{p}_3\cdot\frac{\partial \boldsymbol{u}}{\partial x_3}$$

in the form[6]

$$E = \mu\left\{2\sum_{i=1}^{3}\left(\frac{\partial v_i}{\partial x_i}\right)^2 + \left(\frac{\partial v_2}{\partial x_3}+\frac{\partial v_3}{\partial x_2}\right)^2 + \left(\frac{\partial v_3}{\partial x_1}+\frac{\partial v_1}{\partial x_3}\right)^2 + \left(\frac{\partial v_1}{\partial x_2}+\frac{\partial v_2}{\partial x_1}\right)^2\right\}, \tag{1.86}$$

so that

$$\int_\tau \boldsymbol{u}\cdot\operatorname{div}\boldsymbol{P}d\tau = -\frac{d\Pi_2^*}{dt} - \int_\tau E d\tau .$$

The kinetic energy theorem for the case of a viscous fluid thus takes the form

$$\frac{dT}{dt} = \boldsymbol{K}\cdot\boldsymbol{v}_0 + \boldsymbol{L}\cdot\boldsymbol{\omega} + \varrho\int_\tau \boldsymbol{F}\cdot\boldsymbol{u}d\tau - \frac{d\Pi_2^*}{dt} - \int_\tau E d\tau . \tag{1–87}$$

When all the effective forces applied to the body and to the fluid have a force function U, the above expression reduces to

$$\frac{d}{dt}(T-U) = -\int_\tau E d\tau \tag{1–88}$$

for the energy dissipation rate, since $E \geqslant 0$. Expression (1–88) for E shows that energy is not dissipated in the fluid only when all the components of the deformation rate tensor cancel out:

$$\frac{\partial v_i}{\partial x_i} = 0 \quad (i = 1, 2, 3), \quad \frac{\partial v_2}{\partial x_3} + \frac{\partial v_3}{\partial x_2} = 0 \quad (123), \tag{1–89}$$

which is possible only when the fluid moves as a single solid body.

Let us now clarify the conditions under which it is possible to have such a motion of the fluid subjected to mass forces with a force function U_2. The equation of relative equilibrium of the system will be obtained from Eq. (1–38) by setting $\boldsymbol{u}=0$:

$$\operatorname{grad}\left[U_2 - \frac{p}{\varrho} - \left(\frac{d\boldsymbol{v}_0}{dt}\right)' \cdot \boldsymbol{r} + \frac{1}{2}\omega^2 r^2 - \frac{1}{2}(\boldsymbol{\omega}\cdot\boldsymbol{r})^2\right] = \frac{d\boldsymbol{\omega}}{dt} \times \boldsymbol{r}.$$

This equation can be satisfied only when its right-hand side is a gradient of some function; for this it is necessary and sufficient that

$$\operatorname{curl}\frac{d\boldsymbol{\omega}}{dt} \times \boldsymbol{r} = 2\frac{d\boldsymbol{\omega}}{dt} = 0.$$

Thus, in order for the fluid to move as a single, unchanging system, it is necessary for the vector of the instantaneous angular velocity of the body to remain constant during the entire motion[15]

$$\boldsymbol{\omega} = \mathbf{const}. \tag{1–90}$$

Viscosity forces in a fluid are internal forces; hence their principal vector and principal moment about an arbitrary point are zero. For this reason the integrals of the linear and angular momenta of the system in the form (1–68) –(1–74) are the same for viscous liquids as those for ideal liquids.

2

Elementary Cases of Motion of a Rigid Body Containing a Fluid

The expression "elementary cases of motion of a rigid body with fluid-filled cavities" is here used to refer to instances in which the motion of the fluid in the cavity can be described completely by a finite number of variables. Obviously this is only possible when the fluid fills the entire cavity and thus has no free surface. If the motion of the fluid is potential or homogeneous turbulent motion (quasi-solid), then the motion of the system is described by ordinary differential equations together with Laplace's equation[15,17]. The study of the motion of the system in these cases is appreciably simplified.

In this chapter are presented some results obtained by N. Ye. Zhukovskiy in studying the motion of a rigid body with cavities completely filled with an ideal fluid. The problem of stability is stated, the basic theorem of the method of Lyapunov's functions is given, and the solutions of a number of problems of the stability of motion of a fluid-filled rigid body are presented.

[2-1] IRROTATIONAL MOTION OF A FLUID

We now use Lagrange's theorem for potential motion and assume that the homogeneous, incompressible ideal fluid which completely fills the cavity of the body, and which is subjected to mass potential forces with force function $U_2(x_1, x_2, x_3, t)$ performs vortex-free motion. The velocity potential of the absolute motion of the fluid, expressed in the $Ox_1x_2x_3$ coordinate system, is denoted by $\Phi(x_1, x_2, x_3, t)$; then the fluid velocity vector is

$$\boldsymbol{v} = \operatorname{grad} \Phi . \qquad (2\text{–}1)$$

By virtue of Eq. (1–29), the function Φ should satisfy Laplace's equation $\Delta\Phi = 0$ inside the cavity. The boundary conditions for solving this equation are obtained from condition (1–32): on the walls of the cavity σ the derivatives of the function Φ along the normal should equal the normal components of the velocities of the corresponding points on the walls

$$\frac{d\Phi}{dn} = v_n. \tag{2–2}$$

Thus the determination of the velocity potential reduces to solving Neumann's problem of finding a function satisfying Laplace's equation inside the region τ, on the basis of its normal derivatives specified on its boundary.

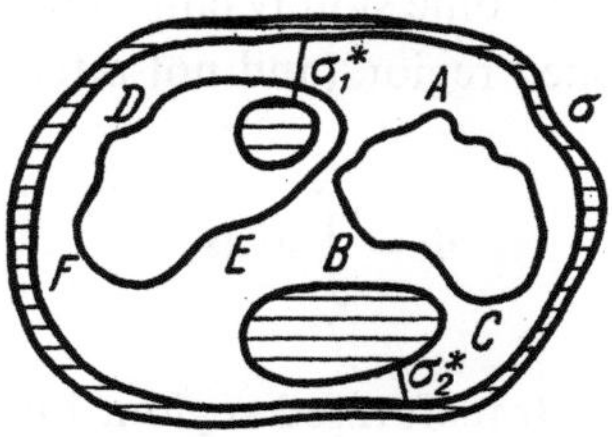

Fig. 7.

If τ is a simply connected region, the problem has a unique solution in the form of a single-valued harmonic function[18]; for a multiply connected region the solution will, in general, be a multivalued function.

We recall that a region is called simply connected if any two points of it can be connected by curves lying entirely within the region. Any two such curves will be mutually reconcilable if they can be made to coincide by continuous deformation without leaving the region. Any closed curve, ABCA, which can be shrunk to a point without leaving the region is called reducible. The region is called k-connected if it is possible to draw between two points within it k and only k mutually irreconcilable paths or, equivalently, if it is possible in the region, without disturbing its connectedness, to set up $k-1$, but not more, barriers. The irreducible curve DEFD is called simply irreducible if it intersects a barrier only once (Fig. 7).*

Let us denote the velocity circulations, taken over the closed, simply irreducible curves l_s $(s=1, \ldots, k-1)$ that can be drawn in the k-connected region, by $\kappa_s = \int_{ls} \mathbf{v} \cdot d\mathbf{s}$. According to Kelvin's theorem, $\kappa_s = \text{const}$ $(s=1, \ldots, k-1)$; we shall define the direction of the curve l_s over which κ_s is taken as the positive direction of the curve. From here on these curves will be called the *principal contours*, while κ_s will be called the *principal circulations* or *cyclic constants*. The value of the circulation for any arbitrary closed curve l that is reducible to any combination of principal contours will be represented as

$$\int_l \boldsymbol{v} \cdot d\boldsymbol{s} = \sum_{s=1}^{k-1} p_s \kappa_s , \tag{2–3}$$

* Corrections made conform to Lamb, Hydrodynamics, page 37, Dover, 1945?

where $p, \ldots, p_{k-1}$ are some integers (positive, negative, or zero). The circulation along a reducible curve is obviously zero. It follows from Eq. (2–3) that in a multiply connected region the velocity potential Φ will be a multivalued function, and when traversing some closed curve the value of Φ will in general differ from its initial value by an amount such as (2–3). Only in the case when all the $\kappa_s = 0$ will Φ be a single-valued function.

Let us set up in a multiply connected cavity the $k-1$ barriers mentioned above. The cavity will thus become simply connected. If we draw some curve l lying inside this transformed region (and not intersecting any of the above barriers), the function

$$\Phi = \int_l \boldsymbol{v} \cdot d\boldsymbol{s}$$

will be single-valued in the transformed region, but its value at two neighboring points on different sides of any sth barrier will differ by $\pm\kappa_s$.

Consequently[17], the irrotational motion of an incompressible fluid in a k-connected cavity of a body will be fully defined if we are given the normal velocity component for each point of the boundary σ of the cavity, as well as the value of the circulation (or the cyclic constant) for each of the $k-1$ independent irreducible closed curves that can be drawn through the cavity. The principal circulations are determined by specifying the irrotational motion of the fluid in the cavity of a body at rest.

Following Zhukovskiy[15], we now break up the motion of the rigid body into translation, with the velocity of any point O, and rotational motion about this point. The velocities of any translational motion have a single-valued potential function $\sum_{i=1}^{3} v_{0i} x_i$, since this motion is imparted to the fluid directly and has no effect on its internal motion. The internal motion is influenced only by the rotational motion of the body about point O, and will remain the same if this point is made to be stationary and the body is rotated about it in such a manner as it rotates about the moving point O.

The velocity potential of a fluid moving relative to a stationary coordinate system, together with a body performing rotational motion, is sought in the form

$$\emptyset = \varphi(x_1, x_2, x_3, t) + \chi(x_1, x_2, x_3), \qquad (2\text{–}4)$$

where inside the cavity the functions φ and χ satisfy Laplace's equation, and at the cavity walls, the conditions

$$\frac{d\varphi}{dn} = (\boldsymbol{\omega} \times \boldsymbol{r}) \cdot \boldsymbol{n}, \quad \frac{d\chi}{dn} = 0. \qquad (2\text{–}5)$$

In addition, the function χ must yield the specified principal circulations κ_s along the principle contours.

If the cavity is simply connected, then $\chi=0$; in the case of a multiply connected cavity

$$\chi = \sum_{s=1}^{k-1} \kappa_s \vartheta_s(x_1, x_2, x_3), \tag{2–6}$$

where $\vartheta_s(x_1, x_2, x_3)$ satisfy Laplace's equation inside the cavity, while their derivatives along the normal at the cavity's walls are zero. The function ϑ_s increases by unity on passing through the sth barrier in the direction of the sth principal contour and varies continuously when passing through all the remaining barriers. If the fluid is stationary when $t=0$, all of the $\kappa_s=0$. The motion of a fluid with potential χ will be called *cyclic motion.*

The function $\varphi(x_1, x_2, x_3)$ will be sought in the form of a linear function of the projections of the instantaneous angular velocity of the body

$$\varphi(x_1, x_2, x_3, t) = \sum_{i=1}^{3} \omega_i \psi_i(x_1, x_2, x_3), \tag{2–7}$$

where the single-valued harmonic functions ψ_i on the walls of cavity σ satisfy the conditions

$$\frac{d\psi_1}{dn} = x_2 n_3 - x_3 n_2 \quad (123). \tag{2–8}$$

It should be noted that representation of the function φ in the form of (2–7) follows naturally from the linearity of Neumann's problem; this was first proposed by Stokes[19]. The functions ψ_i are usually called the Stokes-Zhukovskiy potentials. These functions are determined solely by the geometry of the cavity; they are independent of the motion of the rigid body and can be determined once and for all for cavities of a given shape.

After the functions ψ_i and ϑ_s are determined, it is easy to find the velocity potential of the absolute motion of the fluid:

$$\Phi(x_1, x_2, x_3, t) = \sum_{i=1}^{3} v_{0i} x_i + \varphi(x_1, x_2, x_3, t) \tag{2–9}$$

and the velocity vector of its relative motion:

$$\boldsymbol{u} = \operatorname{grad} \varphi - \boldsymbol{\omega} \times \boldsymbol{r}. \tag{2–10}$$

It is also possible to determine the vector

$$\operatorname{curl} \boldsymbol{u} = -2\boldsymbol{\omega}.$$

It can be seen from this expression that the fluid particles in their relative motion rotate in a direction opposite to that of the body's rotation.

As follows from Eqs. (2–7) and (2–9), the time t is contained in the function Φ only by means of v_{0i} and ω_i. If the body stops after some time has elapsed, then the function Φ will become the function χ, *i.e.*, the fluid inside the cavity will perform the same motion as when the body was at rest.

Let us now calculate the kinetic energy of the fluid with potential $\emptyset$

$$2T_2^{(\theta)} = \varrho \int_\tau \operatorname{grad}^2 \emptyset \, d\tau .$$

To transform this volume integral into an integral over the surface σ, we use Kelvin's formula[17]:

$$\begin{aligned}
&\int_\tau \nabla\varphi\cdot\nabla\varphi' d\tau + \int_\tau \varphi' \Delta\varphi d\tau = \int_\sigma \varphi' \frac{d\varphi}{dn} d\sigma + \sum_{s=1}^{k-1} \kappa_s' \int_{\sigma^{(s)}} \frac{d\varphi}{dn} d\sigma^{(s)} , \\
&\int_\tau \nabla\varphi\cdot\nabla\varphi' d\tau + \int_\tau \varphi \Delta\varphi' d\tau = \int_\sigma \varphi \frac{d\varphi'}{dn} d\sigma + \sum_{s=1}^{k-1} \kappa_s \int_{\sigma^{(s)}} \frac{d\varphi'}{dn} d\sigma^{(s)} ,
\end{aligned} \tag{2–11}$$

which represents the generalization of Green's theorem to the case of cyclic functions φ and φ'. Here $\sigma^{(s)}$ denotes the area of the sth barrier from among the $k-1$ barriers that can be set up in a k-connected cavity without disturbing its connectedness; κ_s and κ_s' are the principal circulations for the above cyclic functions. Here the derivatives $d\varphi/dn$ and $d\varphi'/dn$ on $\sigma^{(s)}$ are taken along the positive direction of the normal to $\sigma^{(s)}$. Equations (2–11) can easily be obtained from Green's first theorem, setting up in the cavity the above barriers and assuming that the above cyclic functions are single-valued in the transformed region.

Then according to Eqs. (2–11)

$$2T_2^{(\theta)} = \varrho \int_\sigma \emptyset \frac{d\emptyset}{dn} d\sigma + \sum_{s=1}^{k-1} \kappa_s \varrho \int_{\sigma^{(s)}} \frac{d\emptyset}{dn} d\sigma^{(s)} .$$

By virtue of the single-valuedness of the functions φ and from the second of boundary conditions (2–5)

$$\int_\sigma \chi \frac{d\varphi}{dn} d\sigma + \sum_{s=1}^{k-1} \kappa_s \int_{\sigma^{(s)}} \frac{d\varphi}{dn} d\sigma^{(s)} = \int_\sigma \varphi \frac{d\chi}{dn} d\sigma = 0 ,$$

as a result of which we find

$$2T_2^{(\theta)} = \varrho \int_\sigma \varphi \frac{d\varphi}{dn} d\sigma + \sum_{s=1}^{k-1} \kappa_s \varrho \int_{\sigma^{(s)}} \frac{d\chi}{dn} d\sigma^{(s)} .$$

Substituting the expressions (2–6) and (2–7) for χ and φ into the above formula and making use of the first of conditions (2–5), we obtain

$$2T_2^{(\theta)} = \omega\cdot\Theta^*\cdot\omega + \sum_{r,s=1}^{k-1} a_{rs}\kappa_r\kappa_s , \tag{2–12}$$

where the symmetric tensor $\boldsymbol{\Theta}^*$ is defined by the components

$$\Theta^*_{ij} = [\psi_i, \psi_j] = \frac{1}{2}\varrho \int\limits_{\sigma} \left(\psi_i \frac{d\psi_j}{dn} + \psi_j \frac{d\psi_i}{dn}\right) d\sigma \quad (i, j = 1, 2, 3), \qquad (2\text{–}13)$$

which depend only on the geometry of the cavity and on the fluid density, while the constant

$$a_{rs} = \varrho \int\limits_{\sigma^{(r)}} \frac{d\vartheta_s}{dn} d\sigma^{(r)} \quad (r, s = 1, \dots, k-1) \qquad (2\text{–}14)$$

represents the vector flux grad ϑ_s through the rth barrier.

Thus the kinetic energy of a fluid with potential $\emptyset$ is composed of the kinetic energy

$$T^* = \tfrac{1}{2}\boldsymbol{\omega}\cdot\boldsymbol{\Theta}^*\cdot\boldsymbol{\omega} \qquad (2\text{–}15)$$

of motion with potential φ and of the constant kinetic energy of cyclic motion with potential χ.

It is important to note that T^* has precisely the same form as the kinetic energy of some rigid body with inertia tensor $\boldsymbol{\Theta}$ moving about a stationary point.

We now present one remarkable theorem that was established by Kelvin[17]: the irrotational motion of an incompressible fluid occupying a multiply connected region has less kinetic energy than any other motion with the same normal velocity components at the boundary and the same values of the total flux through each of the barriers.

Let T be the kinetic energy of some motion of a fluid in a multiply connected cavity with velocities

$$\boldsymbol{v} = \operatorname{grad}\emptyset + \boldsymbol{v}_1,$$

where, according to the conditions

$$\operatorname{div}\boldsymbol{v}_1 = 0, \quad \int\limits_{\sigma^{(s)}} \boldsymbol{v}_1\cdot\boldsymbol{n}\, d\sigma^{(s)} = 0 \quad (s = 1, \dots, k-1)$$

and on the surface σ, $\boldsymbol{v}_1\cdot\boldsymbol{n}=0$. Setting $T_1 = \frac{1}{2}\varrho\int_\tau v^2 d\tau$, we find

$$T = T^* + T_1 + \varrho \int\limits_{\tau} \boldsymbol{v}_1\cdot\operatorname{grad}\emptyset\, d\tau .$$

But

$$\int\limits_{\tau} \boldsymbol{v}_1\cdot\operatorname{grad}\emptyset\, d\tau = \int\limits_{\sigma^*} \emptyset \boldsymbol{v}_1\cdot\boldsymbol{n}\, d\sigma^*,$$

where σ^* denotes the total surface of the walls of a simply connected cavity, which will be obtained by setting up $k-1$ barriers in the multiply connected cavity. By virtue of the condition for $\boldsymbol{v}_1$ on σ, the last integral will extend only to both sides of each of the barriers. Since, on the opposite sides of the element $d\sigma^{(s)}$, the quantity $\boldsymbol{v}_1 \cdot \boldsymbol{n}$ is of opposite sign, while the values of $\emptyset$ differ by κ_s, this integral is equal to

$$\sum_{s=1}^{k-1} \kappa_s \int_{\sigma^{(s)}} \boldsymbol{v}_1 \cdot \boldsymbol{n} d\sigma^{(s)} = 0\,.$$

Thus

$$T = T^* + T_1\,, \tag{2–16}$$

which was to be proven. This theorem is obviously also valid for a simply connected region.

[2-2] MOTION OF A FLUID-CONTAINING RIGID BODY. ZHUKOVSKIY'S THEOREM

Let us now turn to the study of motion of a rigid body in the case of potential motion of the fluid contained in the cavity. Since the homogeneous fluid completely fills the cavity of the body, the mass center of the system is stationary relative to the body, so that this point may be used as the origin of the moving coordinate system $Ox_1x_2x_3$. The equations of motion of the system then take the form of Eqs. (1–40) for the motion of the mass center and the relative motion about the mass center.

It is obvious that $\boldsymbol{G}$, the absolute angular momentum of the system for the mass center, is equal to the angular momentum of the system $\boldsymbol{G}'$ with respect to the mass center. The angular acceleration $\boldsymbol{G}_2'$ of the fluid about the mass center is equal to the geometric sum of the angular momenta when moving with potentials φ and χ. The following vector

$$\boldsymbol{R} = \sum_{s=1}^{k-1} \kappa_s \varrho \int_{\tau} \boldsymbol{r} \times \nabla \vartheta_s d\tau \tag{2–17}$$

represents a linear function of the principal circulations. Obviously, this vector is constant in the above coordinate system.

Let us note at the same time that the vector $\boldsymbol{R}$ is independent of the location of the point O, for which it is defined. In fact, let us take any point O_1 with radius vector $\boldsymbol{r}_1$ relative to point O. The radius vector of a fluid particle relative to the point O_1 is denoted by $\boldsymbol{\varrho}$. Obviously, $\boldsymbol{r} = \boldsymbol{r}_1 + \boldsymbol{\varrho}$. Then

$$\int_{\tau} \boldsymbol{r} \times \nabla \vartheta_s d\tau = \boldsymbol{r}_1 \times \int_{\tau} \nabla \vartheta_s d\tau + \int_{\tau} \boldsymbol{\varrho} \times \nabla \vartheta_s d\tau\,.$$

But

$$\int_\tau \nabla \vartheta_s d\tau = 0 \quad (s = 1, \ldots, k-1),$$

since the velocity of the mass center of the fluid when moving with potential ϑ_s is zero (when the rigid body is at rest).

As a result of Eqs. (2–7) and (2–8), the angular momentum of the fluid with the potential φ

$$\boldsymbol{G}^* = \varrho \int_\tau \boldsymbol{r} \times \nabla \varphi d\tau = \varrho \int_\sigma \boldsymbol{r} \times \varphi \boldsymbol{n} d\sigma$$

is a linear vectorial function of the instantaneous angular velocity vector $\boldsymbol{\omega}$:

$$\boldsymbol{G}^* = \boldsymbol{H}^* \cdot \omega, \tag{2–18}$$

where the tensor $\boldsymbol{\Theta}^*$ is defined in Sec. 1 by its components (2–13).

Thus the vector $\boldsymbol{G}^*$ has exactly the same form as the angular momentum of some rigid body with the inertia tensor $\boldsymbol{\Theta}^*$, where

$$\boldsymbol{G}^* = \mathrm{grad}_\omega T^* .$$

The total angular momentum of the rigid body and the fluid in its cavity is

$$\boldsymbol{G} = \boldsymbol{G}_1 + \boldsymbol{G}^* + \boldsymbol{R} = \mathrm{grad}_\omega (T_1 + T^*) + \boldsymbol{R} . \tag{2–19}$$

Consequently, the equation of motion of the system relative to the mass center becomes

$$\frac{d}{dt}(\boldsymbol{G}_1 + \boldsymbol{G}^*) + \omega \times (\boldsymbol{G}_1 + \boldsymbol{G}^* + \boldsymbol{R}) = \boldsymbol{L} . \tag{2–20}$$

This is also the form of the equation of motion of the system about a stationary point.

If initially, when the rigid body was at rest, the fluid was stationary, then all the $\kappa_s = 0$, and $\boldsymbol{R} = 0$. Remembering Eq. (2–19), we can see that Eq. (2–20) represents in this case the equation of motion of a single rigid body with an ellipsoid of inertia for point O:

$$\boldsymbol{r} \cdot (\boldsymbol{H}^{(1)} + \boldsymbol{H}^*) \cdot \boldsymbol{r} = 1 . \tag{2–21}$$

Such a rigid body can be obtained by replacing the fluid in the body cavity by a rigid body with mass equal to the mass of the fluid and mass center identical to that of the fluid, and the ellipsoid of inertia for point O is expressed as

$$\boldsymbol{r} \cdot \boldsymbol{H}^* \cdot \boldsymbol{r} = 1 . \tag{2–22}$$

This body was called an *equivalent body* by Stokes.

Zhukovskiy established that the ellipsoid of inertia of the equivalent body (2–22) is always contained in the ellipsoid of inertia of the fluid. In fact, let the equation of the ellipsoid of inertia of the fluid for point O have the form

$$F(x_1, x_2, x_3) = 1 .$$

From Eq. (2–10) we find the quantity

$$\boldsymbol{\omega} \times \boldsymbol{r} = \operatorname{grad} \varphi - (\boldsymbol{u} - \operatorname{grad} \chi),$$

which we square, multiply by $\varrho d\tau$ and integrate with respect to τ. We obtain

$$F(\omega_1, \omega_2, \omega_3) = 2T^*(\omega_1, \omega_2, \omega_3) + \varrho \int_\tau [(\boldsymbol{u} - \nabla\chi)^2 - 2\nabla\varphi(\boldsymbol{u} - \nabla\chi)]\, d\tau .$$

Applying Eqs. (2–11) and (1–9), and using of Eqs. (2–5) and (1–32), we find

$$\int_\tau \nabla\varphi \cdot \nabla\chi d\tau = \int_\sigma \varphi \frac{d\chi}{dn} = 0, \quad \int_\tau \boldsymbol{u} \cdot \nabla\varphi d\tau = \int_\sigma \varphi u_n d\sigma = 0,$$

with the result that

$$F(\omega_1, \omega_2, \omega_3) = 2T^*(\omega_1, \omega_2, \omega_3) + \varrho \int_\tau (\boldsymbol{u} - \nabla\chi)^2\, d\tau \tag{2–23}$$

from which it follows that

$$F(x_1, x_2, x_3) > 2T^*(x_1, x_2, x_3) .$$

Let us now take any point $P(x_1, x_2, x_3)$ situated on the ellipsoid of inertia of the fluid. By virtue of the above inequality, we have for this point

$$2T^*(x_1, x_2, x_3) = \boldsymbol{r} \cdot \Theta^* \cdot \boldsymbol{r} < 1 ,$$

which leads us to the conclusion that the point P is situated inside the ellipsoid (2–22).

Let us now consider the central ellipsoid of inertia of the equivalent body; let its equation in the $O_2 x_1^*, x_2^*, x_3^*$ coordinate system, with the origin in the mass center of the fluid and axes parallel to axes x_1, x_2, x_3, have the form

$$F^*(x_1^*, x_2^*, x_3^*) = 1 .$$

According to the Huygens-Steiner theorem, the moment of inertia of a system is equal to the moment about an axis parallel to the given one and passing through the mass center, plus the product of the mass by the square of the distance between the two axes. Hence the moment of inertia of the equivalent rigid body about a line with direction cosines α_i $(i=1, 2, 3)$ passing through point O is

$$2T^*(\alpha_1, \alpha_2, \alpha_3) = F^*(\alpha_1, \alpha_2, \alpha_3) + M_2 [r_2^2(\alpha_1^2 + \alpha_2^2 + \alpha_3^2) - \\ - (x_{21}\alpha_1 + x_{22}\alpha_2 + x_{23}\alpha_3)^2],$$

where $\boldsymbol{r}_2 = \sum_{s=1}^{3} x_{2s}\boldsymbol{i}_s$ is the radius vector of the mass center of the fluid. Replacing α_i by ω_i, we obtain

$$2T^*(\omega_1, \omega_2, \omega_3) = F^*(\omega_1, \omega_2, \omega_3) + M_2(\boldsymbol{r}_2 \times \boldsymbol{\omega})^2.$$

On the other hand, let us divide the motion of the rigid body into translational motion, with the velocity of the mass center O_2 of the fluid equal to $\boldsymbol{\omega} \times \boldsymbol{r}_2$, and rotational motion about this center. As a result of this rotation the fluid will undergo irrotational motion with potential φ', which differs from φ. Using König's theorem, we find

$$2T^* = [\varphi', \varphi'] + M_2(\boldsymbol{\omega} \times \boldsymbol{r}_2)^2.$$

Equating this expression to the preceding, we conclude that

$$F^*(\omega_1, \omega_2, \omega_3) = [\varphi', \varphi'].$$

Consequently the equation of the central ellipsoid of inertia of the equivalent body can be obtained from Eq. (2–22) [see (2–15)] independently of the location of the stationary point.

It follows that, having replaced the fluid by an equivalent body, we completely reproduce its mechanical effect, no matter what the motion of the rigid body. Since the ellipsoid of inertia of the equivalent body contains within itself the ellipsoid of inertia of the fluid, the equivalent body has a smaller moment of inertia than the fluid about any given axis.

Following Zhukovskiy, we shall call a solid body with the equivalent body associated with it a *transformed body*.

Let us now consider the case of $\boldsymbol{R} \neq 0$. Then Eq. (2–20) is identical with the equation of motion about a stationary point of a rigid body with attached rotor. The axis of rotation of this rotor forms angles with the x_1, x_2, x_3 axes, cosines of which are in the ratio of $R_1 : R_2 : R_3$, while the product of the initial velocity of the rotor (with the rigid body remaining stationary) by its moment of inertia about the axis of rotation is equal to $|\boldsymbol{R}|$. Such a mechanical system is called a *hydrostat*.

Thus the effect of a fluid mass having zero initial velocity is identical to the effect of some equivalent rigid body; a fluid mass with an initial velocity in a multiply connected cavity also performs an action that is similar to the action of some rotor attached to the rigid body. This important theorem is due to Zhukovskiy. Particular cases of it were considered by Stokes and Neumann.

We note that if the x_1, x_2, x_3 axes are directed along the principal axes of the ellipsoid (2–21), its equation will have the form

$$\boldsymbol{r}\cdot\boldsymbol{\Theta}_*\cdot\boldsymbol{r} = A_* x_1^2 + B_* x_2^2 + C_* x_3^2 = 1\,,\quad \boldsymbol{\Theta}_* = \boldsymbol{\Theta}^{(1)} + \boldsymbol{\Theta}^*\,,$$

where A_*, B_*, C_* denote the principal moments of inertia of the transformed body for point O, which are equal to the sums of the moments of inertia of the rigid body and of the equivalent body relative to the corresponding coordinate axes. When axes rigidly fixed to the body are selected in this manner, the vector equation (2–20) projected on these axes yields three scalar ordinary differential equations:

$$A_* \frac{d\omega_1}{dt} + (C_* - B_*)\omega_2\omega_3 + \omega_2 R_3 - \omega_3 R_2 = L_1 \quad (A_* B_* C_*;\quad 123)\,. \tag{2–24}$$

In addition to the statements made in Sec. 8 of Chapter 1 concerning the integrals of the equations of motion, it also should be noted that if the ellipsoid of inertia of the transformed body is an ellipsoid of revolution ($A_* = B_*$) the projections of the cyclic angular momentum of the fluid on the x_1 and x_2 axes do not exist ($R_1 = R_2 = 0$), and if the moment of the external forces with respect to the axis of symmetry $L_3 = 0$, then an integral such as (1–75)

$$\omega_3 = \text{const}$$

also exists in the case when the cavity does not have the shape of a body of revolution about the x_3 axis.

We also note that, if the fluid is initially at rest, $R_i = 0$ ($i = 1, 2, 3$) and Eq. (2–24) reduce to Euler's equations of motion for a rigid body about a stationary point.

[2-3] VELOCITY POTENTIALS AND MOMENTS OF INERTIA OF EQUIVALENT BODIES FOR SEVERAL CAVITY SHAPES

Let us now examine the determination of the velocity potentials and ellipsoids of inertia of the equivalent bodies for several cavity shapes. The determination of the former, as was pointed out above, reduces to solving Neumann's problem for Laplace's equation. This boundary-value problem can be reduced to integral equations by introducing Green's function, for the solution of which effective methods are available[21]. However, we shall restrict ourselves to the presentation of some results due primarily to Zhukovskiy[15], as well as to Stokes[19], Lamb[17], and Chetayev[11].

[2-3.1] Ellipsoidal Cavities

Assume that the rigid body rotates about the center of the cavity. Directing the coordinate axes along the axes of the cavity, we represent the equation of the cavity walls in the form

$$f(x_1, x_2, x_3) = \sum_{i=1}^{3} \left(\frac{x_i}{a_i}\right)^2 - 1 = 0, \tag{2–25}$$

where the a_i denote the lengths of the ellipsoid semiaxes. The cosines of the angles that the external normal $\boldsymbol{n}$ to the surface $f(x_1, x_2, x_3) = 0$ makes with the coordinate axes are determined by the expression

$$n_i = \frac{1}{|\nabla f|} \frac{\partial f}{\partial x_i}.$$

Substituting these quantities for the surface (2–25) into conditions (2–8), we obtain boundary conditions for the harmonic functions ψ_i:

$$\frac{d\psi_1}{dn} = \frac{1}{|\nabla f|} x_2 x_3 \left(\frac{1}{a_2^2} - \frac{1}{a_3^2}\right) \quad (123).$$

Seeking the function ψ_1 in the form

$$\psi_1 = a x_2 x_3$$

and substituting into the first of these conditions, we obtain

$$\left(\frac{1}{a_2^2} + \frac{1}{a_3^2}\right) a x_2 x_3 = \left(\frac{1}{a_3^2} - \frac{1}{a_2^2}\right) x_2 x_3,$$

from which we find

$$a = \frac{a_2^2 - a_3^2}{a_2^2 + a_3^2}.$$

The function ψ_1 is determined completely, and the functions ψ_2 and ψ_3 are found by cyclical transposition of indices. Consequently the velocity potential for the fluid moving in an ellipsoidal cavity has the form[17]

$$\varphi(x_1, x_2, x_3, t) = \frac{a_2^2 - a_3^2}{a_2^2 + a_3^2} \omega_1 x_2 x_3 + \frac{a_3^2 - a_1^2}{a_3^2 + a_1^2} \omega_2 x_3 x_1 + \frac{a_1^2 - a_2^2}{a_1^2 + a_2^2} \omega_3 x_1 x_2. \tag{2–26}$$

The equation of the ellipsoid of inertia of the equivalent rigid body has the form of (2–22); the components of the inertia tensor are

$$\Theta_{11}^* = \frac{M_2}{5} \frac{(a_2^2 - a_3^2)^2}{a_2^2 + a_3^2} \quad (123), \quad \Theta_{ij}^* = 0 \quad (i \neq j); \tag{2–27}$$

i.e., the principal axes of inertia of the equivalent body are identical with the axes of the cavity.

Zhukovskiy showed that the streamlines of the relative motion of the fluid represent families of ellipses along which the planes conjugate to the direction

$$\omega_1' : \omega_2' : \omega_3', \; \omega_1' = -\frac{2\omega_1}{a_2^2 + a_3^2} \quad (123),$$

intersect a family of concentric ellipsoids similar to (2–25). He called this kind of fluid motion *elliptical rotation*; its direction is opposite to that of the rigid body.

In the case of a spherical cavity ($a_1 = a_2 = a_3$) rotation of the rigid body about the center of the cavity does not cause motion in the fluid ($\varphi = 0$). The equivalent body is a material point in which the entire mass of the fluid M_2 is concentrated.

Let us make one more remark. If the rigid body rotates about some point O that is not identical with the center of the cavity, the determination of the velocity potential and of the moments of inertia of the equivalent body reduces to the case in which the body rotates about the center of the cavity and the fluid has a velocity potential φ'. In fact, the motion of the body about O can be broken up into a translational motion with the velocity $\boldsymbol{\omega} \times \boldsymbol{r}_2$ of the center of the cavity, and a rotational motion with angular velocity $\boldsymbol{\omega}$ about the center of the cavity. The first motion has the velocity potential $(\boldsymbol{\omega} \times \boldsymbol{r}_2) \cdot \boldsymbol{r}$, and the second has the potential φ', so that the composite motion has the velocity potential

$$(\boldsymbol{\omega} \times \boldsymbol{r}_2) \cdot \boldsymbol{r} + \varphi,$$

where φ is a function into which φ' is transformed on changing from a coordinate system with the origin in the center of the cavity to the coordinate system $Ox_1x_2x_3$. The moments of inertia of the equivalent bodies are then determined by the Huygens-Steiner theorem.

For example, for an ellipsoidal cavity

$$\sum_{i=1}^{3} \frac{(x_i - x_{2i})^2}{a^2} - 1 = 0$$

with the center in point O_2, which is defined by the radius vector $\boldsymbol{r}_2 = \sum_{i=1}^{3} x_{2i} \boldsymbol{i}_i$, on rotation of the body about the origin the velocity potential differs from that given by (2–26) by the component

$$2\left\{\frac{\omega_1}{a_2^2 + a_3^2}(a_3^2 x_{22} x_3 - a_2^2 x_{23} x_2) + \frac{\omega_2}{a_3^2 + a_1^2}(a_1^2 x_{23} x_1 - a_3^2 x_{21} x_3) + \right.$$
$$\left. + \frac{\omega_3}{a_1^2 + a_2^2}(a_2^2 x_{21} x_2 - a_1^2 x_{22} x_1)\right\}.$$

[2-3.2] Cylindrical Cavities

Let us examine a cavity bounded by a cylindrical surface and two planes normal to this surface. We direct the x_3 axis parallel to the directrix of the cylinder, and we select the x_1Ox_2 plane to divide the cavity into halves. Then one of the principal axes of inertia of the equivalent body is directed along the x_3 axis. In fact,

$$[\psi_2, \psi_3] = \varrho \int_\sigma \frac{d\psi_2}{dn} \psi_3 d\sigma = \varrho \int_\sigma \psi_3 (x_3 n_1 - x_1 n_3)\, d\sigma = 0\,,$$

since the function ψ_3 is a function only of x_1 and x_2, $n_3=0$ on the surface of the cylinder, and $n_1=0$ and $n_3=\pm 1$ on the bases of the cylinder, so that the integrand for points with equal x_1 and x_2 will be the same quantity but with opposite sign. Similarly, $|\psi_3, \psi_1|=0$.

Let us first examine the motion of the fluid when the body rotates about the x_3 axis. We replace ψ_3 by a function $f(x_1, x_2)$, setting

$$\frac{\partial \psi_3}{\partial x_1} = \frac{\partial f}{\partial x_2}, \quad \frac{\partial \psi_3}{\partial x_2} = -\frac{\partial f}{\partial x_1}.$$

Obviously, f should be a harmonic function. Determining the relative velocity of the fluid from Eq. (2–10) and integrating the differential equation of the streamlines, we find the paths of relative motion of the fluid:

$$f(x_1, x_2) + \tfrac{1}{2}(x_1^2 + x_2^2) = \text{const}\,. \tag{2–28}$$

Since the contour of the cavity cross section should be one of the lines of this family, the following condition should be satisfied there:

$$f = -\tfrac{1}{2}(x_1^2 + x_2^2) + \text{const}\,.$$

The moment of inertia of the equivalent body about the x_3 axis can be expressed in terms of f:

$$[\psi_3, \psi_3] = [f, f] = 2h\varrho \iint \left[\left(\frac{\partial f}{\partial x_1}\right)^2 + \left(\frac{\partial f}{\partial x_2}\right)^2\right] dx_1 dx_2\,, \tag{2–29}$$

where the integral is taken over the area of the contour and $2h$ is the length of the cylinder's directrix. Thus the determination of f reduces to Dirichlet's problem of finding a harmonic function, given its value at the boundary of a plane region.

Zhukovskiy used two methods to solve this problem. One of them consists in using Fourier series, when the isothermic coordinates corresponding to the specified contour have been found. The second consists in first specifying a harmonic function and selecting its parameters so that the contour obtained has a simple shape. This inverse method of solution was first used by Saint-Venant in the problem of torsion of prisms, which is analogous to the hydrodynamic problem we are considering.

For example, specifying a function

$$f = k(x_1^2 - x_2^2), \quad -1 < 2k < 1, \quad k = \text{const}, \tag{2–30}$$

and applying Eq. (2–28), Zhukovskiy found that the cavities are elliptical cylinders with semiaxes a and b. The moment of inertia of the equivalent body as found from Eq. (2–29)

$$[f, f] = M_2 k^2 (a^2 + b^2), \tag{2–31}$$

while the ratio of the above moment of inertia to the moment of inertia of the fluid is

$$4k^2 = \left(\frac{a^2 - b^2}{a^2 + b^2}\right)^2 .$$

For a circular cylinder ($a=b$) the moment of inertia of the equivalent body is zero. Zhukovskiy used a similar method to determine the velocity potential of a fluid moving in a triangular prism.

The isothermic coordinates for a cylindrical cavity with a circular sector cross section with radius a and angle β are

$$\xi = \ln r, \quad \vartheta = \arctan \frac{x_2}{x_1} \quad (r^2 = x_1^2 + x_2^2), \tag{2–32}$$

if the origin is at the center of the circle and the x_1 axis is directed along one of the extreme radii of the sector. For a cavity of this shape, Stokes[19] found the function f in the form of a Fourier series ($\beta \neq \pi/2$)

$$f = -\frac{r^2}{2}\frac{\cos(\beta - 2\vartheta)}{\cos\beta} + \frac{8a^2}{\pi} \sum_{n=1,3,\ldots}^{\infty} \frac{\sin n\nu\theta}{n[(n\nu)^2 - 4]}\left(\frac{r}{a}\right)^{n\nu}, \quad \nu = \frac{\pi}{\beta}. \tag{2–33}$$

The moment of inertia of the equivalent body is

$$[f, f] = \frac{2M_2 a^2 \nu}{\pi} \times \left(\frac{\tan\beta}{4} - \frac{16\nu}{\pi} \sum_{n=1,3,\ldots}^{\infty} \frac{1}{(n\nu - 2)(n\nu + 2)^2 n\nu}\right). \tag{2–34}$$

This expression makes it possible to determine the moments of inertia of the equivalent body for different angles, provided $\beta \neq \pi/2$.

When $\beta=\pi/2$, the first and second terms of Eq. (2–34) become infinite, but their difference has a finite limit, which is $1/\pi$[15]. Hence,

$$[f, f] = \frac{4M_2 a^2}{\pi^2}\left\{1 - 2 \sum_{n=3,5,\ldots}^{\infty} \frac{1}{n(n-1)(n+1)^2}\right\}. \tag{2–35}$$

Zhukovskiy found that if the cavity is a circular cylinder divided by diametral barriers into two, four, or eight equal parts, then the moment of inertia of the equivalent body is 0.5211, 0.7904 or 0.9077 of the moment of inertia of the fluid. The same method was used in[15] for a cavity with a cross-sectional contour in the form of two confocal ellipses.

Let us now consider the motion of the fluid as a result of rotation of the body about an axis perpendicular to the generatrix of the cylinder. We set

$$\psi_1 = F_1 - x_2 x_3, \quad \psi_2 = F_2 + x_1 x_3, \tag{2–36}$$

where the harmonic functions F_i, according to Eq. (2–8), should satisfy the boundary conditions

$$\frac{dF_i}{dn} = 0 \quad (i = 1, 2) \tag{2–37}$$

on the lateral surface of the cylinder, and the conditions

$$\frac{dF_1}{dn} = \pm 2x_2, \quad \frac{dF_2}{dn} = \mp 2x_1 \tag{2–38}$$

on the bases of the cylinder; here the upper sign pertains to the top base and the lower sign to the bottom.

The moments of inertia of the equivalent body are most conveniently calculated from Eq. (2–23), where one must set $\omega_3=\chi=0$ and replace u_i by the sums of the relative velocities produced by the rotations ω_1 and ω_2, *i.e.*,

$$u_1 = \omega_1 \frac{\partial F_1}{\partial x_1} + \omega_2 \frac{\partial F_2}{\partial x_1},$$

$$u_2 = \omega_1 \frac{\partial F_1}{\partial x_2} + \omega_2 \frac{\partial F_2}{\partial x_2},$$

$$u_3 = \omega_1 \left(\frac{\partial F_1}{\partial x_3} - 2x_2\right) + \omega_2 \left(\frac{\partial F_2}{\partial x_3} + 2x_1\right).$$

Equating the coefficients of ω_1^2 and $\omega_1\omega_2$ in the right and left sides of the equation obtained by this substitution, we obtain

$$[\psi_1, \psi_1] = \varrho \int_\tau (x_2^2 + x_3^2)\, d\tau - \varrho \int_\tau \left[\left(\frac{\partial F_1}{\partial x_1}\right)^2 + \left(\frac{\partial F_1}{\partial x_2}\right)^2 + \left(\frac{\partial F_1}{\partial x_3} - 2x_2\right)^2\right] d\tau,$$

$$[\psi_2, \psi_2] = \varrho \int_\tau (x_1^2 + x_3^2)\, d\tau - \varrho \int_\tau \left[\left(\frac{\partial F_2}{\partial x_1}\right)^2 + \left(\frac{\partial F_2}{\partial x_2}\right)^2 + \left(\frac{\partial F_2}{\partial x_3} + 2x_1\right)^2\right] d\tau,$$

$$[\psi_1, \psi_2] = -\varrho \int_\tau x_1 x_2 d\tau -$$

$$- \varrho \int_\tau \left[\frac{\partial F_1}{\partial x_1}\frac{\partial F_2}{\partial x_1} + \frac{\partial F_1}{\partial x_2}\frac{\partial F_2}{\partial x_2} + \left(\frac{\partial F_1}{\partial x_3} - 2x_2\right)\left(\frac{\partial F_2}{\partial x_3} + 2x_1\right)\right] d\tau.$$

Applying Eq. (1–1) and making use of boundary conditions (2–37) and (2–38), we obtain

$$\begin{aligned} [\psi_1, \psi_1] &= A_2 + 2\varrho \int_{\sigma_0} (F_1^{(1)} - F_1^{(2)})\, x_2 d\sigma_0 - 8\varrho h \int_{\sigma_0} x_2^2 d\sigma_0, \\ [\psi_2, \psi_2] &= B_2 - 2\varrho \int_{\sigma_0} (F_2^{(1)} - F_2^{(2)})\, x_1 d\sigma_0 - 8\varrho h \int_{\sigma_0} x_1^2 d\sigma_0, \qquad (2\text{–}39) \\ [\psi_1, \psi_2] &= 6\varrho h \int_{\sigma_0} x_1 x_2 d\sigma_0 - 2\varrho \int_{\sigma_0} (F_1^{(1)} - F_1^{(2)})\, x_1 d\sigma_0, \end{aligned}$$

where σ_0 denotes the cross-sectional area of the cavity;

$$F_i^{(1)} = F_i|_{x_3 = h}, \quad F_i^{(2)} = F_i|_{x_3 = -h} \quad (i = 1, 2).$$

It is easy to see that when the cavity is symmetrical about the $x_2 x_3$ plane, the function F_1 also has this property. Then the integrands in both integrals in the right-hand side of the expression for $|\psi_1, \psi_2|$ will change sign on moving from the point (x_1, x_2) to the point $(-x_1, x_2)$, with their absolute value remaining constant, with the result that $|\psi_1, \psi_2|$ vanishes; *i.e.*, the coordinate axes will be the principal axes of inertia of the equivalent body.

Zhukovskiy noted that the determination of the functions F_i is extremely difficult and gave the solution of this problem for cavities in the shape of a rectangular parallelepiped and a circular cylinder[15].

Let the cavity be a rectangular parallelepiped. This can be taken as a cylinder that yields, in cross section, a rectangle with sides a and b. We make the x_3 axis coincide with the edge that is $2h$ long, and we direct the x_1 and x_2 axes along the sides a and b of the center cross section. Then the function

F_1 can be found in the form

$$F_1 = bx_3 - \frac{8b^2}{\pi^3} \sum_{n=1,3,\ldots}^{\infty} \frac{1}{n^3} \frac{\sinh \dfrac{n\pi x_3}{b}}{\cosh \dfrac{n\pi h}{b}} \cos \frac{n\pi x_2}{b}. \tag{2–40}$$

For an axis passing through the center of the cavity and parallel to the x axis, the moment of inertia of the equivalent body is found from Eq. (2–39) to be

$$[\psi_1', \psi_1'] = \frac{M_2}{12}(a^2 + 4h^2) - M_2 b^2 \left\{\frac{1}{3} - \left(\frac{2}{\pi}\right)^5 \frac{b}{h} \sum_{n=1,3,\ldots}^{\infty} \frac{1}{n^5} \tanh \frac{n\pi h}{b}\right\}.$$

The series contained in this expression converges rapidly. In the case of a cube ($a=b=2h$),

$$[\psi_1', \psi_1'] = M_2 a^2 \left\{\frac{64}{\pi^5} \sum_{n=1,3,\ldots}^{\infty} \frac{1}{n^5} \tanh \frac{n\pi}{2} - \frac{1}{6}\right\} = 0.0261 M_2 a^2 .$$

Since the moment of inertia of a cube is $1/6\, M_2 a^2$, the moment of inertia of the equivalent body is only 0.1565 of the moment of inertia of the fluid.

For a circular cylindrical cavity with radius a, it is convenient to use cylindrical coordinates r, ϑ, z with their origin in the center of the cavity. In this case,

$$F_1 = 4a^2 \sin \vartheta \sum_{n=1}^{\infty} \frac{1}{\zeta_n(\zeta_n^2 - 1)} \frac{J_1\left(\zeta_n \dfrac{r}{a}\right)}{J_1(\zeta_n)} \frac{\sinh\left(\zeta_n \dfrac{x_3}{a}\right)}{\cosh\left(\zeta_n \dfrac{h}{a}\right)}, \tag{2–42}$$

where $J_1(\zeta)$ denotes a first order and first kind and ζ_n denotes the roots of the equation

$$\frac{dJ_1(\zeta)}{d\zeta} = 0 .$$

The moment of inertia of the equivalent body is found to be

$$[\psi_1, \psi_1] = M_2\left(\frac{a^2}{4} + \frac{h^2}{3}\right) - M_2 a^2 \left\{1 - 8\frac{a}{h} \sum_{n=1}^{\infty} \frac{1}{\zeta_n^3(\zeta_n^2 - 1)} \tanh\left(\zeta_n \frac{h}{a}\right)\right\}. \tag{2–43}$$

N. G. Chetayev[11] has determined the velocity potentials and the moments of inertia of the equivalent bodies for cavities in the shape of a circular cylinder with one plane diaphragm and with a cross piece. For example, for a cavity in the shape of half a circular cylinder, the functions F_1 and F_2 were found in the form

$$F_1 = -\frac{16}{\pi}\sum_{n,m}\frac{\lambda_{nm}\cos n\vartheta\sinh(\lambda_{nm}x_3)\,J_n(\lambda_{nm}r)}{(n^2-1)(\lambda_{nm}^2a^2-1)\cosh(\lambda_{nm}h)\,[J_n(\lambda_{nm}a)]^2}\times$$

$$\times\int_0^a r^2 J_n(\lambda_{nm}r)\,dr \quad (n=0,2,4,\dots),$$

$$F_2 = -4a^2\cos\vartheta\sum_n\frac{1}{\zeta_n(\zeta_n^2-1)}\frac{J_1\left(\zeta_n\dfrac{r}{a}\right)}{J_1(\zeta_n)}\frac{\cosh(\zeta_n x_3)}{\cosh\left(\dfrac{\zeta_n h}{a}\right)}, \tag{2–44}$$

where $J_n(\zeta)$ denotes the Bessel function of nth order and first kind, and $\lambda_{nm}a=\zeta_{nm}$, where ζ_{nm} denote the roots of equation

$$\frac{dJ_n(\zeta)}{d\zeta}=0.$$

The moments of inertia of the equivalent body are

$$[\psi_1,\psi_1]=\frac{M_2}{2}\left\{\frac{h^2}{3}-\frac{3}{4}a^2+\frac{128}{\pi^2a^2h}\times\right.$$

$$\left.\times\sum_{n,m}\frac{\lambda_{nm}\tanh(\lambda_{nm}h)}{(\lambda_{nm}^2a^2-1)(n^2-1)^2}\left(\frac{\int_0^a r^2J_n(\lambda_{nm}r)\,dr}{J_n(\lambda_{nm}a)}\right)^2\right\},$$

$$[\psi_2,\psi_2]=\frac{M_2}{2}\left\{\frac{h^2}{3}+\frac{5}{4}a^2-4\frac{a^3}{h}\sum_n\frac{1}{\zeta_n^3(\zeta_n^2-1)}\tanh\left(\zeta_n\frac{h}{a}\right)\right\},$$

$$[\psi_1,\psi_2]=0. \tag{2–45}$$

[2-3.3] Cavities in the Shape of Bodies of Revolution

Let us consider a cavity in the shape of a body of revolution, directing the x_3 axis along the axis of rotation of the cavity. The rotation of the body about its axis of revolution does not set the fluid into motion, since, for the surface of revolution,

$$\frac{d\psi_3}{dn}=0,\quad \psi_3=\text{const}.$$

Hence

$$[\psi_3,\psi_3]=[\psi_2,\psi_3]=[\psi_3,\psi_1]=0,$$

as a result of which the equivalent body can be represented in the form of a segment of a straight line of homogeneous material situated along the axis of the cavity, and having a fluid mass M_2 and a length

$$\sqrt{12\,\frac{[\psi_1, \psi_1]}{M_2}}.$$

The problem is to determine the motion of the fluid that results from rotation of the body about an axis perpendicular to an axis of the cavity, for example, the x_1 axis.

If we change from Cartesian coordinates x_i to spherical coordinates r, ϑ, φ,

$$x_1 = r\sin\vartheta\cos\varphi\,, \quad x_2 = r\sin\vartheta\sin\varphi\,, \quad x_3 = r\cos\vartheta\,,$$

then Laplace's equation will assume the form

$$\Delta\psi_1 = \frac{\partial}{\partial r}\left(r^2\,\frac{\partial\psi_1}{\partial r}\right) + \frac{1}{\sin\vartheta}\,\frac{\partial}{\partial\vartheta}\left(\sin\vartheta\,\frac{\partial\psi_1}{\partial\vartheta}\right) + \frac{1}{\sin^2\vartheta}\,\frac{\partial^2\psi_1}{\partial\varphi^2} = 0\,.$$

We write the function ψ_1 in the form

$$\psi_1 = \sin\varphi F(r, \vartheta)\,,$$

where the function $F(r, \vartheta)$ is given by the expression

$$\frac{\partial}{\partial r}\left(r^2\,\frac{\partial F}{\partial r}\right) + \frac{1}{\sin\vartheta}\,\frac{\partial}{\partial\vartheta}\left(\sin\vartheta\,\frac{\partial F}{\partial\vartheta}\right) - \frac{F}{\sin^2\vartheta} = 0\,.$$

From Eq. (2–8), we have the condition for the function F on the surface of the cavity

$$\sin\varphi\,\frac{dF}{dn} = x_2 n_3 - x_3 n_2\,.$$

If we construct a meridian cross section for the point on the surface under study and draw an external normal $\boldsymbol{n}$, it is easy to show that

$$x_2 n_3 - x_3 n_2 = \tfrac{1}{2}\sin\varphi\,\frac{dr^2}{ds}\,,$$

as a result of which, on the surface of the cavity,

$$\frac{dF}{dn} = \frac{1}{2}\,\frac{dr^2}{ds}\,. \tag{2–46}$$

Here $ds = -(dr/\sin(\gamma - \vartheta))$ denotes an arc element of a half of the meridian cross section; here, the angle ϑ is measured from the Ox_3 axis, $0 \leqslant \vartheta \leqslant \vartheta_0$, $\vartheta_0 \leqslant \pi$, and γ is the angle between the x_3 axis and the vector $\boldsymbol{n}$. Obviously, for a circle, $dr = 0$, $\gamma = \vartheta$, $ds = r d\vartheta$.

The moment of inertia of the equivalent body is

$$[\psi_1, \psi_1] = \varrho \int_\sigma \frac{d\psi_1}{dn} \psi_1 d\sigma = \varrho \int_0^{2\pi} \sin^2 \varphi d\varphi \int F \frac{dF}{dn} r \sin \vartheta ds .$$

Integrating with respect to φ, we obtain

$$[\psi_1, \psi_1] = \frac{\pi}{2} \varrho \int F r \sin \vartheta dr^2 , \tag{2–47}$$

where the integration is performed over half of the meridian cross section. As with cylindrical cavities, we can either seek the function F on the basis of a specified cavity, or first specify a function that would satisfy Laplace's equation and then appropriately select the parameters of which it is a function to determine the shape of the cavity. Here the given function is usually expanded in terms of spherical functions.

Applying these methods, Zhukovskiy determined the velocity potentials for cavities in the form of cones, hemispheres, and simply connected and annular intersecting hyperboloids of revolution.

For example, for a hemispherical cavity with radius a, the function F was

$$\begin{aligned} F = r x_3 \sin \vartheta - \tfrac{3}{4} r a \vartheta_1^{(1)} + r^2 \sum_{n=3,5,\dots}^{\infty} (-1)^{\frac{n-1}{2}} \frac{2n+1}{(n+3)(n-2)} \times \\ \times \frac{1 \cdot 3 \cdot 5 \cdots (n-2)}{2 \cdot 4 \cdot 6 \cdots (n-1)} \left(\frac{r}{a}\right)^{n-2} \vartheta_n^{(1)} , \end{aligned} \tag{2–48}$$

where $\vartheta_n^{(1)}$ are spherical functions of the first kind that satisfy the equation

$$\frac{1}{\sin \vartheta} \frac{d}{d\vartheta} \left(\sin \vartheta \frac{d\vartheta_n^{(1)}}{d\vartheta} \right) + \left[n(n+1) - \frac{1}{\sin^2 \vartheta} \right] \vartheta_n^{(1)} = 0 .$$

The moment of inertia of the equivalent body is

$$\begin{aligned} [\psi_1, \psi_1] = 3 M_2 a^2 \left\{ \frac{3}{32} - \sum_{n=3,5,\dots}^{\infty} \frac{2n+1}{(n+3)^2 (n+1)(n-2)} \times \right. \\ \left. \times \left(\frac{1 \cdot 3 \cdot 5 \cdots (n-2)}{2 \cdot 4 \cdot 6 \cdots (n-1)} \right)^2 \right\} . \end{aligned} \tag{2–49}$$

Zhukovskiy calculated the above quantity with a precision of 0.0001 and found that

$$[\psi_1, \psi_1] = 0.2387 M_2 a^2 .$$

From this we have the interesting result that if we divide a sphere into two halves by a plane passing through its center, the moment of inertia of the equivalent body will be 0.5968 of the moment of inertia of the fluid, while when the undivided sphere rotates about a diameter the moment of inertia of the equivalent body is zero.

[2-3.4] Multiply Connected Cavities

It is generally more difficult to determine velocity potentials for multiply connected cavities than for simply connected ones. Zhukovskiy studied a doubly connected cavity consisting of an infinitesimally thin closed tube of arbitrary shape and cross section, as well as an annular cavity of revolution. Referring the reader to Ref. [15] for details, we will only point out that, for all annular cavities that are bodies of revolution about the x_3 axis, the velocity potential for a fluid undergoing cyclical motion is

$$\chi = \frac{\kappa}{2\pi}\,\vartheta\,, \quad \vartheta = \arctan\frac{x_2}{x_1}\,. \tag{2–50}$$

In this case

$$v_r = \frac{\partial\chi}{\partial r} = 0\,, \quad v_\vartheta = \frac{1}{r}\frac{\partial\chi}{\partial\vartheta} = \frac{\kappa}{2\pi r} \quad (r^2 = x_1^2 + x_2^2)$$

and since $v = \infty$ when $r = 0$, the cavity should not contain points for which $r = 0$.

The projections of the vector $\boldsymbol{R}$ of the angular momentum of a fluid performing cyclical motion are

$$R_1 = R_2 = 0\,, \quad R_3 = \varrho\int_\tau \frac{d\chi}{d\vartheta}\,d\tau = \frac{M_2\kappa}{2\pi}\,; \tag{2–51}$$

i.e., for annular-shaped cavities of revolution, the initial principal angular momentum is equal to the product of the mass of the fluid and the velocity of circulation, divided by 2π.

In this section we have presented the velocity potentials and moments of inertia of equivalent bodies only for cavities with the most elementary shapes. They do not nearly exhaust all the numerous problems of this kind that have already been solved. Numerical methods become necessary and are apparently most effective for intricately shaped cavities. This question is examined in more detail in Chapter 5.

[2-4] HOMOGENEOUS VORTEX FLOW OF A FLUID

Let us now consider another elementary case of fluid motion—homogeneous vortex, or quasi-solid, motion. This is the name given to the motion when the vector curl $\boldsymbol{v}$ is the same at any given instant for all the fluid particles.

Let the cavity be in the shape of the ellipsoid given by Eq. (2–25) with semiaxes $a_i\,(i=1,2,3)$. The velocity of a homogeneous, incompressible ideal fluid in the cavity can be represented in the form[23,24]

$$\boldsymbol{v} = \operatorname{grad}\Phi + \boldsymbol{\Omega}\times\boldsymbol{r}\,, \tag{2–52}$$

where $\Phi(x_1, x_2, x_3, t)$ is a harmonic function of the position of the point, and $\boldsymbol{\Omega}=\boldsymbol{\Omega}(t)$ is a function of t only.

The kinematic condition (1–32) yields the following boundary condition at the walls of the cavity for the function Φ:

$$\frac{d\Phi}{dn} = \boldsymbol{v}_0\cdot\boldsymbol{n} + [(\boldsymbol{\omega}-\boldsymbol{\Omega})\times\boldsymbol{r}]\cdot\boldsymbol{n}\,.$$

It is easy to check that this condition is satisfied by the harmonic function

$$\begin{aligned}\Phi(x_1, x_2, x_3, t) = \sum_{i=1}^{3} v_{0i}x_i + \frac{a_2^2-a_3^2}{a_2^2+a_3^2}(\omega_1-\Omega_1)\,x_2x_3 + \\ + \frac{a_3^2-a_1^2}{a_3^2+a_1^2}(\omega_2-\Omega_2)\,x_3x_1 + \frac{a_1^2-a_2^2}{a_1^2+a_2^2}(\omega_3-\Omega_3)\,x_1x_2\,.\end{aligned} \tag{2–53}$$

To determine $\boldsymbol{\Omega}(t)$, we use Helmholtz's equation for vortex flow of a fluid (1–39). Since, in the case under study,

$$\operatorname{curl}\boldsymbol{v} = 2\boldsymbol{\Omega}\,,$$

Helmholtz's equation projected on the moving axes yields three scalar equations:

$$\frac{d\Omega_1}{dt} = 2a_1^2\left(\frac{\omega_3\Omega_2}{a_1^2+a_2^2} - \frac{\omega_2\Omega_3}{a_3^2+a_1^2}\right) - 2\,\frac{a_1^2(a_3^2-a_2^2)}{(a_1^2+a_2^2)(a_1^2+a_3^2)}\,\Omega_2\Omega_3 \quad (123)\,. \tag{2–54}$$

The right-hand sides of these equations are independent of the position of the points of the fluid, justifying the representation of the fluid motion in the form (2–52).

It is easy to show that Eq. (2–54) permits the first integral

$$a_2^2a_3^2\Omega_1^2 + a_3^2a_1^2\Omega_2^2 + a_1^2a_2^2\Omega_3^2 = \text{const}\,, \tag{2–55}$$

which expresses Helmholtz's theorem on the constancy of the vortex intensity.

It should be noted additionally that the motion of a fluid in an ellipsoidal cavity can be described by formulas that differ in appearance from (2–52). Zhukovskiy[15] assumed that the motion of the fluid consists of a potential motion, with the potential (2–26), and an elliptical motion, with velocities

$$u_1' = a_1^2(\omega_2'x_3 - \omega_3'x_2) \quad (123)\,,$$

where $\omega_i' = \omega_i'(t)$ are functions of time only. It is obvious that such a motion can be described by Eq. (2–52) if we set

$$\Omega_1 = \tfrac{1}{2}(a_2^2 + a_3^2)\omega_1' \quad (123).$$

Poincaré[25] assumed that the velocities of the fluid particles are linear functions of their position. He called this motion *simple* and pointed out, on the basis of Helmholtz's theorems, that if the motion of the fluid in an ellipsoidal cavity is initially simple, it will remain so all of the time. The motion of the fluid then can be graphically interpreted from the kinematic point of view. Each fluid particle in the ellipsoidal cavity can be made to correspond to some imagined particle with coordinates

$$x_i' = \frac{x_i}{a_i} \quad (i = 1, 2, 3)$$

and velocities relative to the rigid body

$$u_i' = \frac{u_i}{a_i} \quad (i = 1, 2, 3).$$

An ensemble of these imagined particles will fill a sphere Σ

$$\sum_{i=1}^{3} x_i'^2 = 1.$$

In the case of simple motion of the fluid, the imagined particles will move as points of a rigid body; *i.e.*, the motion of the fluid will be represented by rotation of the sphere Σ as a single rigid body. Denoting the projections of the instantaneous angular velocity of the sphere Σ on the moving axes by Ω_i, we find that the projections of the absolute velocity of the fluid on the same axes are defined by the expressions

$$v_1 = \frac{a_1}{a_3}\Omega_2' x_3 - \frac{a_1}{a_2}\Omega_3' x_2 + \omega_2 x_3 - \omega_2 x_2 \quad (123). \qquad (2\text{–}56)$$

Here the projections of the vector $\frac{1}{2}$ curl $\boldsymbol{v}$ are

$$\Omega_1 = \omega_1 + \frac{a_2^2 + a_3^2}{2a_2 a_3}\Omega_1' \quad (123). \qquad (2\text{–}57)$$

Substituting for the function Φ in (2–52) its expression given by (2–53), and replacing Ω_i by Eq. (2–57), we obtain Eq. (2–56). Conversely, on the basis of Eq. (2–56), making use of Eqs. (2–57) and (2–53), we get Eq. (2–52).

It is interesting that if the rigid body is held stationary ($\omega_i = 0$, $v_{0i} = 0$), then (Eq. 2.54) can be transformed, using Eq. (2–57), into

$$a_1^2(a_2^2 + a_3^2)\frac{d\Omega_1'}{dt} = [a_2^2(a_3^2 + a_1^2) - a_3^2(a_1^2 + a_2^2)]\,\Omega_2'\Omega_3' \quad (123). \qquad (2\text{–}58)$$

These expressions have the form of Euler's equations for a solid body with principal moments of inertia proportional to $a_1^2({}_2^2a+a_3^2)$ (123) and moving by inertia about a stationary point.

Let us find the vector of the angular momentum of a fluid performing homogeneous vortex motion. It is easy to calculate that

$$G_2 = \boldsymbol{\Theta}^* \cdot \omega + \boldsymbol{\Theta}' \cdot \boldsymbol{\Omega}, \tag{2–59}$$

where $\boldsymbol{\Theta}^*$ is the inertia tensor of the equivalent body which is defined by its components (2–27), and

$$\boldsymbol{\Theta}' = \boldsymbol{\Theta}^{(2)} - \boldsymbol{\Theta}^*$$

denotes a tensor that is equal to the difference of the tensors of inertia of the fluid and of the equivalent body, with the components

$$A' = \Theta'_{11} = \frac{4}{5} \frac{M_2 a_2^2 a_3^2}{a_2^2 + a_3^2} \quad (A'B'C'; 123), \quad M_2 = \tfrac{4}{3}\pi\varrho a_1 a_2 a_3 . \tag{2–60}$$

Let us consider the motion of a rigid body containing a fluid performing homogeneous vortex motion. We place the origin of the coordinate system $Ox_1x_2x_3$ at the mass center of the rigid body with the fluid. Then the equation of motion of the mass center has the form of the first of Eqs. (1–40). The second—the equation of motion about the mass center—will be written, with the help of (2–59), in the form

$$\boldsymbol{\Theta}_* \cdot \frac{d\omega}{dt} + \boldsymbol{\Theta}' \cdot \frac{d\Omega}{dt} + \omega \times (\boldsymbol{\Theta}_* \cdot \omega + \boldsymbol{\Theta}' \cdot \boldsymbol{\Omega}) = \boldsymbol{L}, \tag{2–61}$$

where the tensor $\boldsymbol{\Theta}_* = \boldsymbol{\Theta}^1 + \boldsymbol{\Theta}^*$ is equal to the sum of the inertia tensors of the rigid and the equivalent bodies. In projecting on the moving axes, we obtain the three scalar equations

$$\begin{gathered} A_* \frac{d\omega_1}{dt} + A' \frac{d\Omega_1}{dt} + (C_* - B_*)\,\omega_2\omega_3 + C'\omega_2\Omega_3 - B'\omega_3\Omega_2 = L_1 \\ (A_*B_*C_*;\; A'B'C';\; 123), \end{gathered} \tag{2–62}$$

which are equivalent to the vector equation (2–61).

Thus the system of equations of motion about the mass center of a rigid body containing a fluid performing homogeneous vortex motion in an ellipsoidal cavity consists of the first-order ordinary differential equations (2–62) and (2–54), which, depending on the effective forces acting on the system, can also be supplemented by equations for force-defining parameters, for example, Eqs. (1–5).

Equation (2–62) is identical with the equations of motion for a rigid body with a tensor of inertia $\boldsymbol{\Theta}_*$ about a stationary point and a rotating gyroscope with a tensor of inertia $\boldsymbol{\Theta}'$ associated with the rigid body. The gyroscope rotates according to Eqs. (2–54) so that the geometry of the system's masses does not change. Consequently, the effect of a fluid performing homogeneous vortex motion is identical to the effect of some equivalent body and a rotating gyroscope, connected to the body, which together form a gyrostat.

We note that in the particular case of a spherical cavity ($a_1 = a_2 = a_3 = a$) the motion of the fluid in the cavity has no effect on the motion of the rigid body. In fact, in this case, as can be seen from Eqs. (2–27),

$$\boldsymbol{\Theta}^* = 0,\ \Theta'_{ii} = \tfrac{2}{5} M_2 a^2 \quad (i = 1, 2, 3),$$

and Eq. (2–54) takes the form

$$\frac{d\Omega}{dt} + \boldsymbol{\omega} \times \boldsymbol{\Omega} = 0 .$$

Then Eq. (2–61) is written as

$$\boldsymbol{\Theta}^{(1)} \cdot \frac{d\omega}{dt} + \boldsymbol{\omega} \times \boldsymbol{\Theta}^{(1)} \cdot \boldsymbol{\omega} = \boldsymbol{L} ,$$

i.e., the system moves as a single rigid body with the tensor of inertia $\boldsymbol{\Theta}^{(1)}$. We recall that the case considered corresponds the to center of the cavity situated at point O. If, however, the center is situated at a point with a radius vector $\boldsymbol{r}_2$, the tensor of inertia is given by $\boldsymbol{\Theta}^{(1)} + M_2(\boldsymbol{E}r_2^2 - \boldsymbol{r}_2\boldsymbol{r}_2)$, where $\boldsymbol{E} = \sum_{s=1}^{3} \boldsymbol{i}_s \boldsymbol{i}_s$ is a unit tensor.

Remark 1

Equations (2–62) and (2–54), defining the motion of a rigid body containing a fluid in homogeneous vortex motion, can be obtained as Poincaré's equation in terms of group variables [25].

Remark 2

Dirichlet [26] and Poincare [25] showed that, if the region occupied by the fluid remains ellipsoidal during the entire motion, and the fluid performs homogeneous vortex motion at some instant of time, it will perform this kind of motion at all times. What are the other shapes of cavities for which homogeneous vortex motion is possible? It follows from Helmholtz's equations that for this it is necessary and sufficient that the velocities of the fluid particles be linear functions of these coordinates. If we restrict ourselves to closed cavities of revolution, it can be shown that such a fluid motion is possible only in an ellipsoidal cavity [27].

[2-5] PROBLEMS OF STABILITY OF MOTION. LYAPUNOV'S AND CHETAYEV'S THEOREMS

The elementary cases of a rigid body with cavities completely filled by ideal fluids performing vortex-free (irrotational) or homogeneous vortex (rotational) motions have many things in common; in particular, in both, the velocities of the fluid particles depend on a finite number of variables. As a result the motion of a rigid body containing a fluid, *i.e.*, a system with an infinite number of degrees of freedom, is completely described by a finite number of variables, and the equations of motion of the system are found to be ordinary differential equations. This makes it possible to state the problem of stability of motion of a solid body containing a fluid in the cases under study as the problem of stability in the sense of Lyapunov for systems with a finite number of degrees of freedom.*

Let us recall some postulates of the theory of stability according to Lyapunov. Let q_j $(j=1,\dots,n)$ denote the Lagrangian coordinates of some holonomic mechanical system, and $\dot{q}_j$ denote the generalized velocities. We now consider some particular solution of the differential equations of motion

$$q_j = f_j(t),$$

which has corresponding to it a specific motion of our system which we define as undisturbed. The functions $f_j(t)$ satisfy the specified initial conditions, *i.e.*,

$$q_{j0} = f_j(t_0), \quad \dot{q}_{j0} = \dot{f}_j(t_0).$$

This motion of the system can be compared, in certain respects, to other motions that are possible for this system while subject to the same forces but under other initial conditions

$$q_{j0} = f_j(t_0) + \varepsilon_j, \dot{q}_{j0} = \dot{f}_j(t_0) + \varepsilon'_j,$$

where the ε_j and ε'_j are some real, numerically sufficiently small constants which are called perturbations.

Let Q_s $(s=1,\dots,k)$ be some specified, continuous, real functions of q_j, $\dot{q}_j$ and t. For undisturbed motion these functions are some known functions $F_s(t)$, and for disturbed motion they are some functions of time and of the disturbances ε_j and ε'_j.

Lyapunov's Definition [28]

Let L_s be arbitrarily specified positive numbers. If, for any L_s, there can be selected positive numbers E'_j and E'_j, no matter how small, in a manner such that for any disturbances ε_j and ε'_j satisfying the conditions

$$|\varepsilon_j| \leqslant E_j, \quad |\varepsilon'_j| \leqslant E'_j,$$

* This irrotational motion of a fluid first attracted the attention of N.G. Chetayev, who also solved the problem of stability of rotational motion of a projectile with a cavity having the form of a circular cylinder, a cylinder with one diaphragm, and a cylinder with a cross piece[11].

and for any $t \geqslant t_0$, the following inequalities are satisfied:

$$|Q_s - F_s| < L_s;$$

then the undisturbed motion is stable with respect to Q_s, while in the opposite case it is unstable.

The question of stability can be resolved by examining the differential equations that are satisfied by the differences

$$z_s = Q_s - F_s.$$

Equations for z_s are called *equations of disturbed motion*; they can be reduced to the normal form

$$\frac{dz_s}{dt} = Z_s(z_1, \ldots, z_k, t) \qquad (s = 1, \ldots, k),$$

where the functions Z_s, which vanish when $z_s = 0$ $(s = 1, \ldots, k)$, will be assumed to be holomorphic functions with coefficients that are known functions of time.

A. M. Lyapunov developed a direct method for the study of stability that consists of finding continuous single-valued functions $V(z_1, \ldots, z_k, t)$ which have certain properties. The total time derivatives of the functions V are also considered; these are taken by virtue of the differential equations of disturbed motion:

$$\dot{V} = \frac{\partial V}{\partial t} + \sum_{s=1}^{k} \frac{\partial V}{\partial z_s} Z_s.$$

The direct method of Lyapunov is based on several theorems, of which we present the following.

Lyapunov's Stability Theorem

If the differential equations of disturbed motion are such that it is possible to find a sign-definite function V, the derivative $\dot{V}$, of which, by virtue of these equations, is either a sign-definite function the sign of which is opposite to that of V, or identically zero, then the undisturbed motion is stable.

Chetayev's Instability Theorem[29]

If the differential equations of disturbed motion are such that it is possible to find a function V, bounded in the region $V > 0$, existing for any $t \geqslant t_0$, and for any variables z_s of magnitude as small as desired, the derivative $\dot{V}$, of which, by virtue of these equations, is positive definite in the above region, then the undisturbed motion is unstable.

The proofs of these theorems are well known[28,29] and will be omitted here.

Functions satisfying the conditions of Lyapunov's stability theorem are called Lyapunov's functions, while those satisfying Chetayev's instability theorem are called Chetayev's functions.

Lyapunov's functions may be effectively constructed by an extensively used method, due to Chetayev, which briefly consists of the following: Let there be known some holomorphic integrals of the equations of disturbed motion

$$V_i = \text{const} \quad (i = 1, \ldots, m).$$

Let us consider the function

$$V = \sum_i \lambda_i V_i ,$$

where the λ_i are constants selected in such a manner that the power series expansion of the function V in terms of z_s, in the vicinity $z_s = 0$ of undisturbed motion, would start with a quadratic (or, in general, homogeneous of even order p) form $V^{(p)}$ of the variables z_s. If it is found that the function V constructed in this manner is sign-definite, then, since $\dot{V} \equiv 0$, it will be a Lyapunov function.

We also note that if among the solutions of equations of disturbed motion, there exist solutions

$$V_\alpha = c_\alpha$$

for which the variable constants c_α, by virtue of the mechanical or geometrical meaning of these solutions, can only be equal to zero, then in this case the function V can also be constructed in the form[30]

$$V = \sum_j \lambda_j V_j + \sum_\alpha \mu_\alpha V_\alpha ,$$

where μ_α are some variables with bounded derivatives $\dot{\mu}_\alpha$. In this case, obviously,

$$\dot{V} = \sum_\alpha \dot{\mu}_\alpha V_\alpha = 0 .$$

If the form of $V^{(p)}$, which starts the power series expansion of the function V in terms of z_s, is sign-definite, then the function V will also be sign-definite in the region of z_s of sufficiently small absolute magnitude; if, however, the form of $V^{(p)}$ is sign-variable, then the function V will also be such. Any odd-order form is obviously a sign-variable function.

The following sign-definiteness criterion, which is due to Sylvester, is valid for quadratic forms ($p = 2$).

The Sylvester Criterion

The real quadratic form

$$\varphi = \sum_{i,j=1}^{k} c_{ij} z_i z_j \quad (c_{ij} = c_{ji})$$

will be positive definite if and only if all of the principal diagonal minors of its discriminant are positive:

$$\begin{vmatrix} c_{11} \dots c_{1r} \\ \cdot \;\; \cdot \;\; \cdot \;\; \cdot \\ c_{r1} \;\dots c_{rr} \end{vmatrix} > 0 \quad (r = 1, \dots, k) .$$

No effective sign-definiteness criteria exist for even-order forms with $p > 2$.

Remark

In the case when the solutions of equations of disturbed motion $V_i = \text{const}$ are not holomorphic functions of z_s, Chetayev's method for constructing Lyapunov functions in the form $V = \sum_i \lambda_i V_i$ is also applicable, if the constants λ_i can be selected in a manner such that the function V has the form

$$V = V^{(p)} + V_* ,$$

where V_* denotes a function that becomes zero when $z = \dots = z_k = 0$ and is of a higher order of smallness than p. In this case, the function V will also be sign-definite if the form $V^{(p)}$ is of definite sign.

Before the appearance of Lyapunov's theorem, the problem of stability was usually solved by examining the linear approximation

$$\frac{dz_s}{dt} = p_{s1} z_1 + \dots + p_{sk} z_k , \quad p_{sr} = \text{const} \quad (s = 1, \dots, k),$$

which are obtained from the equations of disturbed motion by dropping all nonlinear terms with respect to z_s. By virtue of the same fact, the procedure reduces essentially to replacing the problem with another problem, with the solution of which it is in no way related[28]. Having noted this, Lyapunov stated and solved the problem of stability on the basis of the first approximation, *i.e.*, independently of higher-order terms. He proved that if the real parts of all of the roots of the characteristic equation of the first approximation

$$\Delta(\kappa) = \| p_{ij} - \delta_{ij} \kappa \| = 0$$

are negative, then the undisturbed motion is stable. If, however, there exists among the roots κ_s at least one with a positive real part, then the undisturbed motion is unstable, no matter what are the terms higher than first order of smallness.

These theorems leave unclarified the cases when some of the roots of the characteristic equation have negative real parts and others have zero real parts. These cases are critical in the sense that their stability and instability cannot be clarified by examining only the first approximation. In these cases the nonlinear terms in the right-hand sides of the equations of disturbed motion can always be chosen in a manner such that the system is stable or unstable, as desired [29]. In particular, systems for which the law of conservation of energy applies can be stable only in those critical cases when the real parts of all the roots of the characteristic equation are zero. Consequently, in the critical cases the stability conditions obtained by analysis of the roots of the characteristic equation can be regarded only as necessary conditions.

Nevertheless, this method is extensively used, even at present. In cases for which the Lyapunov function has been constructed, the direct method for investigating stability in the critical cases has many advantages over the methods of the theory of small vibrations not only in accuracy and rigorousness, but also in simplicity.

Below we present examples of solutions of problems of the stability of motion of a rigid body containing a fluid. The problems are of interest in themselves from the theoretical and practical points of view; all of them pertain to cases which are critical according to Lyapunov.

[2-6] STEADY SCREW DISPLACEMENT OF A BODY CONTAINING A FLUID

Let us consider a rigid body with a cavity of arbitrary shape that is completely filled with a fluid performing irrotational motion. Let the kinetic energy of the body and the fluid be

$$T = \tfrac{1}{2}M\left(v_{01}^2 + v_{02}^2 + v_{03}^2\right) + \tfrac{1}{2}\left(A_*\omega_1^2 + B_*\omega_2^2 + C_*\omega_3^2\right)$$

and let the system move by inertia. The equations of motion can be taken in the form of Eqs. (1–40); in the given case, $\boldsymbol{K}=\boldsymbol{L}=0$. Obviously the equations of motion have the general solutions

$$T = \text{const}, \quad \boldsymbol{v}_0 \cdot \boldsymbol{G} = \text{const}, \quad \boldsymbol{v}_0 = \mathbf{const}$$

and allow the particular solution

$$v_{01} = v,\ v_{02} = v_{03} = 0, \quad \omega_1 = \omega, \quad \omega_2 = \omega_3 = 0,$$

which corresponds to a steady screw displacement of the rigid body along the x_1 axis, while the fluid in the cavity moves with the velocity potential $\Phi = vx_1 + \omega_1 \psi_1(x_1, x_2, x_3)$. Let us examine the stability of this motion with respect to the six variables v_{0i} and ω_1 $(i=1, 2, 3)$ [Fig. 8].

Setting in the disturbed motion

$$v_{01} = v + z_1, \quad \omega_1 = \omega + z_2$$

and retaining the previous notation for the remaining variables, we write the

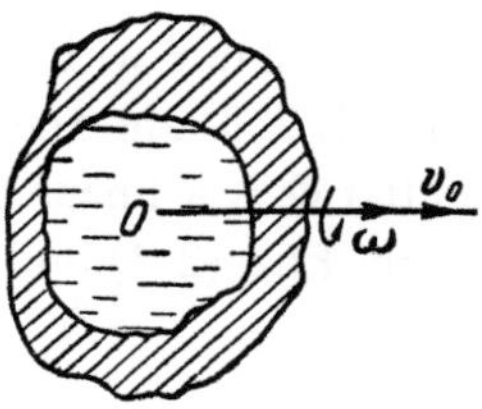

Fig. 8.

first integrals of the equations of disturbed motion corresponding to the above first integrals of the equations of motion:

$$\begin{aligned} V_1 &= M(z_1^2 + 2vz_1 + v_{02}^2 + v_{03}^2) + A_*(z_2^2 + 2\omega z_2) + B_*\omega_2^2 + C_*\omega_3^2 = \text{const}, \\ V_2 &= A_*(\omega z_1 + vz_2 + z_1 z_2) + B_* v_{02}\omega_2 + C_* v_{03}\omega_3 = \text{const}, \\ V_3 &= z_1^2 + 2vz_1 + v_{02}^2 + v_{03}^2 = \text{const}. \end{aligned}$$

We construct a Lyapunov function in the form[31]

$$\begin{aligned} V = V_1 - 2\lambda V_2 + \kappa V_3 + V_3^2 = {} & (M + \kappa + 4v^2)\, z_1^2 - 2A_*\lambda z_1 z_2 + \\ & + A_* z_2^2 + (M + \kappa)\, v_{02}^2 - 2B_*\lambda v_{02}\omega_2 + B_*\omega_2^2 + \\ & + (M + \kappa)\, v_{03}^2 - 2C_*\lambda v_{03}\omega_3 + C_*\omega_3^2 + \\ & + (z_1^2 + v_{02}^2 + v_{03}^2)^2 + 4vz_1(z_1^2 + v_{02}^2 + v_{03}^2), \end{aligned} \tag{2–63}$$

where

$$\lambda = \frac{\omega}{v}, \quad M + \kappa = A\frac{\omega^2}{v^2}.$$

According to Sylvester's criterion, if the following inequalities are satisfied:

$$A_* > B_*, \quad A_* > C_*, \tag{2–64}$$

which means that the x_1 axis should be the minor axis of the central ellipsoid of inertia of the transformed body, when V will be a positive definite function of all of the six variables z_1, z_2, v_{0i}, and ω_i $(i=2, 3)$, while its time derivative, by virtue of the equations of disturbed motion, will be $\dot{V} = 0$. On the basis of Lyapunov's stability theorem we conclude that these inequalities are sufficient conditions for stability of a steady screw motion of a rigid body containing fluid in its cavity with respect to the quantities v_{0i} and ω_i $(i=1, 2, 3)$.

Since in the given case the equations of motion of the system about the mass center have the same form as the equations of inertial motion of a rigid body about a stationary point, the undisturbed motion is stable with respect to ω_i $(i=1, 2, 3)$ if the x_1 axis is the major or minor axis of the central ellipsoid of inertia of the transformed body, and is unstable if the x_1 axis is the central axis of the ellipsoid of inertia[30].

We will make another general remark. From the stability of undisturbed motion with respect to v_{0i} and ω_i $(i=1, 2, 3)$, we can also conclude that the motion of the fluid in the cavity is stable with respect to the projections of the fluid particle velocities v_i $(i=1, 2, 3)$ and the pressure p. In fact, making use of the fact that the potential Φ is a linear function of v_{0i} and ω_i, this conclusion can be obtained from Eqs. (2–1) and (1–79).

[2-7] UNIFORM ROTATION OF A BODY ABOUT A STATIONARY POINT

Let us examine a rigid body having a cavity completely filled with an ideal fluid in irrotational motion. Let the body rotate about a stationary point under the action of forces that have a force function $U(\gamma_3)$. Let us assume that the cavity is multiply connected, while the initial motion of the fluid is such that the projections of the angular momentum of the fluid performing cyclical motion on the moving coordinate axes are $R_1=R_2=0$, and $R_3=R$. For example, if the cavity has the shape of a torus obtained by rotating some closed curve that does not intersect the x_3 axis about this axis, then, as was noted in Sec. 3, $R=M_2\chi/2\pi$ (Fig. 9).

From here on, we assume that the finite continuous function $U(\gamma_3)$ has continuous, bounded first and second derivatives.

It is easy to see that the equations of motion (2–24) and (1–41) in the case

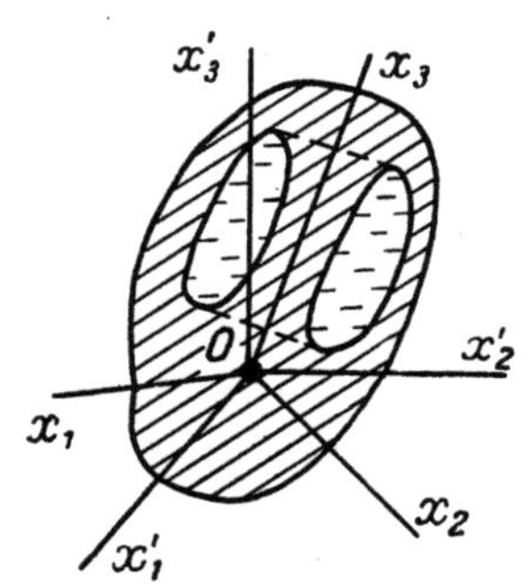

Fig. 9.

under study have the following first integrals: energy integral (1–59); area integral (1–70); geometric integral (1–77); as well as the integral (1–75) stating the constancy of the projections of the instantaneous angular velocity on the x_3 axis, provided that the ellipsoid of inertia (2–21) of the transformed body is an ellipsoid of rotation about this axis, *i.e.*, if

$$A_* = B_* . \tag{2–65}$$

The equations of motion allow the particular solution

$$\omega_1 = \omega_2 = 0 , \quad \omega_3 = \omega ,$$
$$\gamma_1 = \gamma_2 = 0 , \quad \gamma_3 = 1 ,$$

which describes the uniform rotation of the rigid body about the x_3 axis, which coincides with the x_3' axis, and irrotational motion of the fluid with potential $\emptyset = \omega\psi_3(x_1, x_2, x_3) + \chi$. If the cavity is in the shape of a body of revolution about the x_3 axis, then $\psi_3 = \text{const}$.

In the disturbed motion we set

$$\omega_3 = \omega + z_1 ,$$
$$\gamma_3 = 1 + z_2 ,$$

and the previous notation is retained for the remaining variables. The equations of disturbed motion obviously have the first integrals

$$V_1 = A_*\omega_1^2 + B_*\omega_2^2 + C_*(z_1^2 + 2\omega z_1) - 2\Gamma_1 z_2 - \Gamma_2 z_2^2 + \cdots = \text{const} ,$$
$$V_2 = A_*\omega_1\gamma_1 + B_*\omega_2\gamma_2 + C_*(\omega z_2 + z_1 + z_1 z_2) + R z_2 = \text{const} ,$$
$$V_3 = \gamma_1^2 + \gamma_2^2 + z_2^2 + 2z_2 = 0 ,$$

as well as the integral

$$V_4 = z_1 = \text{const}$$

if condition (2–65) is satisfied. The series of dots denote terms containing z_2 to a power higher than the second, and

$$\Gamma_1 = \left.\frac{\partial U}{\partial \gamma_3}\right|_{\gamma_3 = 1} , \quad \Gamma_2 = \left.\frac{\partial^2 U}{\partial \gamma_3^2}\right|_{\gamma_3 = 1} .$$

Let us first consider the case when $A_* = B_*$, and let us construct a Lyapunov function in the form

$$\begin{aligned} V = {} & V_1 + 2\lambda V_2 - (C_*\omega\lambda + R\lambda - \Gamma_1) V_3 - 2C_*(\omega + \lambda) V_4 + \\ & + \frac{C_*(C_* - A_*)}{A_*} V_4^2 + \tfrac{1}{2}\Gamma_2 z_2 V_3 = A_*(\omega_1^2 + \omega_2^2) + \\ & + 2A_*\lambda(\omega_1\gamma_1 + \omega_2\gamma_2) - (C_*\omega\lambda + R\lambda - \Gamma_1)(\gamma_1^2 + \gamma_2^2) + \\ & + \frac{C_*^2}{A_*} z_1^2 + 2C_*\lambda z_1 z_2 - (C_*\omega\lambda + R\lambda - \Gamma_1) z_2^2 + \cdots , \end{aligned} \tag{2–66}$$

where λ is an arbitrary constant. The conditions for positive definiteness of this function yield the inequality

$$A_*\lambda^2 - (C_*\omega + R)\,\lambda - \Gamma_1 < 0\,.$$

The latter is possible if the polynomial on the left-hand side has two different real roots:

$$(C_*\omega + R)^2 + 4A_*\Gamma_1 > 0\,. \tag{2–67}$$

When this condition is satisfied, λ can be selected in a manner such that the function V will become positive definite with respect to all of the variables ω_i, γ_i, z_i $(i=1, 2)$ that it contains; here $\dot{V}=0$. Consequently, according to Lyapunov's theorem, inequality (2–67) is a sufficient condition for stability. We shall now show that it is also necessary for stability. In fact, let us consider Chetayev's function in the form

$$W = \omega_1\gamma_2 - \omega_2\gamma_1 \tag{2–68}$$

and its time derivative, taken by virtue of the equations of disturbed motion,

$$\dot{W} = (1 + z_2)\,(\omega_1^2 + \omega_2^2) - \frac{C_*(\omega + z_1) + R}{A_*}(\omega_1\gamma_1 + \omega_2\gamma_2) - \\ - \frac{1}{A_*}\,\Gamma_1\,(\gamma_1^2 + \gamma_2^2) + \cdots$$

It is easy to see that the function $\dot{W}$ will be positive definite with respect to ω_i, γ_i $(i=1, 2)$ if the inequality

$$(C_*\omega + R)^2 + 4A_*\Gamma_1 < 0 \tag{2–69}$$

is satisfied. It is assumed here that the variable z_2 remains at all times of the order of smallness of the variables ω_i and γ_i, since if it did not the instability of the motion with respect to γ_3 would have been obvious.

When condition (2–69) is satisfied, the undisturbed motion is unstable with respect to ω_i, γ_i $(i=1, 2)$, since then the function W satisfies all of the conditions of Chetayev's instability theorem. Consequently, condition (2–67) is a necessary and sufficient condition for stability of the undisturbed motion under study. From this condition we can also obtain, as particular cases, a number of well-known stability conditions. For example, when $R=0$ and the fluid contained in a multiply connected cavity is initially at rest, or when the cavity is simply connected, or when the cavity contains no fluid, we obtain the condition[12]

$$C_*^2\omega^2 + 4A_*\Gamma_1 > 0\,.$$

For the case of a homogeneous field of gravity forces with force function $U(\gamma_3) = -Mgx_{c3}\gamma_3$, condition (2–67) takes the form[33]

$$(C_*\omega + R)^2 - 4A_*Mgx_{c3} > 0,$$

and when $R=0$ it has the form of Mayyevskiy's condition[29]

$$C_*^2\omega^2 - 4A_*Mgx_{c3} > 0. \qquad (2\text{–}70)$$

For a projectile moving in a flattened trajectory, and acted upon by forces with a force function of the form of (1–43), condition (2–67) assumes the form

$$C_*^2\omega^2 - 4A_*a > 0,$$

as was established by Chetayev for projectiles with solid and fluid contents; in the latter case the cavity was assumed to be in the shape of a circular cylinder with or without a cross piece[11].

Let us now consider the case when $A_* \geqslant B_*$ and construct a Lyapunov function in the form

$$\begin{aligned} V &= V_1 - 2\omega V_2 + (C_*\omega^2 + R\omega + \Gamma_1)\, V_3 + \tfrac{1}{4}\lambda V_3^2 = \\ &= A_*\omega_1^2 - 2A_*\omega\omega_1\gamma_1 + (C_*\omega^2 + R\omega + \Gamma_1)\,\gamma_1^2 + \\ &\quad + B_*\omega_2^2 - 2B_*\omega\omega_2\gamma_2 + (C_*\omega^2 + R\omega + \Gamma_1)\,\gamma_2^2 + \\ &\quad + C_*z_1^2 - 2C_*\omega z_1 z_2 + (C_*\omega^2 + R\omega + \Gamma_1 + \lambda - \Gamma_2)\, z_2^2 + \cdots, \end{aligned} \qquad (2\text{–}71)$$

where the constant

$$\lambda > \Gamma_2 - \Gamma_1 - R\omega.$$

This function will be positive definite if it satisfies the condition

$$(C_* - A_*)\,\omega^2 + R\omega + \Gamma_1 > 0. \qquad (2\text{–}72)$$

Since $\dot{V}=0$ by virtue of the equations of disturbed motion, the above inequality according to Lyapunov's theorem will be a sufficient condition for stability of the undisturbed motion with respect to ω_i and γ_i $(i=1, 2, 3)$.

From this stability condition it is also possible to obtain a number of particular cases. For example, for a homogeneous field of gravitational forces with a force function in the form of (1–42), where $x_{c_1}=x_{c_2}=0$, condition (2–72) takes the form[33]

$$(C_* - A_*)\,\omega^2 + R\omega - Mgx_{c3} > 0.$$

In the case of forces with a force function of the form of (1–43), when $R=0$, condition (2–72) becomes:

$$(C_* - A_*)\,\omega^2 > a. \qquad (2\text{–}73)$$

A condition of this form was obtained by Chetayev as one of the necessary conditions for stability of a projectile in the shape of a circular cylinder with one plane diaphragm[11].

We recall that for cavities in the shape of bodies of revolution, the axial moment of inertia of the equivalent body is equal to zero. The equatorial moment of inertia A_* is determined by the shape and dimensions of the cavity and is proportional to the density ϱ of the fluid. Hence for the same rigid body (A_1, C_1, M_1 constant) and the same conditions of motion (ω constant), greater fluid density, which increases the values of A^* and M_2, has a negative effect on stability conditions (2–70) and (2–73).

[2-8] PERMANENT ROTATION OF A HYDROSTAT MOVING BY INERTIA ABOUT A STATIONARY POINT

Let us examine the stability of the permanent rotation of a rigid body moving by inertia about a stationary point O and having a multiply connected cavity completely filled with a fluid in irrotational motion (Fig. 10). The

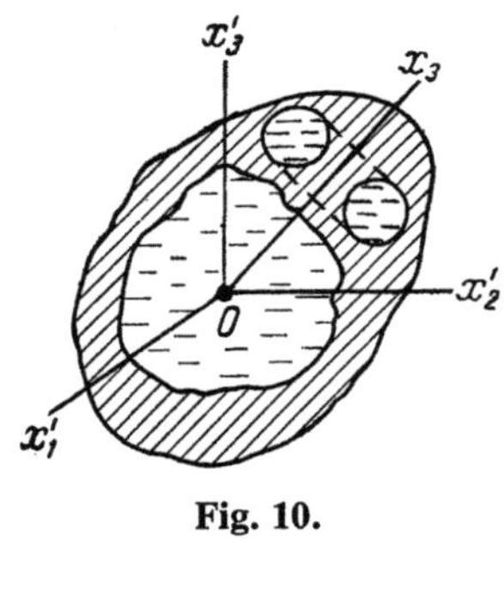

Fig. 10.

equations of motion have the form of (2–24) for $\boldsymbol{L}=0$. They allow energy integrals (1–59) and integrals stating the constancy of angular momentum (1–71). Using these integrals, Zhukovskiy[15] gave an elegant geometric interpretation of the motion of a rigid body. Volterra[34] studied permanent rotations and their stability. The stability in our problem will now be examined, using Lyapunov's direct method.

Of greatest interest are the permanent rotations of a hydrostat about its principal central axes of inertia, which are possible provided the vector of the angular momentum in the cyclical motion of the fluid $\boldsymbol{R}$ is collinear with the principal axis of inertia. The angular velocity ω in this case can have arbitrary constant magnitudes. For example, let $R_1 = R_2 = 0$, and $R_3 = R$. Equations (2–24) allow the particular solution

$$\omega_1 = \omega_2 = 0\,, \quad \omega_3 = \omega\,,$$

which corresponds to permanent rotation of the body about the x_3 axis, and irrotational motion of the fluid with the potential

$$\Phi = \omega\psi_3(x_1, x_2, x_3) + \chi .$$

In the disturbed motion, we set

$$\omega_3 = \omega + z_1 ,$$

while the previous designations are retained for ω_1 and ω_2. The equations of disturbed motion allow the integrals

$$V_1 = A_*\omega_1^2 + B_*\omega_2^2 + C_*(z_1^2 + 2\omega z_1) = \text{const},$$
$$V_2 = A_*^2\omega_1^2 + B_*^2\omega_2^2 + C_*^2(z_1^2 + 2\omega z_1) + 2C_*Rz_1 = \text{const}.$$

Let us consider the function

$$V = V_2 - \left(C_* + \frac{R}{\omega}\right)V_1 + \frac{1}{4\omega^2}\lambda V_1^2 = A_*(A_* - \tilde{C})\,\omega_1^2 + + B_*(B_* - \tilde{C})\,\omega_2^2 + C_*\left(C_*\lambda - \frac{R}{\omega}\right)z_1^2 + \cdots, \tag{2–74}$$

$$\tilde{C} = C_* + \frac{1}{\omega}R, \quad \lambda = \text{const}.$$

It is easy to see that if

$$A_* \gtrless \tilde{C}, \quad B_* \gtrless \tilde{C}, \tag{2–75}$$

where only the upper or only the lower signs apply simultaneously, the constant λ can always be selected so that the function (2–74) is sign-definite. On the basis of Lyapunov's theorem we conclude that, under conditions (2–75), the undisturbed motion is stable. In particular, this conclusion is also valid when $A_* = B_*$.

Let us also consider the function[29]

$$W = \omega_1\omega_2$$

and its time derivative, which, by virtue of the equation of disturbed motion, is

$$\dot{W} = \left(\frac{\tilde{C} - A_*}{B_*}\,\omega_1^2 + \frac{B_* - \tilde{C}}{A_*}\,\omega_2^2\right)\omega + \left(\frac{C_* - A_*}{B_*}\,\omega_1^2 + \frac{B_* - C_*}{A_*}\,\omega_2^2\right)z_1 .$$

We shall assume that during the entire motion the variable z_1 remains of the order of smallness of ω_i, since in the opposite case instability exists with respect to this variable.

Then, if the inequalities

$$\tilde{C} \gtrless A_*, \quad B_* \gtrless \tilde{C} \tag{2–76}$$

are satisfied simultaneously (with only the upper or only the lower signs), $\dot{W}$ will be a sign-definite function of the variables ω_1 and ω_2. On the basis of Chetayev's theorem we conclude that the undisturbed motion is unstable under conditions (2–76).

The quantity $\tilde{C}$ has the dimensions of moment of inertia, and for a given angular velocity ω it can be treated as a "reduced" motion of inertia of the hydrostat for the permanent axis x_3, but only if $\tilde{C}>0$. Introducing the ellipsoid

$$A_* x_1^2 + B_* x_2^2 + \tilde{C} x_3^2 = 1 ,$$

the results obtained can obviously be formulated in the form of the known theorem of stability of permanent rotation of a rigid body with such an ellipsoid of inertia in the Euler-Poinsot case.

It should, however, be remembered that when R and ω are of opposite sign, $\tilde{C}$ may be nonpositive. Then conditions (2–75) with the upper signs, will be satisfied; *i.e.*, the undisturbed motion will be stable.

Thus, if the ratio R/ω is bounded within the limits $A_* - \tilde{C}$ and $B_* - \tilde{C}$, the permanent rotation is unstable, while in the opposite case it is stable with respect to the variables ω_i $(i=1, 2, 3)$. In this case it is also easy to conclude that the undisturbed motion is also stable with respect to disturbances of the principal circulations κ_s if it is stable with respect to ω_i[34].

Let us consider one particular case. Let the moments of inertia A_1, B_1, and C_1 of the rigid body be negligible in comparison with the moments of inertia A^*, B^*, and C^* of the equivalent rigid body. Then the stability conditions (2–75) are (when $R=0$)

$$A^* \gtrless C^* , \quad B^* \gtrless C^* .$$

For example, for a cavity in the shape of an ellipsoid of revolution, when

$$a_1 = a_2 = a , \quad a_3 = c \neq a ,$$

$$A^* = B^* = \frac{4}{5} M_2 \frac{(a^2 - c^2)^2}{a^2 + c^2} , \quad C^* = 0 ,$$

these conditions are always satisfied, independently of the ratio $a/c \neq 1$.

[2-9] STABILITY OF ROTATION OF A RIGID BODY WITH AN ELLIPSOIDAL CAVITY

Let us consider a rigid body with a stationary point O, having an ellipsoidal cavity (2–25) filled with an ideal fluid in homogeneous vortex motion (Fig. 11). The motion and stability of this system were studied by many scientists, starting with Kelvin[35], who experimentally discovered that an oblate ellipsoid is stable while a slightly prolate one is unstable. This problem is considered in detail in Refs. [22–25, and 36–40].

We shall assume that the system is acted upon by forces having a force function of the form $U(\gamma_3)$, in which case the complete system of equations of motion consists of nine ordinary differential equations (2–62), (2–54), and (1–41). This system of equations has the following first integrals: the energy integral (1–59); the area integral (1–70); the geometric integral (1–77); and

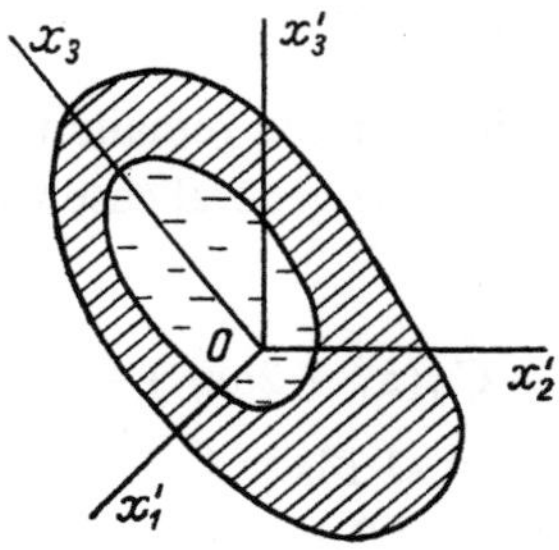

Fig. 11.

the integral (2–55) expressing the constancy of the vortex intensity. When the ellipsoid of inertia of the body for a stationary point and the cavity are in the shape of bodies of revolution about the x_3 axis, *i.e.*, when $A_1 = B_1$ and $a_1 = a_2$, the equations of motion also have the integral (1–75) expressing the constancy of the projection of the instantaneous angular velocity of the body on the x_3 axis.

The equations of motion of the system allow the particular solution

$$\omega_1 = \omega_2 = 0\,, \quad \omega_3 = \omega\,,$$
$$\Omega_1 = \Omega_2 = 0\,, \quad \Omega_3 = \Omega\,,$$
$$\gamma_1 = \gamma_2 = 0\,, \quad \gamma_3 = 1\,.$$

This solution describes uniform rotation of a rigid body about the x_3 axis, made to coincide with the stationary axis x'_3, and also describes the relative elliptical rotation about the same axis with components

$$\omega'_1 = \omega'_2 = 0\,, \quad \omega'_3 = \frac{2(\Omega - \omega)}{a_1^2 + a_2^2}\,.$$

We shall assume that this motion is undisturbed and shall examine its stability with respect to the variables ω_i, Ω, γ_i $(i = 1, 2, 3)$.

We note that in undisturbed motion the quantities ω and Ω can have arbitrary constant values. Of practical interest are the values of Ω in the interval $0<\varepsilon\leqslant\Omega\leqslant\omega$, where ε is an infinitesimal. As was shown experimentally, upon uniform rotation of a body about a stationary axis passing through the center of the cavity, the fluid, provided it has not been disturbed, first performs irrotational motion (in particular, it is at rest), and then is increasingly made to participate in the body's motion by friction forces until it moves with it as a single rigid body. From now on we shall restrict ourselves to the case when $\varepsilon\leqslant\Omega\leqslant\omega$.

In disturbed motion, we set

$$\omega_3 = \omega + z_1, \quad \Omega_3 = \Omega + z_2, \quad \gamma_3 = 1 + z_3,$$

retaining our previous notation for the remaining variables. Obviously the equation of disturbed motion in the general case will have the following first integrals:

$$\begin{aligned} V_1 &= A_*\omega_1^2 + B_*\omega_2^2 + C_*(z_1^2 + 2\omega z_1) + A'\Omega_1^2 + B'\Omega_2^2 + \\ &\quad + C'(z_2^2 + 2\Omega z_2) - 2\Gamma_1 z_3 - \Gamma_2 z_3^2 + \cdots = \text{const}, \\ V_2 &= (A_*\omega_1 + A'\Omega_1)\gamma_1 + (B_*\omega_2 + B'\Omega_2)\gamma_2 + \\ &\quad + C_*(z_1 + \omega z_3 + z_1 z_3) + C'(z_2 + \Omega z_3 + z_2 z_3) = \text{const}, \\ V_3 &= \gamma_1^2 + \gamma_2^2 + z_3^2 + 2z_3 = 0, \\ V_4 &= a_2^2 a_3^2 \Omega_1^2 + a_1^2 a_3^2 \Omega_2^2 + a_1^2 a_2^2 (z_2^2 + 2\Omega z_2) = \text{const}. \end{aligned}$$

We resort to Chetayev's method for constructing Lyapunov's functions. We consider the function

$$\begin{aligned} V &= V_1 - 2\omega V_2 + \lambda_1 V_3 + \lambda_2 V_4 + \tfrac{1}{4}\lambda_3 V_3^2 = \\ &= A_*\omega_1^2 - 2A_*\omega\omega_1\gamma_1 + (A' + \lambda_2 a_2^2 a_3^2)\,\Omega_1^2 - 2A'\omega\Omega_1\gamma_1 + \lambda_1\gamma_1^2 + \\ &\quad + B_*\omega_2^2 - 2B_*\omega\omega_2\gamma_2 + (B' + \lambda_2 a_3^2 a_1^2)\,\Omega_2^2 - 2B'\omega\Omega_2\gamma_2 + \lambda_1\gamma_2^2 + \\ &\quad + C_* z_1^2 - 2C_*\omega z_1 z_3 + (C' + \lambda_2 a_1^2 a_2^2)\, z_2^2 - 2C'\omega z_2 z_3 + \\ &\quad + (\lambda_1 + \lambda_3 - \Gamma_2)\, z_3^2 + \cdots, \end{aligned} \tag{2–77}$$

where, for brevity, we have introduced the notation

$$\lambda_1 = (C_*\omega + C'\Omega)\,\omega + \Gamma_1, \quad \lambda_2 = C'\,\frac{\omega - \Omega}{\Omega a_1^2 a_2^2}$$

for the constant coefficients; the coefficient λ_3 remains as yet arbitrary.

The function (2–77) is a sum of three quadratic forms, which depend on three variables each, and terms of the third and higher orders of smallness with respect to γ_1, γ_2, and z_3, which have been dropped. We shall obtain the conditions for positive definiteness of these forms by using Sylvester's inequalities, which, when $\omega\geqslant\Omega$ and $\lambda_3>\Gamma_2-\Gamma_1$, and making use of the notation for λ_1 and λ_2, reduce to

$$\begin{aligned}&\left[A' + C'\frac{a_3^2}{a_1^2}\left(\frac{\omega}{\Omega} - 1\right)\right][(C_* - A_*)\,\omega^2 + C'\omega\Omega + \Gamma_1] - A'^2\omega^2 > 0\,,\\&\left[B' + C'\frac{a_3^2}{a_1^2}\left(\frac{\omega}{\Omega} - 1\right)\right][(C_* - B_*)\,\omega^2 + C'\omega\Omega + \Gamma_1] - B'^2\omega^2 > 0\,.\end{aligned} \tag{2–78}$$

According to Lyapunov's stability theorem, the above inequalities are sufficient conditions for stability of the undisturbed motion with respect to ω_i, Ω_i, γ_i $(i=1, 2, 3)$. Since under conditions (2–78) the constant quantity Ω is not fixed, they make it possible to clarify the effect of the degree of vorticity of the fluid's motion $(\varepsilon \leqslant \Omega \leqslant \omega)$ on the stability of the undisturbed motion.

In particular, let us see what form the conditions (2–78) will take when $\Omega = \varepsilon$, *i.e.*, when the motion of the fluid in the cavity (2–25) is very close to potential motion, and when $\Omega = \omega$, *i.e.*, when the fluid and the body are rotating as a single rigid body. In the first case, when ε is considered an infinitesimal and $A_* \geqslant B_*$, the conditions (2–78) will obviously be satisfied if

$$(C_* - A_*)\,\omega^2 + \Gamma_1 > \delta(\varepsilon) > 0\,, \tag{2–79}$$

where $\delta(\varepsilon)$ is a small quantity. The condition (2–79) is practically identical with the sufficient condition for stability (2–72), for $R=0$ for rotation of a rigid body having a cavity filled with a fluid in irrotational motion.

In the second case, when $\Omega = \omega$, conditions (2–78) for $A \geqslant B$ reduce to the single inequality

$$(C_* + C' - A_* - A')\,\omega^2 + \Gamma_1 > 0\,. \tag{2–80}$$

This condition obviously coincides with the sufficient condition for stability of rotation of a single rigid body with moments of inertia A, B, and C, which are equal to the sum of the moments of inertia of the body and the fluid contained in its cavity.

When $U=0$, the conditions (2–78) reduce to sufficient conditions for the stability of undisturbed inertial motion of the system. For example, when $\Omega = \omega$, they take the form of (2–80) if we set $\Gamma_1 = 0$ in the latter.

Let us now pass to the stability of undisturbed motion in the case when the x_3 axis is the axis of symmetry of the cavity and the axis of dynamic symmetry of the body, *i.e.*, when

$$a_1 = a_2\,, \quad A_1 = B_1\,.$$

Under these conditions the equations of disturbed motion of the system have, in addition to the previously cited first integrals, the integral

$$V_5 = z_1 = \text{const}\,.$$

It is easy to see that conditions (2–78) for stability of undisturbed motion now reduce to a single inequality

$$[(C_1 - A_*)\,\omega^2 + C'\omega\Omega + \Gamma_1]\left(\frac{a_1^2 - a_3^2}{a_1^2 + a_3^2} + \frac{\omega}{\Omega}\right) - \frac{4a_1^2 a_3^2}{(a_1^2 + a_3^2)^2}\, C'\omega^2 > 0\,.$$

It is also possible to point out other sufficient conditions for stability, for which purpose we now consider the function

$$\begin{aligned} V &= V_1 + 2\lambda V_2 - (C_*\omega\lambda + C'\Omega\lambda - \Gamma_1)\, V_3 + \lambda_1 V_4 - \\ &\quad - 2C_*(\omega + \lambda)\, V_5 + \frac{C_*(C_* - A_*)}{A_*}\, V_5^2 + \tfrac{1}{2}\Gamma_2 z_3 V_3 = \\ &= A_*(\omega_1^2 + \omega_2^2) + 2A_*\lambda(\omega_1\gamma_1 + \omega_2\gamma_2) - (C_*\omega\lambda - \Gamma_1)\,(\gamma_1^2 + \gamma_2^2) + \\ &\quad + \frac{C_*^2}{A_*}\, z_1^2 + 2C_*\lambda z_1 z_3 - (C_*\omega\lambda - \Gamma_1)\, z_3^2 + (A' + \lambda_1 a_1^2 a_3^2)\,(\Omega_1^2 + \Omega_2^2) + \\ &\quad + 2A'\lambda(\Omega_1\gamma_1 + \Omega_2\gamma_2) - C'\omega\lambda(\gamma_1^2 + \gamma_2^2) + \\ &\quad + (C' + \lambda_1 a_1^4)\, z_2^2 + 2C'\lambda z_2 z_3 - C'\omega\lambda z_3^2 + \cdots, \end{aligned} \tag{2–81}$$

where λ is a constant, while $\lambda_1 = -C'(\lambda + \Omega/\Omega a_1^4)$.

For simplicity, the above function can be regarded as a sum of six quadratic forms of two variables each, namely: ω_i, γ_i; Ω_i, γ_i $(i=1, 2)$; z_1, z_3; $z_2\, z_3$. The sum of the first three of these forms in Eq. (2–81) is analogous to Lyapunov's function (2–66); it will be positive definite when condition (2–67), in which R should be assumed to be equal to zero, is satisfied. Then the constant coefficient λ can have any value in the interval

$$\lambda_1 < \lambda < \lambda_2\,, \tag{2–82}$$

where λ_1 and λ_2 are the roots of equation $A_*\lambda^2 + C_*\omega\lambda - \Gamma_1 = 0$. We note that no matter what the sign of $\Gamma_1 \neq 0$, the smaller of the two roots λ_1 is negative when $\omega \neq 0$.

The fourth and fifth quadratic forms in the right-hand side of Eq. (2–81) will be positive definite when we satisfy the following condition for negative λ:

$$\lambda < \frac{a_1^2 + a_3^2}{a_1^2 - a_3^2}\,\Omega\,. \tag{2–83}$$

Finally, as can easily be seen, the last of the quadratic forms in Eq. (2–81) is positive definite when $\lambda < 0$.

Thus, when conditions (2–67) and (2–83) are satisfied, if in the latter we restrict the selection of the negative constant λ by the inequalities (2–82), the function (2–81) will be a positive definite function of all of the variables, thus proving under these conditions, the stability of undisturbed motion with respect to ω_i, $\Omega_i^{\gamma_i}$, $(i=1, 2, 3)$.

Obviously, in the case of very small $\Omega = \varepsilon$, condition (2–83) will be satisfied, as a result of which condition (2–67) is a sufficient condition for stability of undisturbed motion when the fluid performs a motion quite close to potential motion.

In the case when Ω is finite, condition (2–83) is satisfied *a priori* if $a_1 > a_3$. If, on the other hand, $a_1 < a_3$, sufficient conditions for stability of the undisturbed motion will be given by the inequalities (2–67) and

$$\frac{1}{2A_*}\left(-C_*\omega - \sqrt{C_*^2\omega^2 + 4A_*\Gamma_1}\right) < \frac{a_1^2 + a_3^2}{a_1^2 - a_3^2}\,\Omega\,, \tag{2–84}$$

which follow from conditions (2–82) and (2–83).

We note that from the stability of undisturbed motion with respect to ω_i, Ω_i and from Eqs. (2–52), (2–53), and (1–80) follows the stability of the motion of the fluid in the cavity with respect to the projections of the velocities of the fluid particles $v(r, t)$ and the pressure p.

It was pointed out in Sec. 4 that, in a stationary cavity, Eq. (2–58) describing the motion of the fluid has the form of the equations of inertial motion of a rigid body with one stationary point. By analogy with Euler's case, we can conclude that the undisturbed motion of the fluid in the stationary cavity

$$\Omega_1' = \Omega_2' = 0\,, \quad \Omega_3' = \Omega\,,$$

which is in elliptical rotation about the x_3 axis, is stable if the quantity

$$(a_1^2 + a_2^2)\,a_3^2$$

is larger or smaller than the two other quantities that are obtained from the above by cyclical transposition of indices, and that it is unstable if this quantity falls between the two others.

Some particular cases for this problem are considered in Ref. [40]. For the case of motion by inertia about the center of the cavity, when $a_1 = a_2$ and the rigid body is sufficiently thin that its moment of inertia can be disregarded in comparison with that of the fluid, it was shown by constructing Lyapunov functions that the conditions

$$a_1 > a_3 \quad \text{or} \quad a_3 > 3a_1 \tag{2–85}$$

are sufficient for stability of the undisturbed motion with respect to the four variables ω_i, Ω_i $(i=1, 2)$; these conditions were previously obtained in Refs. [24, 37] as necessary conditions for stability. We wish to note that, by constructing Lyapunov functions in a form slightly different from that examined in Ref. [40], it is possible to prove the sufficiency of conditions (2–85) for stability with respect to the six variables ω_i, Ω_i $(i=1, 2, 3)$.

3

Stability of Motion with Respect to a Part of the Variables of a Rigid Body with Cavities Partially Filled with a Fluid

This chapter presents the statement and solution of some problems of stability of the motion of a rigid body with cavities partially or completely filled with an ideal or viscous fluid. These problems deal with the variables defining the motion of such body and with some quantities characterizing the motion of the fluid as a whole[41]. Stated in this way, the problem can be considered as determining the stability of motion of the system with respect to some variables[42], but not with respect to all the variables that characterize the motion of a system having an infinite number of degrees of freedom. This problem can be solved by the Lyapunov methods for systems with a finite number of degrees of freedom.

[3-1] STATEMENT OF THE PROBLEM

Let us consider some free, or constrained, rigid body with a cavity partially or completely filled with an ideal or viscous fluid, and let the effective forces acting on the system be specified.

The position of the rigid body and the fluid, *i.e.*, of a system with an infinite number of degrees of freedom, with respect to the stationary coordinate system $O'x_1'x_2'x_3'$ can be specified by the coordinates of the body q_j $(j=1, \ldots, n)$ and by the absolute x_1, x_2, x_3, or relative x_1', x_2', x_3', coordinates of the fluid particles, which form a continuum. The velocities of points in the system are represented by time derivatives of the system coordinates. These variables satisfy a certain system of equations of motion.

The equations of motion of a rigid body containing fluid can be written in one of the forms presented in Chapter 1, together with the applicable boundary and initial conditions. In the general case, when no assumption is made about the character of motion of the fluid in a cavity of the type considered in Chapter 2, except for self-evident assumptions of continuity of the fluid motion, the equations of motion are a simultaneous system of ordinary nonlinear differential equations and partial differential equations.

The complete system of equations of motion of a fluid-containing rigid body is so complex that finding its general solution still presents unsurmountable difficulties. However, some progress has been made in proving the existence theorems and the uniqueness of solutions to the equations of motion of a rigid body containing an ideal or a viscous fluid[43,44].

On the basis of these studies, and restricting ourselves only to the case of continuous initial conditions and continuous applied forces acting on the system, we shall assume in our discussion that the solutions of the equations exist and are unique.

A much simpler problem is that of finding some particular solutions of this system of equations of motion, which can frequently be found on the basis of certain considerations.

Let us assume that the equations of motion have some particular solution

$$\begin{gathered} q_j = f_j(t), \quad \dot{q}_j = \dot{f}_j(t)\,(j = 1, \ldots, n), \\ v_i = F_i(x_1, x_2, x_3, t), \quad p = F_4(x_1, x_2, x_3, t) \quad (i = 1, 2, 3), \end{gathered} \tag{3–1}$$

satisfying the specified initial conditions

$$\begin{gathered} q_{j0} = f_j(t_0), \; \dot{q}_{j0} = \dot{f}_j(t_0), \\ v_i(x_1, x_2, x_3, t_0) = \varphi_i(x_1, x_2, x_3), \end{gathered} \tag{3–2}$$

where t_0 is the initial time, as well as boundary conditions of the kind of (1–30) - (1–34) or (1–84) and (1–85). The functions $\varphi_i(x_1, x_2, x_3)$ are assumed to be continuous.

The solution (3–1) corresponds to some motion of the rigid body with the fluid. This motion will be taken as undisturbed and the problem of its stability with respect to changes in the initial conditions will be stated. For the disturbed motion, let conditions (3-2) be replaced by

$$\begin{gathered} q_{j0} = f_j(t_0) + \varepsilon_j, \quad \dot{q}_{j0} = \dot{f}_j(t_0) + \varepsilon_j', \\ v_i(x_1, x_2, x_3, t_0) = \varphi_i(x_1, x_2, x_3) + \mathscr{E}_i(x_1, x_2, x_3), \end{gathered}$$

where ε_j and ε_j' are some real constants, and the $\mathscr{E}_i(x_1, x_2, x_3)$ are some real continuous functions satisfying the equation of continuity. These quantities will be called *disturbances*. It is assumed that the disturbed motion of the system is fully defined by specifying the disturbances, since the forces acting on the system are assumed to be constant; the absolute values of the disturbances will be assumed sufficiently small.

The equations of the disturbed motion will be obtained from the initial equations of motion of the system on performing the replacements

$$\begin{aligned} q_j &= f_j(t) + y_j, \quad \dot{q}_j = \dot{f}_j(t) + \dot{y}_j, \\ v_i &= F_i(x_1, x_2, x_3, t) + w_i(x_1, x_2, x_3, t), \\ p &= F_4(x_1, x_2, x_3, t) + p'(x_1, x_2, x_3, t). \end{aligned} \tag{3–3}$$

It is natural to assume the following definition of stability of motion.

If for any disturbances ε_j, ε_j', $\mathscr{E}_i(x_1, x_2, x_3)$ of sufficiently small absolute magnitude, the deviations or variations y_j, $\dot{y}_j$, $w_i(x_1, x_2, x_3, t)$, $p'(x_1, x_2, x_3, t)$ have an absolute magnitude as small as desired for any $t \geqslant t_0$, then the undisturbed motion (3–1) is called stable with respect to q_j, $\dot{q}_j$, v_i, and p. In the opposite case, the motion will be unstable.

The solution of some problems of stability of motion of a fluid-containing rigid body stated in this manner was given in Chapter 2 for elementary cases, when it was assumed that the motion of the ideal fluid in the cavity is potential or homogeneous irrotational motion during the entire time, including t_0.

In the general case, when no special assumptions are made regarding the character of the motion of the fluid, the problem thus stated is extremely difficult. At present the only known methods for solving it in a limited sense are the methods of the theory of small vibrations, which are presented in Part II of this book.

However, these difficulties can be circumvented to a certain extent by somewhat changing the statement of the problem. The point is that, in practice, the problem of prime interest is that of the stability of motion of the rigid body. The stability of motion of the fluid is of interest only to the extent to which this motion affects the stability of motion of the rigid body. Obviously these two aspects are interrelated; *i.e.*, the motion of the body depends on the motion of the fluid, and conversely.

Because of this, it is natural to state the problem of the stability of motion of a fluid-containing rigid body in Lyapunov's sense is with respect to variables that determine the motion of the rigid body and with respect to some values, *i.e.*, the functionals

$$p_s = \int_\tau \Phi_s(x_1, x_2, x_3, v_1, v_2, v_3)\, d\tau\,,$$

which integrally characterize the motion of the fluid. Here $\Phi_s(x_1, x_2, x_3, v_1, v_2, v_3)$ are some real, continuous, bounded functions of their variables, the selection of which depends on the problem at hand. For example, it can be seen from Eqs. (1–26) and (1–27) that, for given applied external forces the motion of the rigid body depends on the linear and angular momenta of the fluid and on the rates at which they change with time, which in turn depend on the motion of the body. Hence in many cases it is expedient to take as the quantities p_s the projections of the vectors of linear and angular momenta of the fluid and its kinetic energy.

Let us note that the p_s are quantities such that the linear and angular momenta, the kinetic energy, the coordinates of the mass center of the fluid, *etc.*, do not completely, but only integrally, characterize the motion of the fluid which is determined by an infinite number of variables. As a result the stability of motion of the fluid with respect to p_s is a conventional stability; *i.e.*, the stability of motion of a continuous medium with respect to a part of the infinite number of variables characterizing its motion.*

When using this approach, the problem of stability of a rigid body containing a fluid, which has an infinite number of degrees of freedom, reduces to the study of stability with respect to a finite number of quantities, q_j, $\dot{q}_j$ and p_s. It is important here that, in case of stability of the system in the above sense, we also have stability of motion of the rigid body with respect to the quantities q_j and $\dot{q}_j$, which define the motion of the body.

The deviations of disturbed motion from undisturbed can in specific cases be measured in terms of the differences in certain quantities that are functions of the motion but in general are different from the coordinates and velocities of the body and from the integral characteristics p_s of the motion of the fluid. In conjunction with this we now consider some given, continuous, real functions Q_r of q_j, $\dot{q}_j$, p_s, and t, which, in general, are functionals and which, by means of p_s depend on the motion and shape of the fluid volume τ. For undisturbed motion, the functions Q_r, after substitution from (3–1), will become some known functions of time $F_r(t)$, and for disturbed motion given by (3–3) [we retain for them the previous designation Q_r] they will be functions of time and of the variables y_j, $\dot{y}_j$, w_i, and p', which, in turn, depend on the disturbances ε_j, ε'_j $\mathscr{E}_i(x_1, x_2, x_3)$. The differences

$$z_r = Q_r - F_r \quad (r = 1, \ldots, k),$$

which are functions of time and of the above disturbances, characterize the deviations of the disturbed motion from the undisturbed.

We assume the following definition of stability. Let L_r be arbitrary positive numbers. If for any L_r, no matter how small, it is possible to select positive numbers E_j, E'_j, and E_i^* in a manner such that for any disturbances ε_j, ε'_j and

* The problem can also be stated in this manner for problems of the stability of motion of a continuous medium if the p_s, the integral characteristics of motion of the medium, are selected in the appropriate manner. In particular, this idea was developed by A.A. Movchan[45] as applied to the stability of an elastic body.

for any continuous functions $\mathscr{E}_i(x_1, x_2, x_3)$ conforming to the continuity equation and satisfying the conditions

$$|\varepsilon_j| \leqslant E_j, \quad |\varepsilon_j'| \leqslant E_j', \quad |\mathscr{E}_i(x_1, x_2, x_3)| \leqslant E_i^*, \tag{3–4}$$

and for any $t_0 \geqslant t$, the inequalities

$$|Q_r - F_r| < L_r, \tag{3–5}$$

are satisfied, then the disturbed motion is stable with respect to Q_r $(r=1, \ldots, k)$, while in the opposite case it is unstable.

[3-2] APPLICATION OF THE METHOD OF LYAPUNOV'S FUNCTIONS TO PROBLEMS OF STABILITY WITH RESPECT TO A PART OF THE VARIABLES

The statement, presented above, of the problem of stability of motion of a rigid body containing a fluid with respect to a part of the variables reduces the problem of the stability of motion of a system with an infinite number of degrees of freedom to the problem of the stability with respect to a finite number of quantities Q_r. This makes it possible to apply to the solution the well-developed methods of the theory of stability for systems with a finite number of degrees of freedom, in particular, the method of Lyapunov's functions.

Each time that it is possible to construct a function V that satisfies all the conditions of Lyapunov's theorem on stability, we can assume that the undisturbed motion is stable with respect to the variables on which the functions V depend. However, in problems of stability with respect to a part of the variables, the cases when the function V depends only on variables with respect to which stability is of interest are apparently few. In turns out, in general, that the functions V depend not only on the variables with respect to which stability is of interest, but also on other variables, and they do not satisfy Lyapunov's theorem on stability. In these cases it is possible to apply the theory of stability with respect to a part of variables which is a modification of the above Lyapunov theorem.

We recall that if the sign-definite function $V(z_1, \ldots, z_n)$ is not explicitly a function of time, and in the region $\sum_{s=1}^{m} z_s^2 \leqslant H$ where $z_{m+1}, \ldots, z_n$ are arbitrary; and $H = \text{const}$, can vanish only when all of the variables $z_r(r=1, \ldots, m \leqslant n)$ are zero, then it is called sign definite with respect to $z_r(r=1, \ldots, m)$. The function V, which is explicitly time-dependent, will be called sign-definite with respect to the variables $z_1, \ldots, z_m$ only under the condition that it is possible to find a nontime-dependent positive definite function $W(z_1,\dots,z_m)$ such that one of the expressions $V-W$ or $-V-W$ is a positive function.

Theorem I

If the differential equations of disturbed motion are such that it is possible to find a function V that is sign-positive with respect to the variables $z_1,\dots,$ $z_1,\dots,z_m$ and if the derivative $\dot{V}$, by virtue of these equations, is a sign-constant function of sign opposite to that of V, or is identically equal to zero, then the undisturbed motion is stable with respect to the above variables.

Proof

Let $V(t, z_1,\dots,z_n)$ be a positive definite function with respect to the variables $z_1,\dots,z_m$, and let its time derivative $\dot{V}$, by virtue of the equations of disturbed motion, be negative or zero. By definition, we will be able to find a nontime-dependent position definite function $W(z_1,\dots,z_m)$ such that $V-W\geqslant 0$.

Let A be some arbitrarily small positive number, smaller than H; let $l>0$ be the exact lower limit of the function W at the sphere (A)

$$\sum_{r=1}^{m} z_r^2 = A\,.$$

The function $V(t_0, z_1,\dots,z_n)$, not being explicitly time-dependent, permits an infinitesimal upper limit, and consequently we will find for l a λ such that for values of z_s satisfying condition $\sum_{s=1}^{n} z_s^2\leqslant\lambda$, the values of the functions $V(t_0, z_1,\dots,z_n)$ will satisfy the inequality

$$V(t_0, z_1,\dots,z_n) < l\,.$$

If the initial values z_{s0} of the variables z_s are selected according to the inequality $\sum_{s=1}^{n} z_{s0}^2\leqslant\lambda$, we conclude from the expression

$$V - V_0 = \int_{t_0}^{t} \dot{V}dt$$

(where $V_0=V(t_0, z_{10},\dots,z_{n0})<l$, and $\dot{V}=0$) that when the z_s are varied, by virtue of the equations of disturbed motion, they will satisfy the conditions

$$W \leqslant V \leqslant V_0 < l\,,$$

and, by virtue of the same fact, also the condition $\sum_{r=1}^{m} z_r^2<A$, since l is the exact lower limit of the function W on sphere (A), which proves the theorem.

Remark

Since the specifics of the differential equation of the disturbed motion played no role in proving Theorem I, it is applicable to systems with a finite as well as with an infinite number of degrees of freedom. In this last case, the function V can, in general, depend on an infinite number of variables or be a functional. Such a function, whether explicitly time-dependent or not, we will call sign-definite with respect to $z_1, \ldots, z_m$ only under the condition that it is possible to find for it a nontime-dependent positive definite function $W(z_1, \ldots, z_m)$ such that one of the expressions $V - W$ or $-V - W$ is negative. The time derivative $\dot{V}$ should obviously be evaluated for the complete system of equations of disturbed motion.

In a number of cases, when constructing functions (or functionals) V that are sign-definite with respect to a part of the variables, it may be convenient to use some inequalities between functions of z_s and functions of y_j, $\dot{y}_j$, and w_i, which are established, by virtue of one or another consideration, for all of the values of the variables in their respective regions of variation.

Since z_s are some functions of the variables t, y_j, $\dot{y}_j$, and w_i that vanish when all these variables are zero, the region of variation of these variables

$$t \geqslant t_0\,, \quad |y_j| \leqslant H\,, \quad |\dot{y}_j| \leqslant H\,, \quad |w_i| \leqslant H \tag{3–6}$$

will have corresponding to it the region

$$t \geqslant t_0\,, \quad |z_s| \leqslant H_1 \tag{3–7}$$

of variation of the variables t and z_s. Here H and H_1 denote some numerically sufficiently small positive constants.

Let it be known that at any $t \geq t_0$, for all values of y_j, $\dot{y}_j$, and w_i in region (3–6), and for the corresponding values of z_s in region (3–7), we have the inequality

$$V(t, z_1, \ldots, z_m) \leqslant \varphi(t, y_1, \ldots, y_n, \dot{y}_1, \ldots, \dot{y}_n, w_1, w_2, w_3)\,, \tag{3–8}$$

where V and φ are real, continuous, bounded, single-valued functions of all the variables that vanish when all the y_j, $\dot{y}_j$, and w_i (and, consequently, also z_s) are zero.

Using Eq. (3–8) it is possible, under specific conditions, to establish the stability of undisturbed motion (3–1) with respect to Q_r. Namely, we have the following theorem[46].

Theorem II

If the differential equations of disturbed motion are such that it is possible to find a function $\varphi(t, y_j, \dot{y}_j, w_i)$ with a time derivative $\dot{\varphi} \leqslant 0$ because of these equations and there exists a positive definite function $V(t, z_1, \ldots, z_m)$ such that the inequality (3–8) applies, then the undisturbed motion is stable with respect to Q_r $(r=1, \ldots, m)$.

Proof

By the definition of a sign-definite function, we will be able to find a non-time-dependent positive definite function $W(z_1, \ldots, z_m)$ such that the inequality

$$V(t, z_1, \ldots, z_m) \geqslant W(z_1, \ldots, z_m)$$

applies. But then, by virtue of the inequality (3–8) the function φ can be considered to be sign-definite with respect to a part of the variables $z_1, \ldots, z_m$, although it is not explicitly expressed in terms of them.

If the time derivative of this function is $\dot{\varphi} \leqslant 0$, it obviously satisfies all of the conditions of Theorem I. Let us, however, give an indirect proof of Theorem II.

Let $A>0$ be an arbitrary number smaller than H_1, and let $l>0$ be the exact lower boundary of the function W on sphere (A)

$$\sum_{r=1}^{m} z_r^2 = A .$$

Let us consider the function $\varphi(t_0, y_j, \dot{y}_j, w_i)$; since it is not explicitly time-dependent, it allows an infinitesimal upper limit, and consequently we will be able to find for the number l numbers $\lambda>0$ and $\lambda_1>0$ such that for all of the values of $y_j, \dot{y}_j$, and w_i satisfying the conditions

$$|y_j| \leqslant \lambda, \quad |\dot{y}_j| \leqslant \lambda, \quad |w_i| \leqslant \lambda, \tag{3–9}$$

and for all the values of z_r corresponding to them and satisfying the conditions $|z_r| \leqslant \lambda_1$, the values of the functions $V(t_0, z_1, \ldots, z_m)$ and $\varphi(t_0, y_j, \dot{y}_j, w_i)$ will satisfy the conditions

$$V(t_0, z_1, \ldots, z_m) \leqslant \varphi(t_0, y_j, \dot{y}_j, w_i) < l .$$

If the initial values of y_j, $\dot{y}_j$, and w_i are selected in accordance with Eq. (3–9), then, from the conditions of the theorem, we have the inequalities

$$W(z_1, \ldots, z_m) \leqslant V(t, z_1, \ldots, z_m) \leqslant \varphi(t, y_j, \dot{y}_j, w_i) < l .$$

We conclude from this that, corresponding to their variations, z_r will satisfy the condition $\sum_{z=1}^{m} z_r^2 < A$, since l is the exact lower limit of the function W on sphere (A), which proves the theorem.

Remark

If, for the conditions in the statement of Theorem II, we have instead of (3–8) the inequality

$$V(t, z_1, \ldots, z_m) \leqslant \varphi(t, y_1, \ldots, y_n, \dot{y}_1, \ldots, \dot{y}_n, w_1, w_2, w_3) + \alpha, \quad (3\text{–}10)$$

where $\alpha > 0$ is some constant, then, for an appropriate selection of y_{j0}, $\dot{y}_{j0}$, and w_{i0}, the following condition will be valid during the entire motion:

$$\sum_{r=1}^{m} z_r^2 < A,$$

where A is a number such that the exact lower limit of the function W on sphere (A) exceeds the number α. In this case the undisturbed motion is bounded with respect to z_r.

Subsequently, in constructing functions satisfying the conditions of the above theorems pertaining to stability with respect to a part of the variables, we shall use the first integrals of the equations of disturbed motion, corresponding to integrals (1–59) and (1–66)–(1–77) of the starting equations of motion, as well as the integral relationship obtained from (1–88) for disturbed motion of a viscous fluid.

[3-3] STABILITY OF MOTION OF A ROTATING BODY FILLED WITH A VISCOUS FLUID

Let us consider a rigid body with a cavity completely filled with viscous fluid, moving about a stationary point O in the force field $U = U(\gamma_3)$. The equations of motion take the form of the second of Eqs. (1–40) and of the Navier-Stokes equations (1–83) together with boundary conditions (1–85). As is easily seen these equations allow the energy relationship (1–88) and the first integrals, *i.e.*, the area integral (1–70) and a geometric integral of the form of (1–77).

The equations of motion have the particular solution

$$\begin{gathered} \omega_1 = \omega_2 = 0, \qquad \omega_3 = \omega, \\ \gamma_1 = \gamma_2 = 0, \qquad \gamma_3 = 1, \\ G_{21} = G_{22} = 0, \quad G_{23} = C_2\omega, \\ u_1 = u_2 = u_3 = 0, \end{gathered}$$

describing uniform rotation about the x_3 axis, coinciding with the x_3' axis of the rigid body and of the fluid in its cavity (Fig. 12).

Let us examine the stability of this motion with respect to the projections of the instantaneous angular velocity of the body ω_i, the projections of the fluid angular momentum G_{2i}, and the direction cosines γ_i of the x_3' axis $(i=1, 2, 3)$[32,47]. For the disturbed motion we set

$$\omega_3 = \omega + z_1\,, \quad G_{23} = C_2\omega + z_2\,, \quad \gamma_3 = 1 + z_3\,,$$

and the previous designations are retained for the remaining variables.

Before attempting to solve the problem of stability, we slightly transform expression (1–13) for the kinetic energy of the fluid. We assume, for simplicity,

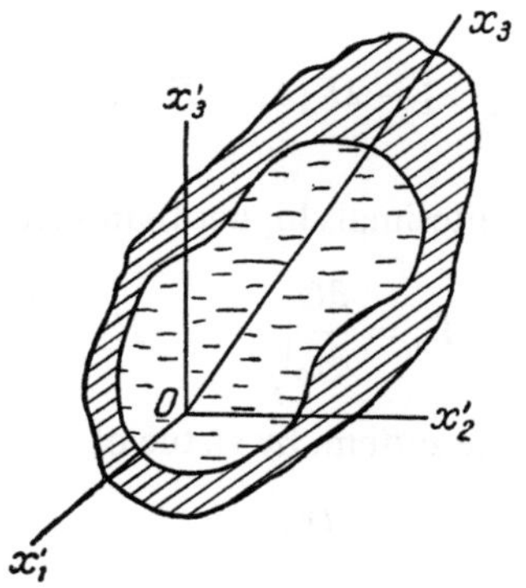

Fig. 12.

that the principal axes of inertia of the body with respect to the fixed point O are also the principal axes of inertia of the cavity. When the system moves, the vector of the angular momentum of the fluid $\boldsymbol{G}_2$ will be some function of time. We replace this vector by the vector $\boldsymbol{\kappa}$, defining it by its projection on the moving axes

$$\kappa_1 = \frac{G_{21}}{A_2}\,, \quad \kappa_2 = \frac{G_{22}}{B_2}\,, \quad \kappa_3 = \frac{G_{23}}{C_2}\,. \tag{3–11}$$

If the fluid velocities $\boldsymbol{v}$ are known, this also defines the vector $\boldsymbol{\kappa}$. Let us also consider the quantity $\boldsymbol{v}_*(\boldsymbol{r}, t)$, defined as

$$\boldsymbol{v}_* = \boldsymbol{v} - \boldsymbol{\kappa} \times \boldsymbol{r}\,. \tag{3–12}$$

Here it is easy to see that, by virtue of Eq. (3–11),

$$\varrho \int_\tau \boldsymbol{r} \times \boldsymbol{v}_* d\tau = 0\,.$$

Using Eqs. (3–11) and (3–12), we can represent the kinetic energy of the fluid in the form

$$2T_2 = \frac{G_{21}^2}{A_2} + \frac{G_{22}^2}{B_2} + \frac{G_{23}^2}{C_2} + \varrho \int_\tau v_*^2 d\tau\,. \tag{3–13}$$

from which immediately follows the validity of the inequality [48]

$$2T_2 I \geqslant G_{21}^2 + G_{22}^2 + G_{23}^2, \quad I = \max(A_2, B_2, C_2). \tag{3–14}$$

Passing to the study of the stability of disturbed motion, we note that

$$\kappa_1 = \kappa_2 = 0, \quad \kappa_3 = \omega, \quad \boldsymbol{v}_* = 0.$$

We consider the function

$$V_1 = A_1\omega_1^2 + B_1\omega_2^2 + C_1(2\omega z_1 + z_1^2) + \frac{1}{A_2} G_{21}^2 + \frac{1}{B_2} G_{22}^2 + \\ + 2\omega z_2 + \frac{1}{C_2} z_2^2 - 2\Gamma_1 z_3 - \Gamma_2 z_3^2 + \varrho \int_\tau v_*^2 d\tau + \cdots,$$

where the dots denote infinitesimals higher than second order with respect to

$$z_3, \quad \text{and} \quad \Gamma_1 = \left.\frac{\partial U}{\partial \gamma_3}\right|_{\gamma_3=1}, \quad \Gamma_2 = \left.\frac{\partial^2 U}{\partial \gamma_3^2}\right|_{\gamma_3=1}.$$

For disturbed motion of the system, by virtue of Eq. (1–88), we will have

$$\frac{dV_1}{dt} \leqslant 0, \tag{3–15}$$

since

$$V_1 = 2(T - U) - C\omega^2 + 2U(1).$$

The equations of disturbed motion also allow the first integrals

$$V_2 = (A_1\omega_1 + G_{21})\gamma_1 + (B_1\omega_2 + G_{22})\gamma_2 + C_1 z_1 + z_2 + \\ + C_1(\omega + z_1) z_3 + (C_2\omega + z_2) z_3 = \text{const},$$
$$V_3 = \gamma_1^2 + \gamma_2^2 + z_3^2 + 2z_3 = 0.$$

We consider the function

$$V = V_1 - 2\omega V_2 + (C\omega^2 + \Gamma_1) V_3 + \tfrac{1}{4}\lambda V_3^2 = \\ = A_1\omega_1^2 - 2\omega(A_1\omega_1 + G_{21})\gamma_1 + \frac{1}{A_2} G_{21}^2 + (C\omega^2 + \Gamma_1)\gamma_1^2 + \\ + B_1\omega_2^2 - 2\omega(B_1\omega_2 + G_{22})\gamma_2 + \frac{1}{B_2} G_{22}^2 + C\omega^2 + (\Gamma_1)\gamma_2^2 + \\ + C_1 z_1^2 - 2\omega(C_1 z_1 + z_2) z_3 + \frac{1}{C_2} z_2^2 + (C\omega^2 + \Gamma_1 - \\ - \Gamma_2 + \lambda) z_3^2 + \varrho \int_\tau v_*^2 d\tau + \cdots, \tag{3–16}$$

where the constant $\lambda > \Gamma_2 - \Gamma_1$.

According to Sylvester's criterion, for positive definiteness of the function (3–16) with respect to the variables ω_i, G_i, and γ_i it is necessary and sufficient to satisfy the inequality

$$(C - A)\,\omega^2 + \Gamma_1 > 0 \quad (A \geqslant B)\,. \tag{3–17}$$

The derivative of the function V, taken by virtue of the equations of disturbed motion, will be nonpositive as a result of inequality (3–15).

Consequently, under condition (3–17), the function (3–16) satisfies all of the conditions of Theorem I on stability with respect to a part of the variables. This proves the stability of the undisturbed motion of a rigid body with a cavity filled with a viscous fluid with respect to the quantities

$$\omega_i,\ G_{2i},\ \gamma_i (i = 1, 2, 3)\,.$$

In addition, stability with respect to $\varrho\int_\tau v_*^2 d\tau$ is also obvious.

We note that if the fluid is ideal, $\dot{V}=0$ and the condition for the stability of undisturbed motion will, as before, have the form of (3–17). Thus the sufficient condition for stability (3–17) is independent of the nature of the fluid in the cavity, which affects only the character of the disturbed motion. Namely, with a viscous fluid, energy will be dissipated according to Eq. (1–88) until the body and the fluid move together as a single rigid body.

It is interesting to compare condition (3–17) for the stability of rotation of a body containing an ideal or viscous fluid with conditions (2–72) when $R=0$, or (2–79) and (2–80), which apply to a body filled with an ideal fluid in irrotational or homogeneous vortex motion. Condition (3–17) is identical with condition (2–80), but differs from conditions (2–72) and (2–79) in that it contains the moment of inertia A_2 and C_2 of the fluid instead of the moments A^* and C^* of the equivalent rigid body.

This is understandable since conditions (2–80) and (3–17) pertain to undisturbed motion, when the fluid and the body move together as one, while conditions (2–72) and (2–79) pertain to irrotational or homogeneous vortex (almost irrotational) motion of the fluid.

For a solid rigid body, when

$$U(\gamma_3) = -\,Mgx_{c3}\gamma_3\,,$$

condition (3–17) for the stability of undisturbed motion becomes

$$(C - A)\,\omega^2 > Mgx_{c3} \quad (A \geqslant B)\,. \tag{3–18}$$

The problem of stability of a symmetrical top with an axially symmetrical cavity completely filled with an ideal fluid performing vortex motion, in its linear formulation, was studied by Sobolev[38], who proved in particular that if

$$(C - A)\,\omega^2 - Mgx_{c3} > 0\,,$$

then some operator characterizing the disturbed motion is bounded. Obviously the condition for the boundedness of this operator is identical to the condition (3–18) for the stability of undisturbed motion with respect to ω_i, G_{2i}, and γ_i $(i=1, 2, 3)$.

We note that for a rigid body containing a fluid and moving by inertia about the mass center of the system ($x_{c_3}=0$), or about a fixed point, condition (3–18) takes the form

$$C > A \geqslant B. \tag{3–19}$$

Consequently, the permanent rotations of a rigid body with a cavity completely filled with an ideal or viscous fluid about the minor axis of the central ellipsoid of inertia of the system are stable with respect to ω_i and G_{2i}.

This result supplements one of Zhukovskiy's theorems about the inertial motion of an unconstrained rigid body with a cavity completely filled with a viscous fluid[15].

Zhukovskiy's Theorem

If some initial velocities are imparted to a body with some cavity filled with a viscous fluid, then its motion will tend to approach a limiting state, in which one of the principal axes of inertia will take the direction of the initial principal angular momentum, and the entire system will rotate about it as a single invariable body with a constant angular velocity, which is obtained by dividing the initial principal angular momentum by the moment of inertia of the system about this axis.

[3-4] STABILITY OF CONSTANT INERTIAL SCREW MOTION OF A BODY CONTAINING A FLUID

This problem was solved in Sec. 6 of Chapter 2 on the assumption that the cavity is filled with an ideal fluid performing irrotational motion. Here we shall forego this assumption and consider an unconstrained rigid body with a cavity completely filled with a viscous fluid, moving by inertia. We shall place the origin of the moving coordinate system $Ox_1x_2x_3$ at the mass center, and we shall direct the coordinate axes along the principal central axes of inertia of the system. The equations of motion will be taken in the form of Eqs. (1–40), together with Eqs. (1–83) and boundary conditions (1–85). In this case the principal vector $\boldsymbol{K}$ and the principal moment $\boldsymbol{L}$ of the applied forces are assumed to be zero.

Obviously the equations of motion allow a relationship of the form of (1–88) and the following first integrals: integral of the constancy of the velocity vector of the mass center (1–69) and of the constancy of the angular momentum of the system about the mass center (1–73). The equations of motion have the particular solution

$$v_{01}=v\,,\quad v_{02}=v_{03}=0\,,\quad \omega_1=\omega\,,\quad \omega_2=\omega_3=0\,,$$
$$G_{21}=A_2\omega\,,\quad G_{22}=G_{23}=0\,,\quad u_1=u_2=u_3=0\,,$$

describing the constant screw motion of the rigid body and the fluid along the x_1 axis. We shall assume this to be our undisturbed motion and we shall examine its stability with respect to v_{0i}, ω_i, and G_{2i} ($i=1, 2, 3$). In disturbed motion, we set

$$v_{01}=v+z_1\,,\quad \omega_1=\omega+z_2\,,\quad G_{21}+A_2\omega=z_3\,,$$

while the previous designation is retained for the remaining variables.

It will be assumed that the coordinate axes are the principal axes of inertia of the rigid body as well as of the fluid. The kinetic energy of the fluid in its motion relative to the mass center will be represented by (3–13).

The equations of the disturbed motion allow the integral relationship

$$\begin{aligned}V_1 &= M(z_1^2+2vz_1+v_{02}^2+v_{03}^2)+A_1(2\omega z_2+z_2^2)+B_1\omega_2^2+\\&\quad+C_1\omega_3^2+2\omega z_3+\frac{1}{A_2}z_3^2+\frac{1}{B_2}G_{22}^2+\frac{1}{C_2}G_{23}^2+\varrho\int_\tau v_*^2 d\tau\leqslant \text{const}.\end{aligned}$$

Here $\dot V_1 \leqslant 0$, and the first integrals are:

$$\begin{aligned}V_2 &= A_1vz_2+A_1\omega z_1+A_1z_1z_2+vz_3+A_2\omega z_1+z_1z_3+\\&\qquad+(B_1\omega_2+G_{22})\,v_{02}+(C_1\omega_3+G_{23})\,v_{03}=\text{const},\\ V_3 &= z_1^2+2vz_1+v_{02}^2+v_{03}^2=\text{const}.\end{aligned}$$

We consider the function

$$\begin{aligned}V &= V_1-2\lambda_1V_2+\lambda_2V_3+V_3^2=\\&=(M+\lambda_2+4v^2)\,z_1^2-2A_1\lambda_1z_1z_2-2\lambda_1z_1z_3+A_1z_2^2+\frac{1}{A_2}z_3^2+\\&\quad+(M+\lambda_2)\,v_{02}^2-2\lambda_1(B_1\omega_2+G_{22})\,v_{02}+B_1\omega_2^2+\frac{1}{B_2}G_{22}^2+\\&\quad+(M+\lambda_2)\,v_{03}^2-2\lambda_1(C_1\omega_3+G_{23})\,v_{03}+C_1\omega_3^2+\frac{1}{C_2}G_{23}^2+\\&\quad+\varrho\int_\tau v_*^2d\tau+(z_1^2+v_{02}^2+v_{03}^2)^2+4vz_1(z_1^2+v_{02}^2+v_{03}^2),\end{aligned}\tag{3–20}$$

where the constants are

$$\lambda_1 = \frac{\omega}{v}, \quad M + \lambda_2 = A\frac{\omega^2}{v^2}.$$

Applying Sylvester's criterion, it is easy to find that the conditions

$$A > B, \quad A > C \tag{3–21}$$

are necessary and sufficient for the positive definiteness of the function (3–20) with respect to the variables $z_i (i=1, 2, 3)$, v_{0s}, ω_s, G_{2s} $(s=2, 3)$. Since the time derivative of V, by virtue of the equations of disturbed motion, is

$$\dot{V} = \dot{V}_1 \leqslant 0,$$

this function satisfies all the conditions of Theorem I on stability with respect to a part of the variables.

Consequently the inequalities (3–21) are sufficient conditions for the stability of screw motion of a rigid body with an ideal or a viscous fluid with respect to v_{0i}, ω_i, G_{2i} $(i=1, 2, 3)$, $\varrho \int_\tau v_*^2 d\tau$. Obviously, conditions (3–21) differ from conditions (2–64) in that the moments of inertia of the equivalent body A^* and C^* in conditions (2–64) have been replaced by A_2 and C_2, the moments of inertia of the fluid.

[3-5] STABILITY OF CIRCULAR MOTION OF AN ARTIFICIAL SATELLITE CONTAINING A FLUID

Let us consider an unconstrained body with a cavity completely filled with a viscous fluid and placed in a central Newtonian force field. This problem simulates, to a given approximation, the problem of the motion of a fluid-containing artificial satellite.

We shall take the stationary gravitational center as the origin of the stationary coordinate system $O'x_1'x_2'x_3'$, and the mass center of the solid body plus the fluid as the origin of the moving coordinate system $Ox_1x_2x_3$ rigidly fixed to the rigid body, and with its axes directed along the principal central axes of inertia. It will be assumed for simplicity that these axes are the principal axes of inertia of the rigid body as well as of the fluid. The position of the rigid body in the stationary coordinate system can be defined by the coordinates of the mass center O and by Euler's angles, or by the cosines of the angles between the stationary and moving axes (Fig. 13).

Introducing the spherical coordinates R, ψ, φ of the system's mass center,

we have
$$x'_{01} = R\cos\psi\cos\varphi\,,\quad x'_{02} = R\cos\psi\sin\varphi\,,\quad x'_{03} = R\sin\psi\,. \tag{3–22}$$

The force function of the gravitational forces can, with a sufficient degree of precision, be represented as[49]

$$U = \mu\frac{M}{R} - \frac{3}{2}\frac{\mu}{R^3}\left(A\beta_1^2 + B\beta_2^2 + C\beta_3^2 - \frac{A+B+C}{3}\right), \tag{3–23}$$

where μ is the gravitational constant, β_i are the cosines of the angles between the line $O'O$ and the x_i axes ($i=1, 2, 3$). The symbol γ_i, is used as before

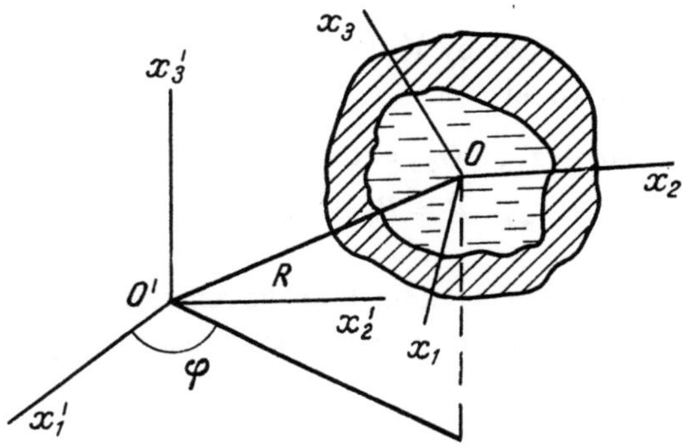

Fig. 13.

to denote the cosines of the angles between the $O'x'$ axis and the x_i axes; $\boldsymbol{k}$ is the unit vector of $\boldsymbol{R}=O'O$.

The equations of motion of the system's mass center and of the motion about the mass center can be written in the form of Eqs. (1–40); the principal vector $\boldsymbol{K}$ and principal moment $\boldsymbol{L}$ of the effective forces are then

$$\boldsymbol{K} = \operatorname{grad}_{\boldsymbol{R}} U\,,\quad \boldsymbol{L} = \operatorname{grad}_{\boldsymbol{k}} U \times \boldsymbol{k}\,.$$

Supplementing Eqs. (1–40) by the Navier-Stokes equations (1–83), together with the boundary conditions (1–85), we obtain a complete system of equations of motion of a rigid body containing a fluid.

It is easy to see that the equations of motion of the system allow an energy relationship of the form of (1–88) and the following first integrals: the area integral (1–70) and geometrical integrals of the form of (1–77). The equations of motion of the system have the particular solution

$$\begin{gathered} R = R_0\,,\quad \psi = 0\,,\quad \varphi = \omega t\,,\quad \omega_1 = \omega_2 = 0\,,\quad \omega_3 = \omega\,,\\ \gamma_1 = \gamma_2 = 0\,,\quad \gamma_3 = 1\,,\quad \beta_1 = \beta_3 = 0\,,\quad \beta_2 = 1\,,\\ G_{21} = G_{22} = 0\,,\quad G_{23} = C_2\omega\,,\quad u_1 = u_2 = u_3 = 0\,, \end{gathered} \tag{3–24}$$

if the constants R_0 and ω satisfy the condition

$$MR_0^3\omega^2 = \mu\left(M + \frac{3}{2}\frac{A - 2B + C}{R_0^2}\right). \tag{3-25}$$

This solution corresponds to the motion mass center of the system in a circular orbit with constant angular velocity ω in the $x'_1O'x'_2$ plane with radius R_0 and with its center at point O'; here the x_1 axis is directed along the tangent to the orbit, the x_2 axis is directed along the line $O'O$, and the x_3 axis is parallel to the x'_3 axis. We shall take this as our undisturbed motion and consider its stability with respect to the variables $R, \dot{R}, \psi, \dot{\psi}, \dot{\varphi}, \omega_i, G_{2i}, \beta_i, \gamma_i$ $(i=1, 2, 3)$.

This problem was solved by Kolesnikov[50] by the method of Sec. 3. In the disturbed motion we set

$$\omega_3 = \omega + z_1, \quad G_{23} = C_2\omega + z_2, \quad \gamma_3 = 1 + z_3,$$
$$\beta_2 = 1 + z_4, \quad R = R_0 + z_5, \quad \dot{\varphi} = \omega + \dot{z}_6,$$

while the previous notation is retained for the remaining variables.

The equations of disturbed motion allow the relationship

$$\begin{aligned}
V_1 = {} & A_1\omega_1^2 + B_1\omega_2^2 + C_1(z_1^2 + 2\omega z_1) + \frac{1}{A_2}G_{21}^2 + \frac{1}{B_2}G_{22}^2 + \\
& + \frac{1}{C_2}z_2^2 + 2\omega z_2 + 3\frac{\mu}{R_0^2}(A\beta_1^2 + Bz_4^2 + 2Bz_4 + C\beta_3^2) + \\
& + M\left[\dot{z}_5^2 + R_0^2\left(\dot{\psi}^2 - \omega^2\psi^2 + \dot{z}_6^2 + 2\omega\dot{z}_6 + 2\frac{\omega^2}{R_0}z_5 + 4\frac{\omega}{R_0}z_5\dot{z}_6\right)\right] + \\
& + \left(M\omega^2 - 2\frac{\mu M}{R_0^3} + 18\frac{\mu B}{R_0^5} - 6\frac{\mu(A + B + C)}{R_0^5}\right)z_5^2 + \\
& + \frac{\mu}{R_0^2}\left(2M - 9\frac{B}{R_0^2} + 3\frac{A + B + C}{R_0^2} - 18\frac{B}{R_0^2}z_4\right)z_5 + \\
& + \varrho\int_\tau v_*^2 d\tau + \cdots \leqslant \text{const}.
\end{aligned}$$

Here $\dot{V}_1 \leqslant 0$, and the first integrals are

$$\begin{aligned}
V_2 = {} & (A_1\omega_1 + G_{21})\gamma_1 + (B_1\omega_2 + G_{22})\gamma_2 + C_1z_1 + C_1z_1z_3 + \\
& + z_2 + z_2z_3 + C\omega z_3 + M(\omega z_5^2 - R_0^2\omega\psi^2 + 2R_0z_5\dot{z}_6 + \\
& + R_0^2\dot{z}_6 + 2R_0\omega z_5) + \cdots = \text{const},
\end{aligned}$$
$$V_3 = \beta_1^2 + \beta_3^2 + z_4^2 + 2z_4 = 0,$$
$$V_4 = \gamma_1^2 + \gamma_2^2 + z_3^2 + 2z_3 = 0.$$

Here and below the dots denote terms of not less than the third order of smallness.

Using Chetayev's method, we construct a function in the form

$$V = V_1 - 2\omega V_2 - 3B\frac{\mu}{R_0^3}V_3 + C\omega^2 V_4 + \lambda_1 V_2^2 + \lambda_2 V_3^2 + \lambda_3 V_4^2 =$$

$$= A_1\omega_1^2 - 2A_1\omega\omega_1\gamma_1 + C\omega^2\gamma_1^2 - 2\omega G_{21}\gamma_1 + \frac{1}{A_2}G_{21}^2 +$$

$$+ B_1\omega_2^2 - 2B_1\omega\omega_2\gamma_2 + C\omega^2\gamma_2^2 - 2\omega G_{22}\gamma_2 + \frac{1}{B_2}G_{22}^2 +$$

$$+ C_1(1 + C_1\lambda_1)z_1^2 - 2C_1\omega(1 - C\lambda_1)z_1z_3 + (C\omega^2 + C^2\lambda_1\omega^2 + 4\lambda_3)\times$$

$$\times z_3^2 - 2\omega(1 - C\lambda_1)z_2z_3 + \left(\frac{1}{C_2} + \lambda_1\right)z_2^2 + 4\lambda_2 z_4^2 +$$

$$+ M\Bigg[\dot{z}_5^2 + R_0^2\dot{\psi}^2 + R_0^2\omega^2\psi^2 + \left(2\frac{\mu}{R_0^3} - 5\omega^2 + 4\lambda_1 MR_0^2\omega^2\right)z_5^2 +$$

$$+ R_0^2(1 + \lambda_1 MR_0^2)\dot{z}_6^2 + 4\lambda_1 MR_0^3\omega z_5\dot{z}_6 + 4\lambda_1 R_0 C\omega^2 z_3 z_5 +$$

$$+ 4\lambda_1 R_0 C_1\omega z_1 z_5 + 4\lambda_1 R_0\omega z_2 z_5 + 2\lambda_1 R_0^2 C\omega z_3\dot{z}_6 + 2\lambda_1 R_0^2 C_1 z_1\dot{z}_6 +$$

$$+ 2\lambda_1 R_0^2 z_2\dot{z}_6\Bigg] - 18\frac{\mu B}{R_0^4}z_4z_5 + 2\lambda_1 C_1 z_1 z_2 +$$

$$+ 3\frac{\mu}{R_0^3}[(A - B)\beta_1^2 + (C - B)\beta_3^2] + \varrho\int_\tau v_*^2 d\tau + \cdots, \tag{3–26}$$

where λ_i $(i=1, 2, 3)$ are some constants.

Applying Sylvester's criterion, it is easy to satisfy ourselves that the constants λ_i can be selected so that V will be a positive definite function with respect to the variables ω_i, G_{2i}, $\gamma_i (i=1, 2)$, β_1, β_3, ψ, $\dot{\psi}$, $z_s (s=1, \ldots, 5)$, $\dot{z}_5$, $\dot{z}_6$ if we satisfy the conditions

$$C > A > B. \tag{3–27}$$

Since the time derivative of V, by virtue of the equations of disturbed motion, is

$$\dot{V} = \dot{V}_1 \leqslant 0,$$

this function, when conditions (3–27) are satisfied, satisfies the requirements of Theorem I concerning stability with respect to a part of the variables.

Consequently, conditions (3–27) are sufficient for the stability of circular motion of an artificial satellite with cavities completely filled with an ideal or viscous fluid, with respect to the eighteen quantities

$$\omega_i,\ G_{2i},\ \gamma_i,\ \beta_i\ (i = 1, 2, 3),\quad R,\ \dot{R},\ \psi,\ \dot{\psi},\ \dot{\varphi},\ \varrho\int_\tau v_*^2 d\tau.$$

Conditions (3–27) for the stability of motion of a rigid body containing a fluid have precisely the same form as the sufficient conditions for the circular motion of one rigid body with moments of inertia A, B, C in a central Newtonian force field[49].

It is interesting that, both in permanent rotation and inertial screw motion and in a central Newtonian force field, the rotation of the system about the minor axis of the central ellipsoid of inertia of the system, which is assumed to be triaxial, ensures stability of the undisturbed motion according to conditions (3–19), (3–21), and (3–27).

[3-6] STABILITY OF ROTATIONAL MOTIONS OF A FLUID-FILLED PROJECTILE

We now consider a projectile with a cavity partially (*i.e.*, with a bubble) or completely filled with an ideal fluid. We assume for simplicity that the central ellipsoid of inertia of the rigid body is an ellipsoid of revolution and that the cavity has the shape of a body of revolution about the axis of dynamic symmetry of the rigid body; we disregard the surface tension of the fluid (Fig. 14).

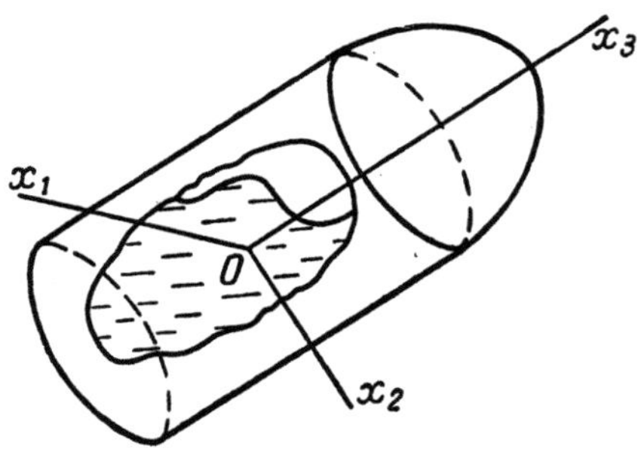

Fig. 14.

When the fluid completely fills the body, the origin O of the moving coordinate system will be taken at the mass center of the system, and the x_1, x_2, and x_3 axes will be directed along the principal axes of inertia of the rigid body for point O. If the fluid partially fills the cavity, the origin will be placed at the mass center of the rigid body, and the coordinate axes will be directed along its principal central axes of inertia. In both cases the x_3 axis will coincide with the axis of rotation of the central ellipsoid of inertia of the body and its cavity.

To examine something more specific, we shall consider the motion of the mass center in a straight line with constant velocity. This case is known for moderately sized segments of flat trajectories of projectiles. For the projectile to move properly it is necessary to ensure that its rotation about the mass center is stable. We shall assume that the rotation is affected only by a tilting couple of the aerodynamic forces, which have the force function (1–43). Without attempting to justify this assumption we only note that the smaller the volume of the air bubble in the fluid filling the cavity, the closer it will be to reality.

Equations (1–57) and (1–58) for the relative motion of the system on the above assumptions have the following first integrals: energy integral (1–66), area integral (1–72), geometric integral (1–77), and integral (1–75) expressing the constancy of the projection of the body's instantaneous angular velocity on the x_3 symmetry axis.

Among the actual relative motions of a body containing an ideal fluid are uniform rotation of the body with angular velocity ω about the x_3 axis, which in this case is collinear with the x_3 axis, and steady-state motion of the fluid in the cavity. Here the projections of the angular momentum of the fluid on the x_1 and x_2 axes are zero, and the projections on the x_3 axis are equal to some constant G. Since the fluid is assumed to be ideal, and the cavity has the shape of a body of revolution, the possible steady-state motions when the rigid body rotates about the x_3 axis, include, in particular, the two involving equilibrium with respect to the $Cx_1''x_2''x_3''$ coordinate system such that $G=0$, and uniform rotation together with the rigid body about the x_3 axis with angular velocity ω; here

$$G = C_2^{(0)}\,\omega ,$$

where $C_2^{(0)}$ denotes the axial moment of inertia of the fluid when it rotates uniformly about the x_3 axis. In addition to these limiting cases, it is also possible to consider intermediate cases when $0<G<C\,\omega_2^{(0)}$.

Thus Eqs. (1–57) allow the particular solution

$$\omega_1 = \omega_2 = 0\,, \quad \omega_3 = \omega\,, \quad G_{21} = G_{22} = 0\,, \quad G_{23} = G\,,$$
$$\gamma_1 = \gamma_2 = 0\,, \quad \gamma_3 = 1\,, \quad v'_{01} = v'_{02} = v'_{03} = 0\,.$$

The motion described by this particular solution will be considered our undisturbed motion, and we shall examine its stability with respect to $\omega_i, G_{2i}, \gamma_i, v'_{0i}$ $(i=1, 2, 3)$. When the fluid fills the cavity completely, the equalities $v'_{0i}=0$ $(i=1, 2, 3)$ are obvious.

In the disturbed motion we set

$$\omega_3 = \omega + z_1\,, \quad G_{23} = G + z_2\,, \quad \gamma_3 = 1 + z_3\,,$$

and the previous notation is retained for the remaining variables. The equations of the disturbed motion have the following first integrals:

$$V_1 = M_1(v_{01}'^2 + v_{02}'^2 + v_{03}'^2) + A_1(\omega_1^2 + \omega_2^2) + C_1(2\omega z_1 + z_1^2) + \\ + \varrho \int_\tau (v_1'^2 + v_2'^2 + v_3'^2)\, d\tau - 2T_2'^{(0)} + 2az_3 = \text{const}.$$

$$V_2 = (A_1\omega_1 + G_{21})\gamma_1 + (A_1\omega_2 + G_{22})\gamma_2 + C_1 z_1 + z_2 + \\ + C_1(\omega + z_1) z_3 + (G + z_2) z_3 = \text{const},$$

$$V_3 = \gamma_1^2 + \gamma_2^2 + 2z_3 + z_3^2 = 0,$$

$$V_4 = z_1 = \text{const},$$

where $T_2'^{(0)}$ denotes the kinetic energy of the fluid in the undisturbed relative motion. We also consider the function

$$H_1 \equiv M_1(v_{01}'^2 + v_{02}'^2 + v_{03}'^2) + A_1(\omega_1^2 + \omega_2^2) + C_1(2\omega z_1 + z_1^2) + \\ + \frac{1}{I}(G_{21}^2 + G_{22}^2 + 2Gz_2 + z_2^2) + 2az_3,$$

where I is the greatest of the moments of inertia of the fluid in the cavity at any instant of time relative to the x_1, x_2, x_3 coordinate axes. By virtue of the inequality (3–4), we have

$$H_1 + \frac{1}{I}G^2 - 2T_2'^{(0)} \leqslant V_1 = \text{const}. \tag{3–28}$$

It is easy to see that the constant $-\alpha \equiv (1/I)\, G^2 - 2T_2'^{(0)} \leqslant 0$. To satisfy ourselves that this is so, it is sufficient to consider the limiting cases. If in the steady motion the fluid is at rest, then $G = T_2'^{(0)} = \alpha = 0$. If, however, the fluid in the above motion moves as a single rigid body with angular velocity ω, then

$$G = C_2^{(0)}\omega, \quad T_2'^{(0)} = \tfrac{1}{2}C_2^{(0)}\omega^2,$$

$$-\alpha = \frac{1}{I}C_2^{(0)2}\omega^2 - C_2^{(0)}\omega^2 = C_2^{(0)}\omega^2\left(\frac{1}{I}C_2^{(0)} - 1\right) \leqslant 0.$$

In this case α is equal to zero if $C_2^{(0)}$ is the greatest of the moments of inertia of the fluid, *i.e.*, $I = C_2^{(0)}$.

We construct the function

$$V = H_1 + 2\lambda V_2 - (C_1\omega\lambda + G\lambda + a) V_3 - 2C_1(\omega + \lambda) V_4 + \\ + \frac{C_1(C_1 - A_1)}{A_1} V_4^2 = A_1\omega_1^2 + 2\lambda(A_1\omega_1 + G_{21})\gamma_1 - \\ - (C_1\omega\lambda + G\lambda + a)\gamma_1^2 + \frac{1}{I}G_{21}^2 + A_1\omega_2^2 + 2\lambda(A_1\omega_2 + G_{22})\gamma_2 - \\ - (C_1\omega\lambda + G\lambda + a)\gamma_2^2 + \frac{1}{I}G_{22}^2 + \frac{C_1^2}{A_1}z_1^2 + 2\lambda(C_1 z_1 + z_2) z_3 - \\ - (C_1\omega\lambda + G\lambda + a) z_3^2 + \frac{1}{I}z_2^2 + M_1(v_{01}'^2 + v_{02}'^2 + v_{03}'^2) + \\ + 2\left(\frac{1}{I}G + \lambda\right) z_2, \tag{3–29}$$

where λ is some constant. The function V is the sum of the quadratic form of the variables $\omega_i, \gamma_i, G_{2i}$ $(i=1, 2)$, z_j, v'_{0j} $(j=1, 2, 3)$ and of a function which is linear with respect to the variable z_2. According to Sylvester's criterion, for positive definiteness of the quadratic part of the function V it is necessary and sufficient to select λ so as to satisfy the inequality

$$(A_1 + I)\,\lambda^2 + (C_1\omega + G)\,\lambda + a < 0. \tag{3–30}$$

This is possible if the polynomial contained in the left-hand side of this expression has two different real roots λ_1 and λ_2, *i.e.*, if

$$(C_1\omega + G)^2 - 4(A_1 + I)\,a > 0. \tag{3–31}$$

When this condition is satisfied, λ can be selected arbitrarily from the interval $\lambda_1 < \lambda < \lambda_2$; if, in addition, we have

$$\left(\frac{1}{I}G + \lambda\right) z_2 \geqslant 0, \tag{3–32}$$

then the function V will be positive definite with respect to all the variables it contains.

By virtue of (3–28),

$$V \leqslant \varphi = V_1 + 2\lambda V_2 - (C_1\omega\lambda + G\lambda + a)\,V_3 - 2C_1(\omega + \lambda)\,V_4 + \\ + \frac{C_1(C_1 - A_1)}{A_1} V_4^2 + 2T_2'^{(0)} - \frac{1}{I}G^2,$$

where, obviously, $\dot{\varphi}=0$. Consequently, when conditions (3–31) and (3–32) are satisfied, the function V will satisfy all the requirements of Theorem II with respect to a part of the variables if $\alpha=0$. These conditions will thus be sufficient conditions for stability (or boundedness, if $\alpha \neq 0$) of the undisturbed motion with respect to the quantities ω_i, G_{2i}, γ_i, v'_{0i} $(i=1, 2, 3)$.

We note that if we set

$$\lambda = -\frac{1}{I}G,$$

then condition (3–32) will be satisfied, while inequality (3–30) will take the form

$$\left(C_1\omega - A_1\frac{G}{I}\right)\frac{G}{I} - a > 0. \tag{3–33}$$

When $(1/I)\,G=\omega$, the above expression becomes

$$(C_1 - A_1)\,\omega^2 - a > 0. \tag{3–34}$$

It is also easy to see that if the motion of the fluid completely filling the cavity is irrotational with a velocity potential Φ, then $z_2=0$ and condition

(3–32) is satisfied. In this case the only sufficient condition for stability of the undisturbed motion with respect to the variables ω_i, G_{2i}, and γ_i is the inequality (3–31), in which we should set $G=0$. As can be seen, it is more rigorous than the necessary and sufficient stability condition (2–67) corresponding to this case.

[3-7] STABILITY OF THE EQUILIBRIUM OF A FLUID-CONTAINING PENDULUM

Let us consider a massive rigid body with a stationary point O, having a cavity partially filled with a homogeneous massive fluid. We shall assume that the cavity surface is closed and everywhere convex and that the vector of its normal varies continuously on passing from point to point.

The axes of the Ox_1, x_2, x_3 coordinate system are directed along the principal axes of inertia of the body. The force function of the gravity forces will have the form (1–42) if the stationary axis Ox_3' is directed vertically upward.

We take the equations of motion in the form of the second of equations (1–40) and of the Euler or Navier-Stokes equations together with the applicable boundary conditions. Obviously the equations of motion in the given case allow the energy integral (1–59) if the fluid is ideal, or the energy dissipation equation (1–88) if the fluid is viscous, as well as the area integral (1–70) and the geometric integral in the form of (1–77).

The equations of motion have the particular solution

$$\begin{gathered}\omega_1 = \omega_2 = \omega_3 = 0\,, \quad G_{21} = G_{22} = G_{23} = 0\,, \quad \gamma_1 = \gamma_2 = 0\,, \\ \gamma_3 = 1\,, \quad u_1 = u_2 = u_3 = 0\,,\end{gathered} \tag{3–35}$$

if the center of gravity of the system is situated on the x_3 axis. This solution corresponds to equilibrium of the body and the fluid, the free surface of which is the plane $x_3 = \bar{x}_3$.

The problem of the stability of the equilibrium of a pendulum was examined in Refs. [51–54, 32].

Let us consider the stability of the equilibrium (3–85) with respect to

$$\omega_i,\ G_{2i},\ \gamma_i\,.$$

In disturbed motion we set

$$\gamma_3 = 1 + z_1\,,$$

while the previous notation is retained for the remaining variables. The equations of disturbed motion have the integral relationship

$$\tfrac{1}{2}V_1 = T_1 + T_2 + M_1 g x_{13} z_1 + \Pi_2 - M_2 g x_{23}^0 \leqslant \text{const}\,,$$

where x_{13} and x_{23}^0 are the coordinates on the x_3 axis of the centers of gravity of the body and the fluid (in equilibrium); here $\dot{V}_1 \leqslant 0$, and the first integral is

$$V_2 = \gamma_1^2 + \gamma_2^2 + z_1^2 + 2z_1 = 0 .$$

By virtue of inequality (3–14),

$$2T_2 \geqslant \frac{1}{I}(G_{21}^2 + G_{22}^2 + G_{23}^2), \tag{3–36}$$

where I denotes the largest of the principal moments of inertia of the massive fluid at any time instant.

Let us turn to the expression for the potential energy of the fluid:

$$\Pi_2 = M_2 g x'_{23} ,$$

where x'_{23} is the coordinate, on the x'_3 axis of the center of gravity of the massive fluid, in the disturbed position. According to Torricelli's principle[1,29] it is obvious that Π_2, the potential energy of the massive fluid in the disturbed position, is not less than the potential energy Π_* of the same mass of this fluid that fills the volume τ_* bounded by the wall of the cavity in its instantaneous position under study and the horizontal plane

$$x_1\gamma_1 + x_2\gamma_2 + x_3\gamma_3 = c , \tag{3–37}$$

i.e.,

$$\Pi_2 \geqslant \Pi_* .$$

The right-hand side of Eq. (3–37) is determined from the condition that the volume of the massive fluid is constant and, obviously, it depends on γ_i. Together with the region τ_*, we shall consider the region τ_0, which is bounded by the walls of the cavity and is "frozen" in by the plane $x_3 = \bar{x}_3$. Since the volumes τ_* and τ_0 are identical,

$$\iint_Q dx_1 dx_2 \int_{\bar{x}_3}^{x_3} dx_3 = 0 ,$$

where Q denotes the part of the plane $x_3 = \bar{x}_3$ that is bounded by the walls of the cavity. From this condition we find

$$c = x_1^1\gamma_1 + x_2^1\gamma_2 + \bar{x}_3\gamma_3 .$$

Here

$$x_i^1 = \frac{1}{S_Q}\iint_Q x_i dx_1 dx_2 \quad (i = 1, 2), \quad S_Q = \iint_Q dx_1 dx_2$$

denote the coordinates of the center of inertia of the area Q. Consequently, Eq. (3–37) becomes

$$(x_1 - x_1^1)\,\gamma_1 + (x_2 - x_2^1)\,\gamma_2 + (x_3 - \bar{x}_3)\,\gamma_3 = 0 .$$

Obviously this plane intersects the plane $x_3 = \bar{x}_3$ along a straight line passing through the center of inertia of the area Q. Let us now evaluate

$$\Pi_* = \varrho g \int_{\tau^*} (x_1\gamma_1 + x_2\gamma_2 + x_3\gamma_3)\, d\tau\,.$$

It is easy to calculate that

$$\Pi_* - \Pi_2^{(0)} = \varrho g \iint_Q dx_1 dx_2 \int_{\bar{x}_3}^{x_3} (x_1\gamma_1 + x_2\gamma_2 + x_3\gamma_3)\, d\tau =$$

$$= \frac{\varrho g}{2\gamma_3} \left\{ (x_1^1\gamma_1 + x_2^1\gamma_2)^2 S_Q - \iint_Q (x_1\gamma_1 + x_2\gamma_2)^2\, dx_1 dx_2 \right\}, \tag{3–38}$$

where $\Pi_2^{(0)} = M_2 g \overset{0}{x}_{23}\gamma_3$ denotes the potential energy of the massive fluid "frozen" in the volume τ_0.

Replacing the variables

$$x_1 - x_1^1 = \xi_1\,, \quad x_2 - x_2^1 = \xi_2\,,$$

we obtain

$$\iint_Q (x_1\gamma_1 + x_2\gamma_2)^2\, dx_1 dx_2 =$$

$$= (x_1^1\gamma_1 + x_2^1\gamma_2)^2\, S_Q + \iint_Q (\xi_1^2\gamma_1^2 + \xi_2^2\gamma_2^2 + 2\xi_1\xi_2\gamma_1\gamma_2)\, d\xi_1 d\xi_2\,.$$

Consequently, Eq. (3–38) takes the form

$$\Pi_* = M_2 g \overset{0}{x}_{23}\gamma_3 - \frac{\varrho g}{2\gamma_3}(A_*\gamma_2^2 + B_*\gamma_1^2 + 2F_*\gamma_1\gamma_2)\,, \tag{3–39}$$

where

$$A_* = \iint_Q \xi_2^2 d1_1 d\xi_2\,, \quad B_* = \iint_Q \xi_1^2 d\xi_1 d\xi_2\,, \quad F_* = \iint_Q \xi_1\xi_2 d\xi_1 d\xi_2$$

denote the moments of inertia and the centrifugal moment of inertia of the area segment Q relative to axes parallel to the x_i axes and passing through the center of inertia of the area segment.

Thus the inequality $\Pi_2 \geqslant \Pi_*$ becomes

$$\Pi_2 \geqslant M_2 g \overset{0}{x}_{23}\gamma_3 - \frac{\varrho g}{2\gamma_3}(A_*\gamma_2^2 + B_*\gamma_1^2 + 2F_*\gamma_1\gamma_2)\,. \tag{3–40}$$

Let us consider the function

$$H_1 = A_1\omega_1^2 + B_1\omega_2^2 + C_1\omega_3^2 + \frac{1}{I}(G_{21}^2 + G_{22}^2 + G_{23}^2) +$$

$$+ 2(M_1 x_{13} + M_2 \overset{0}{x}_{23})\, g z_1 - (A_*\gamma_2^2 + B_*\gamma_1^2 + 2F_*\gamma_1\gamma_2)\, \varrho g + \cdots.$$

On the basis of inequalities (3–36) and (3–40), we conclude that

$$V_1 \geqslant H_1 . \tag{3–41}$$

We construct the function

$$\begin{aligned} V = H_1 - Mgx_{c3}^0 V_2 &= A_1\omega_1^2 + B_1\omega_2^2 + C_1\omega_3^2 + \\ &+ \frac{1}{I}(G_{21}^2 + G_{22}^2 + G_{23}^2) - (Mx_{c3}^{(0)} + A_*\varrho)\, g\gamma_2^2 - \\ &- (Mx_{c3}^{(0)} + B_*\varrho)\, g\gamma_1^2 - 2F_*g\varrho\gamma_1\gamma_2 - Mgx_{c3}^0 z_1^2 + \cdots , \end{aligned} \tag{3–42}$$

where

$$Mx_{c3}^0 = M_1 x_{13} + M_2 x_{23}^0 .$$

By virtue of inequality (3–41) it is obvious that,

$$V \leqslant \varphi = V_1 - Mgx_{c3}^{(0)} V_2 .$$

If we satisfy the inequality

$$Mx_{c3}^0 + A_*\varrho < 0 \quad (A_* \geqslant B_*), \tag{3–43}$$

then, according to Sylvester's criterion, the function V is positive definite and satisfies all of the conditions of Theorem II. Consequently, inequality (3–43) is a sufficient condition for stability of the equilibrium (3–35) of a pendulum containing a fluid with respect to the variables $\omega_i, G_{2i}, \gamma_i$ $(i=1,2,3)$.

4

Stability of Steady Motion of Rigid Bodies with Fluid-Filled Cavities

In this chapter we will consider the stability of steady motion, in particular, of equilibrium of rigid bodies with cavities partially or completely filled with ideal or viscous fluids. This study is based on the definition of stability and on the ideas developed by Lyapunov in the theory of stability of equilibrium shapes of a rotating fluid[48, 55]. It has been proved that the problem of the stability of steady motion or of the equilibrium of a rigid body containing a fluid reduces to the problem of minimizing a certain expression[56, 57] a solution to which is given[54, 58]. The results are used for solving a number of problems of stability of motion of bodies containing a fluid.

[4-1] EQUATIONS OF EQUILIBRIUM AND OF STEADY MOTIONS

Let us consider a perfectly rigid body which has a simply connected cavity of arbitrary shape filled by an incompressible, homogeneous ideal fluid. We shall simultaneously consider the cases of the full and the partially full cavity. When partially filled, the fluid inside the cavity has a free surface, on which there should be satisfied either conditions (1–30) for the hydrodynamic pressure and (1–31) for the contact angle if the surface tension forces are taken into account, or condition (1–34) if surface tension can be disregarded, as well as the kinematic condition (1–33). For fluid particles in contact with the cavity walls, the normal velocity components should be equal to the normal velocity components of the corresponding points of the cavity's walls, according to condition (1–32). We shall consider an unconstrained body, as well as a body with imposed steady constraints. The position of the body in space will be defined by its Lagrangian coordinates q_j $(j=1,\ldots,n)$. We assume that the specified forces applied to the body and the mass forces acting on the fluid particles have respectively force functions $U_1(q_1,\ldots,q_r)$ and $U_2(x_1', x_2', x_3')$ which are not explicitly time dependent, and, in addition, that $r \leqslant n$. As before, we shall assume that the motion of the body is continuous and the fluid moves as a continuum, so that the coordinates of any fluid particle are continuous functions of their initial values and of time.

Under the above conditions, we obtain from the theorem on kinetic energy [or from the differential equations of motion of the system (1–45) and (1–46)] the energy integral

$$T + \Pi = h\,, \tag{4–1}$$

where the potential energy of the forces applied to the system is

$$\Pi = -\,U_1 - \varrho \int_\tau U_2 d\tau + \Pi_2^*\,,$$

and h is a constant of integration.

The potential energy Π of the system depends in general on the coordinates q_j of the body and on the configuration of the fluid inside the cavity; *i.e.*, it is a functional. When the fluid completely fills the cavity of a given shape, the potential energy of the system is a function of $\Pi(q_1, \ldots, q_r)$.

[4-1.1] Equations of Equilibrium

According to the principle of permissible displacements, the equation

$$\delta\Pi = 0 \tag{4–2}$$

is the condition of equilibrium of a rigid body containing a fluid in a cavity.

Changing from the absolute coordinates x_1', x_2', x_3' of the fluid particles to the relative coordinates x_1, x_2, x_3, the force function of the mass forces can be represented in the form $U_2(x_1, x_2, x_3, q_1, \ldots, q_n)$, retaining the previous designation U_2. If the fluid fills the cavity completely, then, obviously, $U_2 = U_2(q_1, \ldots, q_n)$. Making use of Eq. (1–31) for the contact angle θ, we now write Eq. (4–2) in the explicit form:

$$\sum_{j=1}^{n} \frac{\partial\Pi}{\partial q_j}\delta q_j - \sum_{i=1}^{3} \varrho \int_\tau \frac{\partial U_2}{\partial x_i}\delta x_i d\tau + \alpha \int_S \left(\frac{1}{R_1} + \frac{1}{R_2}\right)\delta\zeta dS = 0\,. \tag{4–3}$$

From this, because of the independence of δq_j and δx_i, follows the equations of equilibrium of a rigid body with a fluid:

$$\frac{\partial\Pi}{\partial q_j} = 0 \quad (j = 1, \ldots, n)\,. \tag{4–4}$$

When the fluid has a free surface, the remaining part of Eq. (4–3) gives an equation of the free surface under equilibrium. In fact, applying the Gauss-

Ostrogradskiy formula and making use of the incompressibility condition, we obtain

$$\sum_{i=1}^{3} \varrho \int_{\tau} \frac{\partial U_2}{\partial x_i} \delta x_i d\tau - \alpha \int_{S} \left(\frac{1}{R_1} + \frac{1}{R_2} \right) \delta\zeta dS = \\ = \int_{S} \left[\varrho U_2 - \alpha \left(\frac{1}{R_1} + \frac{1}{R_2} \right) \right] \delta\zeta dS = 0 \,. \tag{4–5}$$

By virtue of the constancy of the fluid volume, the function $\delta\zeta$ should satisfy the condition

$$\int_{S} \delta\zeta dS = 0 \,,$$

while in other respects it is arbitrary, with the result that we obtain from Eq. (4–5) the following equation for the free surface of the fluid under equilibrium:

$$\varrho U_2 - \alpha \left(\frac{1}{R_1} + \frac{1}{R_2} \right) = \text{const} \,. \tag{4–6}$$

Of course, Eqs. (4–4) and (4–6) can be obtained directly from Eqs. (1–45) and (1–46).

[4-1.2] Equations of Steady Motion

Let us now consider the case when the constraints imposed on the body permit rotation of the entire system as a single rigid body about some stationary straight line, while the forces acting on the system do not produce a moment about this line. In this case the kinetic and potential energies of the system are obviously independent of q_n, the angle of rotation of the body about the above line. Then there exists an area integral for the plane orthogonal to the above line. Taking this line as the x'_3 axis of the stationary coordinate system, we write the area integral in the form

$$G_{x'_3} = k = \text{const} \,, \tag{4–7}$$

where $G_{x'_3}$ denotes the projection of the system's angular momentum on the x'_3 axis.

In addition to the stationary and moving coordinate systems, we introduce the coordinate system $O'\xi_1\xi_2 x'_3$, which rotates about the x'_3 axis at some angular velocity ω. Denoting, as before, by $\boldsymbol{v}$ the vector of the absolute velocity of some point of the body or fluid that has the radius vector $\boldsymbol{r}'$ relative to the point O', and by $\boldsymbol{\omega}$ the velocity vector of the same point relative to the coordinate system $O'\xi_1\xi_2x_3'$, we have, according to the velocity addition theorem

$$v = w \times r' + w .$$

Here the kinetic energy and the projection of the system's angular momentum on the x_3' axis can be represented in the form

$$T = T^{(1)} + \omega G^{(1)}_{x'_3} + \tfrac{1}{2}\omega^2 I, \quad G_{x'_3} = G^{(1)}_{x'_3} + \omega I. \tag{4–8}$$

Here

$$T^{(1)} = \tfrac{1}{2}\sum_\nu m_\nu w_\nu^2 \qquad G^{(1)}_{x'_3} = \sum_\nu m_\nu (\xi_{1\nu} w_{2\nu} - \xi_{2\nu} w_{1\nu})$$

denote, respectively, the kinetic energy and the projection on the x_3' axis of the kinetic moment of the system in its motion relative to the coordinate system $O'\xi_1\xi_2x_3'$, and

$$I = \sum_\nu m_\nu (x'^2_{1\nu} + x'^2_{2\nu})$$

is the moment of inertia of the system with respect to the x_3' axis. In all of these formulas the summation is taken over all the points of the system, the masses of which are denoted by $m_\nu(\nu = 1, 2, \ldots)$. The angular velocity of the coordinate system $O'\xi_1\xi_2x_3'$ can be specified arbitrarily; we shall select it so that the projection of the angular momentum of the system in its relative motion on the x_3' axis (*i.e.*, $G^{(1)}_{x_3}$) is equal to zero which, by virtue of (4–7) and (4–8) is equivalent to the equation[55]

$$\omega I = k . \tag{4–9}$$

When ω is selected in this manner, the energy integral (4–1) can be presented as

$$T^{(1)} + \frac{1}{2}\frac{k^2}{I} + \Pi = h . \tag{4–10}$$

[4-1.3] Remark

The energy integral can also be represented in a form similar to (4–10) by generalizing Routh's method for ignoring cyclical coordinates. In fact, if we take the angular velocity ω as being equal to the angular velocity of the rigid body about the x_3 axis, the angle between the x_1' and ξ_1 axes can be taken as the angle q_n. The cyclical coordinate q_n has corresponding to it the first integral (4–7)

$$\frac{\partial L}{\partial \dot{q}_n} = G^{(1)}_{x'_3} + \omega I = k$$

of the equations of motion (1–45), where the Lagrange function is $L = T - \Pi$. From this we find

$$\dot{q}_n = \omega = \frac{k - G^{(1)}_{x'_3}}{I}$$

and we eliminate this quantity from Routh's function $R = L - \omega k$. As a result, we will obtain instead of (1–45) equations of motion in the form of Routh's equations [29]

$$\frac{d}{dt}\frac{\partial R}{\partial \dot{q}_j} - \frac{\partial R}{\partial q_j} = 0 \quad (j = 1, \dots, n - 1),$$

which allow the energy integral

$$R_2 - R_0 = h,$$

where R_i is the part of the function $R = R_2 + R_1 + R_0$ of power i that is homogeneous with respect to the velocities. It is easy to check that in this case

$$R_2 = T^{(2)} = T^{(1)} - \frac{1}{2I} G^{(1)^2}_{x'_3}, \quad R_0 = -\Pi - \frac{1}{2}\frac{k^2}{I},$$

so that the energy integral will have the form

$$T^{(2)} + \frac{1}{2}\frac{k^2}{I} + \Pi = h,$$

which is similar to (4–10). It can be shown [59] that $T^{(2)}$ is a quadratic function of the velocities $\boldsymbol{\omega}_\nu$; here $T^{(2)} > 0$, if not all of the $\boldsymbol{\omega}_\nu = 0$.

On the above assumptions concerning the constraints imposed on the rigid body and on the applied forces, among the actual motions of the system under study are uniform rotations of the entire system as a single rigid body about the stationary x'_3 axis. Then the system will be in equilibrium with respect to the coordinate system $O'\xi_1\xi_2x'_3$ as it rotates about the x'_3 axis with the angular velocity ω of uniform rotation of the system.

For the actual motion of the system, we have the general equation of dynamics

$$\sum_\nu (m_\nu \ddot{\boldsymbol{r}}'_\nu - \boldsymbol{F}_\nu) \cdot \delta \boldsymbol{r}'_\nu = 0, \tag{4–11}$$

expressing the D'Alembert-Lagrange principle, where $\ddot{\boldsymbol{r}}'_\nu$ is the acceleration of the νth point of the system.

When the entire system moves uniformly as a single rigid body about the x_3 axis with angular velocity ω,

$$\ddot{x}'_1 = -\omega^2 x'_1, \quad \ddot{x}'_2 = -\omega^2 x'_2, \quad \ddot{x}'_3 = 0,$$

as a result of which Eq. (4–11) takes the form

$$\tfrac{1}{2}\omega^2\delta\sum_{\nu} m_\nu(x_{1\nu}'^2 + x_{2\nu}'^2) + \sum_{\nu} \boldsymbol{F}_\nu \cdot \delta \boldsymbol{r}_\nu' = 0$$

or

$$\delta U = 0, \tag{4–12}$$

if we introduce the notation

$$U = \tfrac{1}{2}\omega^2 I - \Pi. \tag{4–13}$$

But expression (4–13) can be regarded as a force function of the specified forces and of the centrifugal inertia forces. According to the principle of permissible displacements, Eq. (4–12) is the condition for equilibrium of the system with respect to the $O'\xi_1\xi_2x_3'$ coordinate system, which is rotating with a constant angular velocity ω.

We now introduce an expression for the altered potential energy of the system

$$W = \frac{1}{2}\frac{k_0^2}{I} + \Pi, \tag{4–14}$$

where k_0 is the value of the area constant k when the entire system moves uniformly as a single rigid body about the x_3' axis with angular velocity ω. The variation of this function on the possible displacement of the system is

$$\delta W = -\frac{1}{2}\frac{k_0^2}{I_0^2}\delta I + \delta \Pi,$$

where I_0 is the value of I for steady motion.

Equating the expression δW to the expression

$$\delta U = \tfrac{1}{2}\omega^2 \delta I - \delta \Pi$$

and making use of the fact that $\omega I_0 = k_0$, we conclude that Eq. (4–12) is equivalent to

$$\delta W = 0. \tag{4–15}$$

Thus, in the case of steady motion of the system, the expression for W has the extremum (steady-state) value.

According to definition (4–14), the altered potential energy of the system W depends on those of the body's coordinates $q_1, \ldots, q_{n-1}$ on which Π and I depend, and the configuration of the fluid, as well as the value of the constant k_0; *i.e.*, it is a functional. When the cavity is completely filled with fluid, W is the function $W(q_1, \ldots, q_{n-1})$.

As is easy to see, condition (4–15) yields the equations

$$\frac{\partial W}{\partial q_j} = -\frac{1}{2}\frac{k_0^2}{I_0^2}\frac{\partial I}{\partial q_j} + \frac{\partial \Pi}{\partial q_j} = 0 \quad (j = 1, \ldots, n-1) \tag{4–16}$$

for the coordinates q_j ($j = k, \ldots, n$—1) of a rigid body performing steady motion, as well as equations for the pressure in the fluid, which make it possible to obtain the equation

$$\varrho U_2 + \tfrac{1}{2}\varrho\omega^2 (x_1'^2 + x_2'^2) - \alpha\left(\frac{1}{R_1} + \frac{1}{R_2}\right) = \text{const} \tag{4–17}$$

of its free surface in this motion, if the fluid does not completely fill the cavity. Of course, these equations could have been obtained directly from the equations of motion of the system.

The quantity k_0 can be regarded as a parameter, and this makes it possible to apply to our mechanical system the results of the general theory of "equilibrium" of material systems with potential energy that is a function of a parameter[29].

For a specified value of k_0, Eqs. (4–16) and (4–17) determine the coordinates of the rigid body and the shape of the free surface of the fluid in steady motion. For continuous variation of the parameter k_0, the real roots of Eqs. (4–16) are

$$q_j^{(s)} = \varphi_j^{(s)}(k_0) \tag{4–18}$$

and the modes of relative equilibrium of the fluid corresponding to them will vary. The functions $\varphi_j^{(s)}(k_0)$ are assumed to be continuous functions of k_0.

In n-dimensional space $(q_1, \ldots, q_{n-1}, k_0)$, Eqs. (4–18) define a real curve, different points of which correspond to the different possible steady motions. Individual branches of this curve may intersect in some points when at least two real roots of Eqs. (4–16) coincide. These points are called bifurcation points; Eqs. (4–16) do not have a unique solution in their vicinity.

Remark

When $\omega=0$ (correspondingly, $k_0=0$), $W=\Pi$ and Eqs. (4–16) take the form of Eqs. (4–4), while Eq. (4–17) takes the form of Eq. (4–6). Consequently, the equilibrium case $\omega=0$ can be regarded as a particular case of steady motion, which obviously also takes place when T and Π are explicit functions of the angle q_n of rotation of the body about the x_3' axis.

The constants contained in the right-hand sides of Eqs. (4–6) or (4–17) are determined from the known values of the radii of curvature R_i and the force function U_2 for any point of the free surface of the fluid[8].

From differential geometry we know a formula for the average curvature of a surface[60]:

$$\frac{1}{R_1} + \frac{1}{R_2} = \frac{EN - 2FM + GL}{EG - F}, \tag{4–19}$$

where E, F, G and L, M, N are coefficients of the first and second differential Gaussian forms for the surface. Consequently, Eqs. (4–6) and (4–17) are, in general, nonlinear partial differential equations of the free surface of the fluid in equilibrium or steady motion, the configuration of which is determined by integrating these equations. The integrals of Eqs. (4–6) and (4–17) should satisfy boundary conditions of the type of (1–31):

$$\cos\theta = \frac{\alpha_2 - \alpha_1}{\alpha}$$

if the free surface intersects with the walls of the cavity. Experience shows that the contact angle θ can be either acute or obtuse, depending on the nature of the media in contact. For example[2], for nonwettable walls θ is obtuse and $\alpha_1 > 0$, while for wettable walls θ is acute. In this case it should be assumed that $\alpha_1 < 0$, if the surface tension coefficient at the rigid body–air interface, as is usually done, is assumed to be $\alpha_2 = 0$. In the case when $|\alpha_1/\alpha| > 1$, there is no line of intersection of the free fluid surface with the cavity walls, and the fluid spreads over the entire cavity surface.

Far from the cavity walls, the configuration of the free surface at equilibrium depends on the ratio between the surface tension forces and the mass forces. For example, in a gravity field the configuration of the free surface is determined[16] by the magnitude of the capillary constant $a = \sqrt{2\alpha/g\varrho}$.

It should be noted that integration of Eqs. (4–6) or (4–17) is usually difficult[61,80]. This problem will be considered in more detail in Chapter 7.

If the surface tension forces can be disregarded, $\alpha = 0$, and Eqs. (4–6) and (4–17) become the final equations of the free surface. Thus, Eq. (4–17) takes on the form

$$f_0(x_1', x_2', x_3') = U_2 + \tfrac{1}{2}\omega^2(x_1'^2 + x_2'^2) = c_0 . \tag{4–20}$$

The magnitude of the constant c_0 in the right-hand side of this equation is determined by the amount of fluid in the cavity.

In closing, we note that the conclusions of this section pertaining to the absolute motion of the system also remain valid for the motion of a rigid body, containing a fluid about their common mass center. In fact, in Sec. 8 of Chapter 1 we pointed out the conditions under which there exist an energy integral (1–66) and an area integral (1–72) for the relative motion of the system. If, then, relative equilibrium or relative steady motion of the body plus fluid is possible, it is determined from conditions similar to Eqs. (4–2) and (4–15).

[4-2] STATEMENT OF THE PROBLEM OF STABILITY OF STEADY MOTION

Let us examine some steady motion of a rigid body containing a fluid, corresponding to a specified value of the area constant k_0. Without any loss of generality, we assume that the roots of Eqs. (4–16) for the given value of k_0 are $q_j=0$ $(j=k,\ldots,n-1)$. Then the fluid has the relative equilibrium shape, which is bounded, on partial filling, by the free surface S_0 defined by Eq. (4–17) and by the walls that the fluid is touching. It is assumed from here on that the direction of the normal to the surface S_0 varies continuously over all the surface and that the curvature of its principal normal cross sections is everywhere finite.

We examine the stability of this steady motion (or equilibrium) of a rigid body with a fluid-filled cavity by using the energy integral (4–10). Our mechanical system has $n+\infty$ degrees of freedom and we must decide on the definition of the stability of motion of this system.

When the fluid completely fills the cavity of the body, the term *stability of motion* will be used to denote stability in the sense of Lyapunov[28] with respect to those noncyclical coordinates $q_1,\ldots,q_{n-1}$ of the body on which the potential energy Π and the moment of inertia I of the system depend explicitly, with respect to the generalized velocities $q_1',\ldots,q_n'$, and with respect to the kinetic energy $T_2^{(1)}$ of the fluid.

When the cavity is partially filled and the fluid has a free surface, the situation is more complex. As was made clear by Lyapunov[48, 55] the general energy integral for a fluid is found to be inadequate for discovering the behavior of motion, after disturbance from rest (absolute or relative), that is considered the sign of stable equilibrium in the mechanics of systems with a finite number of degrees of freedom. Lyapunov established that the difficulty vanishes only when the stable form of the equilibrium is defined as that for which, after a sufficiently small disturbance has been imparted to the fluid, its configuration remains different from this equilibrium shape by an amount as small as desired, at least until very thin thread- or sheet-shaped projections form on the fluid surface. These projections may have large linear dimensions, but small volumes, and thus they can carry only small quantities of energy. We shall adopt this definition and, following Lyapunov, present some concepts that are applicable to our problem. We shall compare the configuration f_0 of the relative equilibrium of the fluid to its configuration f at any instant of the disturbed motion, without attending to the motion of the fluid particles proper, but making use of the fluid's kinetic energy in the considerations of magnitude. The configuration f is bounded by the free surface S of the fluid and by the cavity walls with which the fluid is in contact at the given moment. For a disturbed motion that is sufficiently close to the undisturbed motion, in the coordinate system $Ox_1x_2x_3$ fixed to the body, the configurations f_0 and f differ only by the free surfaces S_0 and S. Because of the incompressibility of the fluid, the volume of the configuration f is obviously equal to the volume of the configuration f_0.

Consider any point P on the surface S and nearby point P_0 on the surface S_0 (Fig. 15). The straight line segment between these two points is denoted by n, which will be considered positive or negative depending on whether the point P is situated outside or inside the configuration f_0. In addition to changes in the position of the point P on the surface S and, consequently, also of the point P_0 on the surface S_0, the distance PP_0 will also change, and for some position of the point P this distance will become the greatest of all those possible at the given instant of time. This maximum of the distance

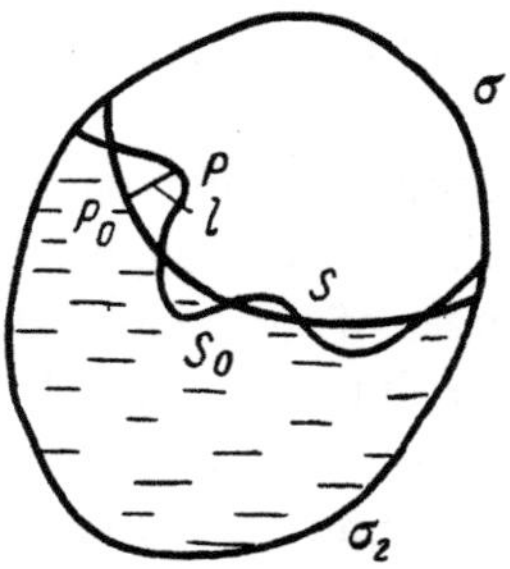

Fig. 15.

PP_0 was called by Lyapunov the *departure of the disturbed from the undisturbed fluid surface*; we shall denote this quantity by l. We now introduce the deviation ∇ of the fluid's configuration f from the equilibrium configuration f_0, taking as the latter the volume of the part of the configuration f that is outside of the configuration f_0 or, which is the same thing, the volume of the part of the configuration f_0 that is outside of f.

If we consider all of the possible surfaces of the fluid, the departures of which from the equilibrium surface are equal to the given quantity l, then ∇ for them will obviously have all the values between zero and some maximum in the form $l\psi(l)$, where $\psi(l)$ is a positive function, all the values of which do not exceed some specified limit. When l does not exceed any specified number A, the function $\psi(l)$ will allow a nonzero minimum. The minimum of the deviation ∇ for the specified departure l will, however, always be zero [55].

Any single-valued and continuous function of l that takes on only values that, for the same l, will be acquired by ∇, and that vanishes only when $l=0$, is called the *possible deviation*, after Lyapunov.

When the surface tension forces are taken into account, in addition to the above quantities we shall also consider the slope of the disturbed surface relative to the undisturbed surface. When the departure l is sufficiently small, the magnitude n of the line segment PP_0 can be regarded as some function $n(u, v)$, where u, v are position parameters of the equilibrium surface S_0. The derivatives n'_u and n'_v describe the slope of the disturbed surface with respect to the undisturbed surface at each point. For the given time instant, these quantities (or one of them) take on their maximum values at some point. We shall denote the maximum $|n'_u|$, $|n'_v|$ by Δ and we shall call it the *slope of the disturbed surface relative to the undisturbed surface*. In continuous motion of the body with the fluid departure l, the deviation ∇ and the slope Δ will be continuous functions of time.

Let us now note the following definition of stability of motion of our system if the cavity is not completely filled. After imparting initial disturbances to the system, we examine its disturbed motion. If the initial departure and the slope that the free fluid surface makes with the equilibrium surface and the initial relative velocities of fluid particles, as well as the initial deviation and relative velocities of the body, can be chosen sufficiently small so that the absolute magnitudes of the coordinates q_j and the velocities $\dot{q}_j$, the kinetic energy $T_2^{(1)}$ of the fluid, its slope Δ, and the departure l of its surface from the equilibrium surface remain smaller than some specified limits, no matter how small, during the entire motion, or at least until the deviation of the fluid configuration from that of equilibrium does not become smaller than some specified possible deviation, no matter how small, then the motion of the system under study is stable, while in the opposite case it is unstable.

Thus, if for any arbitrary positive numbers L_1 and L_2, no matter how small, it is possible to select positive numbers λ_1 and λ_2 in a manner such that, for any initial values of the coordinates* q_{j0} and generalized velocities $\dot{q}_{j0}$ of the body, the departure l_0, the slope Δ_0, the deviation ∇ and the relative fluid velocities $\boldsymbol{w}_0$ satisfying the conditions

$$|q_{j0}| \leqslant \lambda_1, \quad |\dot{q}_{j0}| \leqslant \lambda_1, \quad |l_0| \leqslant \lambda_1, \quad |\boldsymbol{w}_0| \leqslant \lambda_1, \quad \Delta_0 < \lambda_2, \quad \nabla_0 > \varepsilon l_0, \tag{4–21}$$

and for any $t \geqslant t_0$, or at least as long as

$$\nabla > \varepsilon l, \tag{4–22}$$

there are satisfied the inequalities

$$|q_j| < L_1, \quad |l| < L_1, \quad |\Delta| < L_1, \quad |\dot{q}_j| < L_2, \quad |T_2^{(1)}| < L_2, \tag{4–23}$$

then the undisturbed steady motion of the rigid body with a fluid is stable, while in the opposite case it is unstable. Here ε denotes a positive number that is smaller than the minimum of the function $\psi(l)$ under the condition $|l| \leqslant L_1$, while εl can be regarded as the possible deviation of the fluid. We

* These initial data cannot be entirely arbitrary, but should be such that the ensuing motion is continuous.

note that if the fluid completely fills the cavity the conditions pertaining to departure, deviation, and slope should be dropped.

The above definition of stability is general with surface tension taken into account. When this tension can be disregarded, it is simply necessary to drop from the definition the condition related to the slope. On the other hand, when surface tension is taken into account, it is impossible for projections of any precribed size to appear on the free surface, since this involves a prescribed increase in the potential energy. In this case there remains only the hypothetical possibility of the appearance of thread-shaped projections (which, however, is highly unlikely because of physical considerations).

The above is a natural generalization of the definition of stability of equilibrium configurations of a rotating fluid to the case of a rigid body containing a fluid with surface tension. This definition has a serious shortcoming from the logical point of view; *i.e.*, it contains condition (4–22), which cannot be checked. Nevertheless, in our opinion, the above definition has important practical significance. There is an extensive class of technical and engineering problems for which this definition of stability is expedient.

Further on we shall need the concept of the minimum of the expression W. When W is a function $W(q_1, \ldots, q_n - 1)$, its minimum for a specified value of k_0 will denote an isolated minimum of the function with respect to the variables $q_1, \ldots, q_{n-1}$ on which it explicitly depends. If the fluid does not completely fill the cavity, following Lyapunov[48], we shall assume the following definition of the minimum of the expression W.

If for the steady motion under study, when $q_j = 0 (j = 1, \ldots, n-1)$, $l = \Delta = \nabla = 0$, W_0 is a minimum of the expression W, then there exists a sufficiently small positive number E such that for all the systems of values of the body's coordinates q_j, departure l, slope Δ, and deviation ∇ satisfying

$$|q_j| \leqslant E, \quad |l| \leqslant E, \quad |\Delta| \leqslant E, \quad \nabla > \varepsilon l,$$

where ε is a positive number smaller than the minimum of the function $\psi(l)$ under condition $|l| \leqslant E$, all the values assumed by the difference $W - W_0$ will remain positive, becoming zero only when

$$q_j = 0 \quad (j = 1, \ldots, n-1), \quad l = 0, \quad \nabla = 0.$$

We note that for any specified value of l the difference $W - W_0$ can be made as small as desired by selecting a position of the body and a configuration of the fluid for which $|q_j|$, Δ, and ∇ are sufficiently small. But the limiting case, when for $l \neq 0$ all the $q_j = 0 \, (j = 1, \ldots, n-1)$, Δ, ∇, and consequently also $W - W_0$ become zero, is obviously impossible if we consider only configurations that the fluid can take on. To eliminate this inconvenience we have introduced the condition $\nabla > \varepsilon l$[48].

[4-3] SOME THEOREMS CONCERNING THE STABILITY OF STEADY MOTIONS

[4-3.1] Ideal Fluid

We now take up the examination of stability of steady motions (and equilibria) of a rigid body with a fluid in the sense specified in Sec. 2. In the case when the fluid has a free surface, we decide always to select the departure l smaller than the smallest of all of the radii of curvature of the principal normal cross sections of the equilibrium surface.

Theorem I

If, for steady motion of a rigid body with a fluid-filled cavity, the expression

$$W = \frac{1}{2}\frac{k_0^2}{I} + \Pi$$

has an isolated minimum W_0, then the undisturbed motion is stable.

Proof

Let us take the system out of the steady motion under study, imparting to it some sufficiently small initial deviations and velocities; here the slopes between the disturbed and undisturbed surfaces are also considered sufficiently small. Left to itself, our system will continue to move with the energy integral (4–10), which we rewrite in the form

$$T^{(1)} + W + \frac{1}{2}\frac{k^2 - k_0^2}{I} = T^{(10)} + W^{(0)} + \frac{1}{2}\frac{k^2 - k_0^2}{I^{(0)}}, \qquad (4\text{–}24)$$

where the superscript (0) denotes the initial value of the applicable quantity and k is the area constant for the disturbed motion.

Let A be some nonzero arbitrarily small positive number not exceeding the number L_1, which in any case will be assumed smaller than the number E that defines the region in which W is a minimum. We denote by W_1 the smallest possible value that W can take on if the departure l, slope Δ, or one of the coordinates q_j has an absolute value equal to A, and if the remaining quantities and the deviation ∇ satisfy the conditions

$$|q_j| \leqslant A, \quad |l| \leqslant A, \quad |\Delta| \leqslant A, \quad \nabla \geqslant \varepsilon l.$$

Since, by assumption, W for steady motion has a minimum W_0, we will have the inequality

$$W_1 > W_0.$$

But, selecting l, Δ, and $|q_j|$ sufficiently small, while $\nabla > \varepsilon l$, W can be made different from W_0 as by little as desired. Let us select an A sufficiently small so that we satisfy the expression

$$|W_1 - W_0| < L_2. \tag{4–25}$$

The initial values of q_j, Δ, and l can be taken sufficiently small that the initial value $W^{(0)}$ of W is smaller than W_1

$$W^{(0)} < W_1. \tag{4–26}$$

Selecting the initial position of the system in this manner, we shall assume that the initial values of q_j and the initial configuration of the fluid satisfy the inequalities

$$|q_j| < A, \quad |l| < A, \quad |\Delta| < A, \quad \nabla > \varepsilon l.$$

No matter what the initial position and initial configuration of the fluid, the initial velocities of the points of the system can be selected in a manner such that the constants

$$\tfrac{1}{2}|k^2 - k_0^2|, \quad T^{(10)}$$

are as small as convenient. We select constants sufficiently small to satisfy the inequality

$$\tfrac{1}{2}(k^2 - k_0^2)\left(\frac{1}{I^{(0)}} - \frac{1}{I}\right) + T^{(10)} + W^{(0)} < W_1 \tag{4–27}$$

for all values that I can assume when the conditions

$$|q_j| \leqslant A, \quad |l| \leqslant A \tag{4–28}$$

are satisfied. The quantities λ_1 and λ_2, which are contained in the definitions of stability (4–21) and which determine the region of initial disturbances, will be assumed to be selected in such a manner that, when conditions (4–21) are satisfied, the inequality (4–27) will be satisfied for all of the values of I under conditions (4–28). For these initial conditions, during the entire ensuing time of motion, as long as inequalities (4–28) are satisfied, we shall, according to the energy integral (4–24), have the inequality

$$T^{(1)} + W < W_1, \tag{4–29}$$

from which it follows that $W < W_1$. This inequality will be satisfied at least until $|q_j|$, $|l|$, and $|\Delta|$ do not exceed A. But the initial values of q_j, l, and Δ are, by assumption, smaller than A, and the initial deviation $\nabla > \varepsilon l$, and since q_j, l, Δ, and ∇ vary continuously with time, $|q_j|$, $|l|$, and $|\Delta|$ cannot exceed A without first being equal to it. On the other hand, the equalities

$$|q_j| = A, \quad |l| = A, \quad |\Delta| = A$$

because of Eq. (4–29), are obviously impossible when $\nabla > \varepsilon l$.

Making use of (4–25), it follows from Eq. (4–29) that $|T^{(1)}| < L_2$, on the basis of which we conclude that

$$|\dot{q}_j| < L_2 \; (j = 1, \dots, n-1), \quad |T_2^{(1)}| < L_2 .$$

Consequently, if the system moves in a continuous manner, so that q_j, l, Δ, and ∇ vary continuously with time, then, starting with the initial instant of time, we will have the expressions

$$|q_j| < L_1 , \quad |\dot{q}_j| < L_2 , \quad |l| < L_1 , \quad |\Delta| < L_1 , \quad |T_2^{(1)}| < L_2 , \quad \nabla > \varepsilon l ,$$

and each of them will continue to be satisfied as long as the last of them is satisfied, which proves the theorem.

We should note that, if the fluid fills the cavity completely, the conditions imposed on l, Δ, and ∇ become superfluous and, under conditions of the theorem's statement, for any $t \geqslant t_0$ the following inequalities are satisfied:

$$|q_j| < L_1 ; \quad |\dot{q}_j| < L_2 ; \quad |T_2^{(1)}| < L_2 .$$

It is obvious here that the function W for a rigid body with a completely fluid-filled cavity has exactly the same form as that for a single rigid body with a moment of inertia I with respect to the x_3' axis and potential energy Π of the applied forces. Consequently, the conditions for stability of a body containing a fluid in this case will be identical to the stability conditions for the corresponding rigid body.

Remark 1

The conclusion that we obtain on the basis of this theorem for the fluid with a free surface is incomplete, since stability is ensured only as long as the relationship $\nabla > \varepsilon l$ is satisfied. However, as was emphasized by Lyapunov[55], the energy equation by itself cannot yield much more when operating with the definition of stability we have assumed. It is possible to obtain more definite conclusions if, in order to describe the difference between the disturbed and undisturbed configuration of the fluid, we introduce instead of the departure l some other quantities that can become zero only for the undisturbed configuration. As was noted by Lyapunov, who was following Duhem, it is possible to take the deviation ∇ as such a quantity and to prove, by analogy with Theorem I, that if for the steady motion of a body with a cavity partially filled with a fluid W has a minimum W_0 in the sense that $W - W_0 > 0$ for all of values of $|q_j|$, Δ, and ∇ that are not simultaneously equal to zero and that are smaller than a specified constant limit, then, for sufficiently small disturbances, $|q_j|$, $|\dot{q}_j|$ $(j = 1, \dots, n-1)$, Δ, ∇, and $|T(\tfrac{1}{2})|$ will remain throughout the motion smaller than the specified constants no matter how small they are.

Remark 2

If the energy and area integrals for a rigid body with a fluid-filled cavity moving about the mass center of the system exist, and if the system rotates uniformly about the mass center, than Theorem I is also applicable to this relative motion. In this sense, Theorem I is a generalization of Lyapunov's theorem on the stability of the equilibrium configuration of a homogeneous rotating fluid, the particles of which are attracted to one another according to Newton's law[48, 55].

In fact, if our system consists of only one gravitational fluid,

$$W = \frac{1}{2}\frac{k_0^2}{I} - \frac{\mu}{2}\iint \frac{d\tau d\tau'}{r},$$

and if the fluid's equilibrium configuration corresponds to a minimum of W, this equilibrium configuration is stable.

In general, Theorem I is the generalization of Routh's theorem to the case of a rigid body containing a fluid, which is known in the mechanics of systems with a finite number of degrees of freedom.

As a corollary of Theorem I, when $k_0 = 0$ we obtain Lagrange's theorem on the stability of the equilibrium of a rigid body containing a fluid.

Theorem II

If, for the equilibrium position of a rigid body with a fluid-filled cavity, the potential energy of the system Π has an isolated minimum Π_0, then the equilibrium position is stable.

It should be noted that Lagrange's theorem for a rigid body containing a fluid was first proven in its linear formulation by N.N. Moiseyev[62], defining stability as the boundedness of principal vibrations, while in its nonlinear formulation it was proven in Ref. [56].

Let us now consider the behavior of the equilibrium of a rigid body containing a fluid when the potential energy of the system does not have a minimum in the equilibrium position. For systems with a finite number of degrees of freedom, converses of Lagrange's theorem were established by Lyapunov[28] and Chetayev[29]. An analogous theorem can also be proven for a rigid body containing a fluid.

Theorem III

If, in a vicinity (as small as desired) of the equilibrium position of a rigid body containing a fluid the potential energy $\Pi = \Pi^{(2)} + \Pi^{(3)} + \cdots$ can take on negative values and if then the signs of Π and $2\Pi^{(2)}+3\Pi^{(3)}+\cdots$ are determined by the quadratic terms $\Pi^{(2)}$, then this equilibrium position is unstable.

Proof

Let us assume that in the equilibrium position of the rigid body $q_j=0$ $(j=1,\ldots,n-1)$, the fluid occupies the region τ_0 and the coordinates of the fluid particles are $x_i=x_{i0}$ $(i=1, 2, 3)$. In the equilibrium position, let the potential energy be $\Pi=0$ and not a minimum. Then there exists near the equilibrium position a region where $\Pi<0$. The radius vector of the displacement of a fluid particle from the equilibrium to the disturbed position is denoted by $\Delta \boldsymbol{r}=\boldsymbol{r}-\boldsymbol{r}_0$.

We now consider the functional[29]

$$V = -(T+\Pi)\left\{\sum_{j=1}^{n}\frac{\partial L}{\partial \dot{q}_j}q_j+\sum_{i=1}^{3}\varrho\int_\tau \frac{\partial \mathscr{T}}{\partial u_i}\Delta x_i d\tau\right\}.$$

In an area sufficiently near the equilibrium position, *i.e.*, in the region of the coordinates q_j, Δx_i, velocities $\dot{q}_j$, u_i, and slope Δ, the absolute values of which are small, we isolate region (C), which exists on our assumptions for any q_j, Δx_i, $\dot{q}_i$, u_i, and Δ the magnitude of which is as small as desired. The region (C) is defined by the simultaneous inequalities

$$T+\Pi<0\,,\quad \sum_{j=1}^{n}\frac{\partial L}{\partial \dot{q}_j}q_j+\sum_{i=1}^{3}\varrho\int_\tau \frac{\partial \mathscr{T}}{\partial u_i}\Delta x_i d\tau>0\,.$$

In region (C) the functional V is obviously bounded, *i.e.*, $V<N$, where N is some positive number.

By virtue of the equations of disturbed motion (1–45) and (1–46), the time derivative of V has the form

$$\dot{V} = -(T+\Pi)\left\{\sum_{j=1}^{n}\left(\frac{\partial L}{\partial q_j}q_j+\frac{\partial L}{\partial \dot{q}_j}\dot{q}_j\right)+\sum_{i=1}^{3}\varrho\int_\tau\left[\frac{\partial \mathscr{T}}{\partial u_i}u_i+\right.\right.$$
$$\left.\left.+\left(\frac{\partial \mathscr{T}}{\partial x_i}+\frac{\partial U_2}{\partial x_i}-\frac{1}{\varrho}\frac{\partial p}{\partial x_i}\right)\Delta x_i\right]d\tau\right\}.$$

Making use of the fact that

$$\sum_{j=1}^{n}\frac{\partial L}{\partial \dot{q}_j}\dot{q}_j+\sum_{i=1}^{3}\varrho\int_\tau \frac{\partial \mathscr{T}}{\partial u_i}u_i d\tau=2T\,,$$

which was established in Sec. 8 of Chapter 1, we represent the expression for $\dot{V}$ in the form

$$\dot{V} = -(T + \Pi)\left\{2T + \sum_{j=1}^{n} \frac{\partial T}{\partial q_j} q_j + \sum_{i=1}^{3} \varrho \int_{\tau} \frac{\partial \mathcal{T}}{\partial x_i} \Delta x_i d\tau + \right.$$

$$\left. + \sum_{j=1}^{n} \frac{\partial U}{\partial q_j} q_j + \sum_{i=1}^{3} \varrho \int_{\tau} \left(\frac{\partial U_2}{\partial x_i} - \frac{1}{\varrho}\frac{\partial p}{\partial x_i}\right) \Delta x_i d\tau \right\}.$$

Since the constraints imposed on the rigid body are assumed not to be time dependent, the kinetic energy of the system T is positive definite with respect to $\dot{q}_j$ and u_i. For q_j and Δx_i with sufficiently small absolute magnitudes, the expression

$$2T + \sum_{j=1}^{n} \frac{\partial T}{\partial q_j} q_j + \sum_{i=1}^{3} \varrho \int_{\tau} \frac{\partial \mathcal{T}}{\partial x_i} \Delta x_i d\tau$$

by continuity will also be positive definite with respect to q_j and u_i.

Expanding the potential energy Π in a series in the vicinity of the equilibrium position, we obtain

$$\Pi = \Pi^{(2)} + \Pi^{(3)} + \cdots,$$

where $\Pi^{(k)}$ denotes the kth order terms with respect to q_j and Δx_i. Then, as is easy to check, we have

$$\sum_{j=1}^{n} \frac{\partial U}{\partial q_j} q_j + \sum_{i=1}^{3} \varrho \int_{\tau} \left(\frac{\partial U_2}{\partial x_i} - \frac{1}{\varrho}\frac{\partial p}{\partial x_i}\right) \Delta x_i d\tau = -2\Pi^{(2)} - 3\Pi^{(3)} + \cdots$$

Since the values of the total energy $T+\Pi$ in region (C) are negative, by definition of this region, and since, by the statement of the theorem, the signs of Π and $2\Pi^{(2)}+3\Pi^{(3)}+\cdots$ are determined by the second-order terms, then in region (C)

$$\Pi^{(2)} < 0.$$

Consequently, the time derivative of the functional V

$$\dot{V} = -(T + \Pi)\left\{2T + \sum_{j=1}^{n} \frac{\partial T}{\partial q_j} q_j + \sum_{i=1}^{3} \varrho \int_{\tau} \frac{\partial \mathcal{T}}{\partial x_i} \Delta x_i d\tau - 2\Pi^{(2)} + \cdots\right\}$$

will be positive definite with respect to $\dot{q}_j$, u_i, q_j, and Δx_i in region (C). Selecting the initial disturbances q_{j0}, $\Delta x_i^{(0)}$, $\dot{q}_{j0}$, u_{i0}, Δ_0 in this region as small as desired and such as to satisfy conditions (4–21), so that the initial value $V_0 < 0$, we conclude from the equation

$$V = V_0 + \int_{t_0}^{t} \dot{V} dt$$

that the inequality $V < N$ will become invalid with time. Nonconformance to this inequality attests to nonconformance to conditions (4–23), which attests to instability of the equilibrium position, and proves the theorem. In its linear statement, this theorem was proved by Kreyn[78].

It should be noted that Theorem II is also applicable to relative equilibrium of a rigid body with a fluid-filled cavity with respect to a coordinate system rotating about a fixed axis at a constant angular velocity.

Let us assume for definiteness that, in addition to forces that are derivatives of the force function Π, our system is also acted upon by nonconservative forces that yield a moment N, directed along the x'_3 axis, the magnitude of which is such that during the entire motion the angular velocity ω of the rigid body moving about the x'_3 axis remains constant. In this case, instead of the energy integral (4–1) and the area integral (4–7), we will have the expressions[59]

$$d(T + \Pi) = N\omega dt , \quad \frac{dG_{x'_3}}{dt} = N ,$$

from which it follows that during the entire motion

$$T + \Pi - \omega G_{x'_3} = \text{const} .$$

Again introducing the coordinate system $O'\xi_1\xi_2x'_3$, rotating about the x'_3 axis with constant angular velocity ω, and applying Eqs. (4–8), we represent this integral in the form

$$T^{(1)} + \Pi - \tfrac{1}{2}\omega^2 I = \text{const} . \tag{4–30}$$

The positions of relative equilibrium of the rigid body containing a fluid are determined by condition (4–12), which yields the equations of relative equilibrium (4–16), where k_0^2/I_0^2 should be replaced by ω^2, and Eq. (4–17) for the free surface of the fluid in relative equilibrium. By repeating almost verbatim the proof of Theorem I, it is easy to see that the following theorem is valid.

Theorem IV

If, for the relative equilibrium position of a rigid body with a fluid, the expression

$$W_* = \Pi - \tfrac{1}{2}\omega^2 I \tag{4–31}$$

has an isolated minimum, then this relative equilibrium position is stable.

Since the equations of relative equilibrium and of steady motion are identical when $k_0 = I_0\omega$, we are in a position to compare the positions of relative equilibrium of our system, rotating with a constant angular velocity ω, with the steady motion in the presence of area integral (4–7). Then the following theorem is valid.

Theorem V

If, for some relative equilibrium position, W_* has a minimum, then for the corresponding steady motion W also has a minimum.

In fact, let W_{*0} be the minimum of expression W_*, i.e., in a sufficiently small vicinity of the relative equilibrium position

$$\Pi - \Pi_0 - \tfrac{1}{2}\omega^2 (I - I_0) > 0 .$$

Let us assume that W has no minimum for the corresponding steady motion, *i.e.*, in a sufficiently small vicinity of it there are points in which

$$\frac{1}{2} k_0^2 \left(\frac{1}{I_0} - \frac{1}{I} \right) - \Pi + \Pi_0 \geqslant 0 .$$

Replacing k_0 by $I_0\omega$ in this expression and adding this to the first inequality, we get

$$-\frac{1}{2} \frac{\omega^2}{I} (I - I_0)^2 > 0 ,$$

which is impossible. Consequently, if the relative equilibrium position of a rigid body containing a fluid in the case when $\omega = \text{const}$ is stable, then the corresponding steady motion in the case when $G_{x'3} = \text{const}$ is also stable[59].

For a specified value of the parameter k_0 (or ω), let the expression for W (or W_*) have a minimum; *i.e.*, let the steady motion (relative equilibrium) be stable. We shall now continuously vary $k_0(\omega)$ so that the roots of Eqs. (4–16) will describe some branch C of the curve of "equilibria." Then the expression for W (W_*) will also vary continuously, and for all the points of Curve C for which $W(W_*)$ retains its minimum, the steady motions (relative equilibria) will be stable. A change in stability on this branch can only take place at bifurcation points[29].

Consequently, if a specified value of $k_0(\omega)$ has corresponding to it a steady motion (relative equilibrium) for which $W(W_*)$ has an isolated minimum, then for any other value of $k(\omega)$ sufficiently close to $k_0(\omega)$ there is a steady motion (relative equilibrium) sufficiently close to the first for which $W(W_*)$ also has a minimum.

[4-3.2] Effect of Viscosity

Up to now, in this chapter we have assumed that the fluid filling the cavity is ideal. Let us now consider the motion of a rigid body containing a cavity filled with a viscous fluid. When potential forces are acting on the system, its total mechanical energy does not remain constant, but is dissipated as a result of the action of viscous forces. The result is that, instead of the energy integral (4–1) we have the inequality

$$T + \Pi \leqslant h .$$

The area integral (4–7) occurs under the same conditions as for an ideal fluid, since viscous forces, being internal, do not yield a moment with respect to any axis. Introducing, as in Sec. 1, the coordinate system $O'\xi_1\xi_2x'_3$ rotating about the x'_3 axis, and performing the operations specified there, we obtain, instead of the energy integral (4–24), the inequality

$$T^{(1)} + W + \frac{1}{2}\frac{k^2 - k_0^2}{I} \leqslant T^{(10)} + W^{(0)} + \frac{1}{2}\frac{k^2 - k_0^2}{I^{(0)}} \tag{4–32}$$

which changes nothing in the proof of Theorem I. This remark is obviously applicable to the other theorems as well.

Thus, Theorems I–V remain valid for a viscous fluid. Moreover, the results then obtained are even more rigorous. For example, Theorem I is then replaced by the following theorem.

Theorem VI

If, for the steady motion of a rigid body with a cavity filled with a viscous fluid, the expression for W has an isolated minimum, then this motion is stable, while any disturbed motion sufficiently close to it, will approach, in the limit, the steady motion of the system as a single rigid body, if the inequality $\nabla > \varepsilon l$ is not violated.

The validity of the first part of this theorem is obvious on the basis of the remark made just previously. Let us now prove the second part.

Consider some disturbed motion that, during the initial instant of time, is sufficiently close to the undisturbed motion, and let the inequality $\nabla > \varepsilon l$ apply continuously as long as $|l|$ does not exceed L_1. Then the disturbed motion will always be sufficiently close to the steady undisturbed motion. According to Eq. (1–88), the total mechanical energy of the system in its disturbed motion is dissipated until the fluid and the rigid body start moving as a single rigid body. It is consequently possible to make only one of two assumptions: either the total mechanical energy of the system will constantly decrease and the system will finally come to rest, or its motion, in the limit, will become one of rotation of the entire system as a single rigid body corresponding to the extremum of the expression $\frac{1}{2}(k^2)/I+\Pi$. The first assumption, when $G_{x'_3}\neq 0$, is impossible by virtue of the existence of the area integral (4–7), leaving only the second assumption[15], which proves the theorem.

Remark

Theorem IV can also be reinforced in a similar manner for a viscous fluid if $\omega=\text{const}$ during the entire motion.

Theorem VII

If, in the vicinity of isolated steady motion of a rigid body containing a cavity filled with a viscous fluid, the expression for W can take on negative values, then this motion is unstable.

Proof

Let the roots of Eqs. (4–16) for the motion under study be $q_j=0$ ($j=1,\dots,n-1$) and $W_0=0$. We shall assume that there exists a positive number L_1 sufficiently small that for all of the values of q_j and of the departure l satisfying the conditions

$$|q_j|\leqslant L_1\,,\quad |l|\leqslant L_1\,, \tag{4–33}$$

the expression for W has no extremum, except for the single one corresponding to $q_j=0$ and $l=0$. This assumption also means that the steady motion is isolated. Since, for the latter, W does not have a minimum, there exists in the region (4–33) a region where $W<0$. Then in the region where the absolute magnitudes of q_j, $\dot{q}_j$, and l are small, it is possible to isolate a region, existing under our assumptions for coordinates and velocities, the absolute magnitude of which is as small as desired, and which is defined by the expression

$$T^{(1)}+W<0\,.$$

We select our initial disturbances in precisely this region and, in addition, in a manner such that the area constant k remains equal to k_0. Left to itself, the system will move in accordance with Eq. (4–32), which, under the above conditions, takes on the form

$$T^{(1)}+W\leqslant T^{(10)}+W^{(0)}<0\,.$$

Let us assume, contrary to the statement of the theorem, that the undisturbed motion is stable. By definition, this means that during the entire time, or at least until $\Delta>\varepsilon l$, the conditions (4–23) are satisfied. When this is true, it is obviously possible to denote a positive number L, depending on

L_1 and L_2, which from the above limits the absolute magnitude of the mechanical energy of the system

$$|T^{(1)} + W| < L. \tag{4–34}$$

But in the region defined by Eqs. (4–23), by definition, nowhere, except for the undisturbed motion under study, are Eqs. (1–89) identically satisfied. Consequently, the energy of the system will be continuously dissipated and will increasingly differ from its initial value; finally, its absolute value will exceed L, contradicting condition (4–34). This means that the system will move out from region (4–23), which proves the theorem.

In a similar manner it is possible to prove the instability of the isolated position of the absolute (or relative) equilibrium of a rigid body containing a viscous fluid if the expression for Π (or W_*) does not have a minimum*.

[4-4] THE PROBLEM OF THE MINIMUM

According to the theorems proven in Sec. 3, the problem of stability of steady motion (or of equilibrium) of a rigid body with cavities partially or completely filled with fluids reduces to the problem of the minimum of the altered potential energy of the system W (or of the potential energy Π). In the case when the fluid fills the cavity completely, W is a function of a finite number of variables q_j ($j=1,\ldots,n-1$) and the conditions of the minimum for this function are easy to find.

When the cavity is only partially filled, W is a functional that depends on the configuration of the volume τ and of the free surface S of the fluid, and finding conditions of the minimum for it is much more difficult. Since the property of the minimum is local in the sense that the values of W in the given point are compared to the values in all the points sufficiently close to the given point, including those infinitesimally close to the given point, for a rigorous solution of the problem of the minimum, excluding certain particular cases, it is possible to restrict oneself to an examination of the second-order infinitesimals. Hence, in solving this problem, the methods of the theory of small vibrations can be used, if the displacement $n(u, v)$ of the free surface from the equilibrium position is represented in the form of a series in terms of a system of eigenfunctions corresponding to the boundary-value problem. These were precisely the methods used in solving the problem of the minimum of the potential energy of a massive pendulum with a cavity filled with a liquid[51, 52] and of the minimum of the changed potential energy for a massive

* A different proof of theorems similar to VI and VII, which uses the idea of a theorem due to Barbashin and Krasovskiy[63], was suggested by Pozharitskiy[64].

rigid body with a cavity filled with a weightless fluid[57] (these problems will be examined below).

However, the calculations then turn out to be cumbersome, since it is necessary to operate with infinite series, multiply them, take their integrals over certain regions, *etc.* It is more convenient to use another method for solving the problem of the minimum, based on the fact that the potential energy of a massive fluid in a vessel displaced from the equilibrium position has a minimum if the free surface of the fluid is horizontal[54]. We shall present here the solution[58] of the problem of the minimum for W, using the above method and disregarding the surface tension forces.

[4-4.1] Reduction to the Problem of a Minimum of a Function of a Finite Number of Variables

Let us consider, in the vicinity of steady motion of a rigid body containing a fluid, for which $q_j=0$ ($j=1, \ldots, n-1$), the region of varying q_j:

$$|q_j| \leqslant H \quad (j = 1, \ldots, n-1), \tag{4–35}$$

where $H>0$ is a sufficiently small constant quantity.

Let q_j be the coordinates of some fixed point in region (4–35). Let us clarify what should be the configuration of the free surface of the fluid, for which, for the given specified q_j the expression for W should have an extremum. To solve this problem, we now find the first variation of W for the above q_j and equate it to zero:

$$\delta W = -\varrho \int_\tau \left[\frac{k_0^2}{I^2} (x_1' \delta x_1' + x_2' \delta x_2') + \sum_{i=1}^{3} \frac{\partial U_2}{\partial x_i'} \delta x_i' \right] d\tau = 0, \tag{4–36}$$

where I is the moment of inertia of the system about the x_3' axis for the given q_j and for the sought configuration of the free surface of the fluid.

Making use of the equation of continuity and of the boundary conditions (1–32)–(1–34) for the fluid, we find immediately that Eq. (4–36) is possible if and only if the free surface of the fluid has the form

$$f(x_1', x_2', x_3') \equiv \frac{k_0^2}{2I^2} (x_1'^2 + x_2'^2) + U_2(x_1', x_2', x_3') = c, \tag{4–37}$$

where the constant c is determined by the quantity of fluid in the given cavity.

For steady motion, when $q_j=0$, Eq. (4–37) becomes Eq. (4–20) of the free surface in this motion.

Since the fluid presents no resistance to tension, the forces acting on those particles situated on the free surface are directed into the fluid. Hence in steady motion the fluid should be situated to that side of the surface (4–20) where the function $f_0(x_1', x_2', x_3') > c_0$. The simply connected region that the fluid occupies in this motion will be denoted by D_0; for all its points, $f_0 \geqslant c_0$.

From the possible positions that the fluid can take when $q_j \neq 0$, we select only those where everywhere inside the fluid $f(x_1', x_2', x_3') > c$. This region, which is bounded by the surface (4–37) and the walls σ_1 of the cavity, we denote by D.

Let us now clarify the behavior of the minimum of W for the fixed q_j from region (4–35) when the fluid fills region D. Here we shall assume that c does not have the extremum value from among all the values that the function f_0 takes in the region of the surface $f_0 = c_0$. The locus of the points of intersection of the surface (4–20) with the walls σ_1 of the cavity is represented by some closed curve s, which is the boundary of the free surface of the fluid in steady motion.

Let us imagine a unit vector of the normal $n_1(P)$ to the surface (4–20) in point P of curve s, directed in the direction of $f_0 < c_0$, and a unit vector of the normal $n_2(P)$, directed into the cavity. We shall assume that the angle $\theta(P)$ formed by these vectors varies continuously between the fixed limits $0 < \theta_1 \leqslant \theta(P) \leqslant \theta_2 < \pi$ when the point P traverses curve s.

Let us also consider the two-parameter family of surfaces

$$f = \frac{k_0^2}{2(I_0 + \Delta I)^2}(x_1'^2 + x_2'^2) + U_2(x_1', x_2', x_3') = c_0 + \Delta c\,, \qquad (4\text{–}38)$$

which are continuous functions of ΔI and Δc, assumed to be of sufficiently small absolute magnitude. We shall assume that the unit vector of the normal n_1 to the surfaces (4–38) is a continuous function of $x_1', x_2', x_3', \Delta I$, and Δc in a sufficiently small vicinity of the curve s and that the vector n_2 is a continuous function of the points of the cavity surface in the same vicinity.

On the basis of the above assumptions we can claim that, for sufficiently small H and Δc, the surfaces belonging to the family (4–38) continuously intersect the walls of the cavity of the solid body in the position q_j that is being considered[54]. The loci of the points of intersection of these surfaces for different ΔI and Δc, which are closed curves, transform continuously into curve s as $\sum_{j=1}^{n-1} |q_j|^2 + |\Delta I|^2 + |\Delta c|^2 \to 0$. Consequently, for any numerically sufficiently small q_j, ΔI, and Δc, there is a simply connected region D' bounded by the surface of the cavity walls that is deflected into position q_j, and by the surface (4–38); this region does not contain points $f < c_0 + \Delta c$ and transforms continuously into the region D_0 as

$$\sum_{j=1}^{n-1} |q_j|^2 + |\Delta I|^2 + |\Delta c|^2 \to 0\,.$$

It is obvious on the basis of the above that the constant H, which defines the region (4–35) can always be selected sufficiently small such that for any q_j from region (4–35) the surface (4–37) should belong to the family (4–38); then ΔI and Δc, which are determined from the condition of constancy of volume of the fluid, will be continuous functions of q_j that vanish when $q_j=0$, while the departure l of the surface (4–37) from the surface (4–20) inside the cavity will not exceed H.

For certain fixed q_j from region (4–35) we now consider region D, bounded by the walls of the cavity in the deflected position and by the surface (4–37), where the constant $c=c_0+\Delta c$ is determined. Let $d<c_0+\Delta c$ be some constant such that the region $D'(d)$, which is bounded by the surface of the cavity's walls and by the surface $f=d$, and which continuously transforms into region D as $d \to c_0+\Delta c$, does not contain points $f<d$.

The existence of such a region $D'(d)$ is obvious, on the basis of what was said above, by values of d that are sufficiently close to c_0. However, in many cases it is possible to take the value of d sufficiently removed from c_0, and if the cavity is closed it may turn out that any possible positions of the fluid in the cavity belong to the region $D'(d)$, where d depends only on the magnitude of H. For example, for the case of a gravity force field this is valid for any closed surface that is everywhere convex.

Let us consider any possible position of the fluid that entirely fills region $D'' \subset D'(d)$, and let us find the change in W for fixed q_j when the fluid passes from region D to region D''. We then have

$$W'' - W = -\varrho \int\limits_{D''} f\,d\tau + \varrho \int\limits_{D} f\,d\tau + \frac{k_0^2}{2I_0^3}(\Delta I)^2 + \cdots$$

This difference is positive if region D' differs from region D. In fact, the difference of the first two terms is the change in the potential energy of the fluid in the force field

$$f = \frac{k_0^2}{2I^2}(x_1'^2 + x_2'^2) + U_2(x_1', x_2', x_3')$$

upon transfer of some part of the fluid from the region $f \geqslant c_0+\Delta c$ to the region where $f<c_0+\Delta c$; here the potential energy of the fluid obviously increases. The third term is positive. Thus when the fluid is located in region D we find a minimum of the changed potential energy of the fluid with respect to all its possible positions in region $D'(d)$. Consequently, we have the following lemma.

Lemma

For fixed values of q_j $(j=1,\dots,n-1)$ the expression for W has a minimum if the free surface of the fluid is defined by Eq. (4–37).

As a corollary we note that for the case of equilibrium of the rigid body with the fluid, when $k_0=0$, the expression $W=\Pi$ for fixed q_j has a minimum if the free surface of the fluid is expressed by the equation

$$U_2(x_1', x_2', x_3') = \text{const}.$$

For any given set of values of q_j from region (4–35), we can find corresponding to it some rigid body, which we shall call "transformed," consisting of the given rigid body and a solidified fluid with free surface (4–37).

Then, according to our lemma, W for the transformed body has a minimum compared with all those possible for the free fluid surfaces sufficiently close to (4–37).

Theorem VIII

In order for W in steady motion of a rigid body with a fluid-containing cavity to have a minimum, it is necessary and sufficient that W have a minimum when $q_j=0$ for the transformed rigid bodies in the region (4–35).

Proof

Let the expression for W for all the possible transformed rigid bodies in the region (4–35) have a minimum W_0 when $q_j=0$; then $W-W_0>0$. For a body containing a fluid the above difference will, according to the lemma, be more positive, which proves the sufficiency.

Let us now prove the necessity. Let us assume that the expression for W for a rigid body containing a fluid has a minimum when $q_j=0$. This means that for all of the possible systems of values of q_j, the departure l, and the deviation ∇ such that $|q_j|\leqslant H, |l|\leqslant H$, $\nabla\geqslant \varepsilon l$, all of the values assumed by the difference $W-W_0$ will remain positive, vanishing only when $q_j=0$, $l=0$, and $\nabla=0$. Consequently, this difference will also be positive for values of the departure given by $|l|<H$, which characterize the transition in the fluid's configuration from that undisturbed motion to that defined by Eq. (4–37) for any q_j from region (4–35). This also means that W has a minimum W_0 for the transformed bodies when $q_j=0$, which proves the theorem.

Thus, the problem of the minimum of W reduces to the problem of the minimum of a function of a finite number of variables, which is the expression for W for a rigid body with a fluid bounded by walls σ_1 of the cavity and by the free surface (4–37).

Setting $k_0=0$, we obtain the necessary and sufficient condition for the existence of a minimum of the potential energy of the system[54]. An analogous theorem for the case of potential energy of the gravity forces was first established, in a different manner, in Ref. [62].

We should note that if the configuration of the cavity and the mass forces acting on the fluid are such that any possible positions of the fluid in the cavity are in the region $D'(d)$ discussed above, then the minimum of W for steady motion of a rigid body containing a fluid ensures, according to Remark 1 to Theorem I, stability with respect to q_j, $\dot{q}_j$, ∇, and T_2^1 during the entire motion.

[4-4.2] Evaluating the Second Variation

Let us find the change in W for the transformed rigid body when changing from the position corresponding to steady motion of the system when $q_j=0$ to disturbed position in the region (4–35). This transition can be thought of as consisting of two stages: 1) displacement into a disturbed position of the entire system as a single rigid body; 2) deformation of the fluid configuration by imposing on its free surface a layer τ_1, the volume of which is zero, in a configuration with a free surface (4–37).

Then the increment of W will be written in the form

$$\Delta W = \Delta_1 W + \Delta_2 W \tag{4–39}$$

where Δ_1 is the increment on displacement into disturbed motion of the entire system as a single rigid body and Δ_2 is the increment on subsequent deformation of the free surface of the fluid into the surface (4–37). Similarly,

$$\Delta I = \Delta_1 I + \Delta_2 I .$$

Subsequently we shall restrict ourselves to calculating ΔW to within second-order infinitesimals inclusively; *i.e.*, we shall calculate the second variation $\delta^2 W$, the positive definiteness of which is sufficient for the existence of a minimum of the functional W.

To within second-order infinitesimals in q_j, we will have

$$\begin{aligned} \Delta_1 W &= \frac{1}{2} \sum_{i,j=1}^{n-1} \left(\frac{\partial^2 W}{\partial q_j \partial q_i} \right)_0 q_i q_j + \cdots , \\ \Delta_2 W &= -\varrho \int_{\tau_1} \left[\frac{1}{2} \omega^2 (x_1'^2 + x_2'^2) + U_2(x_1', x_2', x_3') \right] d\tau + \\ &\quad + \frac{\omega^2}{2I_0} [(\Delta_2 I)^2 + 2\Delta_1 I \Delta_2 I] + \cdots , \end{aligned} \tag{4–40}$$

where the subscript 0 means that the quantity is calculated for the undisturbed position of the system. The quantities $\Delta_1 I$ and $\Delta_2 I$ here should be determined to within first-order infinitesimals. It is obvious that

$$\Delta I_1 = \sum_{j=1}^{n-1} \left(\frac{\partial I}{\partial q_j}\right)_0 q_j + \cdots .$$

To calculate $\Delta_2 W$ it is convenient to introduce a moving coordinate system $x_1 x_2 x_3$ that is rigidly fixed to the rigid body, the x_3 axis of which in the undisturbed motion of the system is made to coincide with the x_3' axis.

The integrand in the expression for $\Delta_2 W$, which is transformed into the variables x_1, x_2, x_3 is denoted

$$\Phi(x_1, x_2, x_3, q_j) = [U_2(x_1', x_2', x_3') + \tfrac{1}{2}\omega^2(x_1'^2 + x_2'^2)]_{x'_i \to x_i,\, q_j} .$$

The equation of the free surface (4–20) of the solidified fluid, in terms of the variables x_1, x_2, x_3, has the form

$$\Phi(x_1, x_2, x_3, 0) = \tfrac{1}{2}\omega^2(x_1^2 + x_2^2) + U_2(x_1, x_2, x_3) = c_0 . \qquad (4\text{–}41)$$

Let us assume that the above equation can be uniquely solved with respect to one of the variables x_1, x_2, x_3; for definiteness, let it be with respect to x_3), for which it is sufficient that the continuous derivative $(\partial\Phi/\partial x)_0$ should not vanish at any point of the surface (4–41). This requirement is inconsequential and is introduced only to simplify the formulas. We denote by Q the region of the plane (x_1, x_2) bounded by the projection onto this plane of the closed curve s, which is the locus of the points of intersection of the surface (4–41) with the walls σ of the cavity.

Surface (4–37), in terms of the moving axes, will take the form

$$\Phi_1(x_1, x_2, x_3, q_j) = c , \qquad (4\text{–}42)$$

where the constant $c = c_0 + \Delta c$ is determined from the condition of equality of the fluid volumes with free surfaces (4–41) and (4–42). The last condition is equivalent to the requirement that the volume of the layer τ_1 undergoing deformation must be equal to zero:

$$\int_{\tau_1} d\tau = 0$$

which, in the first approximation, is equivalent to

$$\iint_Q dx_1 dx_2 \int_{x_{30}}^{x_{31}} dx_3 = 0 ,$$

where x_{30} and x_{3_ω} denote, respectively, the values of the variable x_3 for points on the surfaces (4–41) and (4–42). Replacing x_3 by a new variable $\mu=\Phi(x_1,x_2,x_3,q_j)-c_0$, we shall represent the last equation, with the same accuracy, in the form

$$\iint_Q \left(\frac{\partial x_3}{\partial \Phi}\right)_0 (\mu_1-\mu_0)\,dx_1dx_2=0\,, \tag{4–43}$$

where, to within first-order infinitesimals,

$$\begin{aligned}\mu_0 &= \Phi(x_1,x_2,x_{30},q_j)-c_0=\sum_{j=1}^{n-1}\left(\frac{\partial \Phi}{\partial q_j}\right)_0 q_j+\cdots,\\ \mu_1 &= \Phi(x_1,x_2,x_{31},q_j)-c_0=\Delta c+\frac{\omega^2}{I_0}(x_1^2+x_2^2)\,\Delta I+\cdots,\end{aligned} \tag{4–44}$$

since, in the first approximation, the functions $\Phi(x_1, x_2, x_3, q_j)$ and $\Phi_1(x_1, x_2, x_3, q_j)$ differ only by the component $\omega^2/I_0(x_1^2+x_2^2)\,\Delta I$.

Substituting the values of μ_0 and μ_1 into (4–43), we obtain a linear equation relating $\Delta_2 I$ to Δc. The second similar equation is obtained by calculating, in the first approximation, the expression

$$\Delta_2 I=\varrho\int_{\tau_1}(x_1'^2+x_2'^2)\,d\tau=\varrho\iint_Q\left(\frac{\partial x_3}{\partial \Phi}\right)_0(x_1^2+x_2^2)(\mu_1-\mu_0)\,dx_1dx_2\,. \tag{4–45}$$

The above equation, together with Eq. (4–43) and making use of Eq. (4–44), takes the form

$$\begin{gathered}\iint_Q\left(\frac{\partial x_3}{\partial \Phi}\right)_0\left[\Delta c+\frac{\omega^2}{I_0}(x_1^2+x_2^2)\Delta I-\sum_{j=1}^{n-1}\left(\frac{\partial \Phi}{\partial q_j}\right)_0 q_j\right]dx_1dx_2=0\,,\\ \varrho\iint_Q\left(\frac{\partial x_3}{\partial \Phi}\right)_0(x_1^2+x_2^2)\left[\Delta c+\frac{\omega^2}{I_0}(x_1^2+x_2^2)\Delta I-\sum_{j=1}^{n-1}\left(\frac{\partial \Phi}{\partial q_j}\right)_0 q_j\right]dx_1dx_2=\Delta_2 I\end{gathered} \tag{4–46}$$

and makes it possible, as is easy to see, to uniquely determine $\Delta_2 I$ and Δc as linear functions of q_j. We note that if

$$\begin{gathered}\iint_Q\left(\frac{\partial x_3}{\partial \Phi}\right)_0\mu_0 dx_1dx_2=\iint_Q\left(\frac{\partial x_3}{\partial \Phi}\right)_0(x_1^2+x_2^2)\,\mu_0 dx_1dx_2=0\,,\\ \left(\frac{\partial I}{\partial q_j}\right)_0=0\quad (j=1,\dots,n-1)\,,\end{gathered} \tag{4–47}$$

then $\Delta_1 I=0$, $\Delta_2 I=0$, $\Delta c=0$ in the first approximation.

We thus have

$$\Delta_2 W = -\frac{1}{2}\varrho \iint\limits_Q \left(\frac{\partial x_3}{\partial \Phi}\right)_0 (\mu_1^2 - \mu_0^2)\, dx_1 dx_2 + \frac{\omega^2}{2I_0}\left[(\Delta_2 I)^2 + 2\Delta_1 I \Delta_2 I\right] + \cdots. \tag{4–48}$$

According to Eqs. (4–39) and (4–40), ΔW is represented in the form of a sum of the quadratic form of the variables $q, \ldots, q_{n-1}$:

$$\begin{aligned} W^{(2)} = \frac{1}{2}\sum_{i,j=1}^{n-1}\left(\frac{\partial^2 W}{\partial q_i \partial q_j}\right)_0 q_i q_j + \frac{\omega^2}{2I_0}\left[(\Delta_2 I)^2 + 2\Delta_1 I \Delta_2 I\right] - \\ -\frac{1}{2}\varrho \iint\limits_Q \left(\frac{\partial x_3}{\partial \Phi}\right)_0 (\mu_1^2 - \mu_0^2)\, dx_1 dx_2 \end{aligned} \tag{4–49}$$

and higher-order infinitesimals. The conditions for positive definiteness of the quadratic form $W^{(2)} = \frac{1}{2}\delta^2 W$ will be the conditions for the minimum of W for a rigid body with a fluid-filled cavity in a force field with potential energy Π. In the case when $W^{(2)}$ can take on negative values, W does not have a minimum for the steady motion under study.

If, on the other hand, $W^{(2)}$ is a positive, sign-constant form, the problem of the minimum of W can be solved only by examining terms of order of smallness higher than two. We shall not dwell on this case.

In examining the stability of the equilibrium position of a rigid body containing a fluid, the formula for the second variation of the potential energy of the system can obviously be obtained from Eq. (4–49) for $W^{(2)}$, setting $\omega = 0$ in the latter. We thus have

$$\Pi^{(2)} = \frac{1}{2}\sum_{i,j=1}^{n}\left(\frac{\partial^2 \Pi}{\partial q_i \partial q_j}\right)_0 q_i q_j - \frac{1}{2}\varrho \iint\limits_Q \left(\frac{\partial x_3}{\partial \Phi}\right)_0 (\mu_1^2 - \mu_0^2)\, dx_1 dx_2\,, \tag{4–50}$$

where

$$\Phi(x_1, x_2, x_3, q_j) = U_2(x_1', x_2', x_3')|_{x_i' \to x_i,\, q_j}.$$

The quantities μ_0 and μ_1 are then determined from

$$\mu_0 = \sum_{j=1}^{n}\left(\frac{\partial \Phi}{\partial q_j}\right)_0 q_j + \cdots, \quad \mu_1 = \Delta c + \cdots, \tag{4–51}$$

which are analogous to Eqs. (4–44).

In the study of the stability of the position of relative equilibrium of a rigid body containing a fluid in a coordinate system $O\xi_1\xi_2 x_3'$ rotating about the x_3' axis, we have the following formula for the second variation of W_*, which is defined by Eq. (4–31):

$$W_*^{(2)} = \frac{1}{2} \sum_{i,j=1}^{n} \left(\frac{\partial^2 W_*}{\partial q_i \partial q_j} \right)_0 q_i q_j - \frac{1}{2} \varrho \iint_Q \left(\frac{\partial x_3}{\partial \Phi} \right)_0 (\mu_1^2 - \mu_0^2)\, dx_1 dx_2 . \tag{4–52}$$

Here, as in Eq. (4–49),

$$\Phi(x_1, x_2, x_3, q_j) = [U_2 + \tfrac{1}{2}\omega^2 (x_1'^2 + x_2'^2)]_{x'_i \to x_i,\, q_j},$$

and μ_0 and μ_1 are defined by Eqs. (4–51).

Remark

The above solution of the problem of the minimum case disregards the surface tension forces of the fluid. When these forces are taken into account, the problem becomes more complex. In fact, if these forces are large in comparison with the mass forces (or if the latter do not act at all), the configuration of the free surface is determined primarily (or exclusively) by the former, and it is generally not possible to reduce the problem to determining the minimum of a function of a finite number of variables. The conditions for the minimum of W in this case can be found by solving the appropriate boundary-value problem. On the other hand, it is intuitively clear that if the surface tension forces are small in comparison with the mass forces, and the dimensions of the vessel are sufficiently large, the effect of the surface tension forces will be small. We shall not consider this problem in more detail here.

[4-5] STABILITY OF MOTION OF AN ARTIFICIAL SATELLITE MOVING IN A CIRCULAR ORBIT

As the first example of an application of the theory of stability of steady motion of a rigid body containing a fluid presented in this chapter, we again consider the problem of the stability of motion of an artificial satellite, moving in a circular orbit, with a cavity filled with a fluid (as was considered in Sec. 5, Chapter 3). Retaining the notation used there we note that when the fluid fills the cavity completely, W can be represented as a function of the variables $R, \psi, \gamma_1, \gamma_2, \beta_1, \beta_3$. In fact, according to the Huygens-Steiner theorem, the moment of inertia of the system about the stationary $O'x_3'$ axis is

$$I = MR^2 \cos^2\psi + A\gamma_1^2 + B\gamma_2^2 + C\gamma_3^2 .$$

Eliminating γ_3 from this expression, and eliminating β_2 from (3–23) using integrals such as (1–77), we represent W in the form

$$W = \frac{1}{2} k_0^2 [MR^2 \cos^2 \psi + (A - C) \gamma_1^2 + (B - C) \gamma_2^2 + C]^{-1} -$$

$$- \mu \frac{M}{R} + \frac{3}{2} \frac{\mu}{R^3} \left[(A - B) \beta_1^2 + (C - B) \beta_3^2 + \frac{1}{3} (2B - A - C) \right]. \quad (4\text{–}53)$$

The variables ψ, β_1, β_3, λ_1, λ_2 are, however, not independent but are related by the expression

$$\chi \equiv \beta_1 \gamma_1 + \sqrt{1 - \beta_1^2 - \beta_3^2}\, \gamma_2 + \beta_3 \sqrt{1 - \gamma_1^2 - \gamma_2^2} - \sin \psi = 0 .$$

Determination of the roots of Eq. (4–16) for steady motion reduces to finding the absolute extrema of the function

$$W_1 = W + \lambda \chi ,$$

where λ is some constant. We have the equations

$$\frac{\partial W_1}{\partial R} = - \frac{k_0^2}{I^2} MR \cos^2 \psi + \mu \frac{M}{R^2} - \frac{9}{2} \frac{\mu}{R^4} \times$$

$$\times \left[(A - B) \beta_1^2 + (C - B) \beta_3^2 + \frac{1}{3} (2B - A - C) \right] = 0 ,$$

$$\frac{\partial W_1}{\partial \psi} = \left(\frac{k_0^2}{I^2} MR^2 \sin \psi - \lambda \right) \cos \psi = 0 ,$$

$$\frac{\partial W_1}{\partial \beta_1} = 3 \frac{\mu}{R^3} (A - B) \beta_1 + \lambda [\gamma_1 - \beta_1 \gamma_2 (1 - \beta_1^2 - \beta_2^2)^{-1/2}] = 0 , \quad (4\text{–}54)$$

$$\frac{\partial W_1}{\partial \beta_3} = 3 \frac{\mu}{R^3} (C - B) \beta_3 +$$

$$+ \lambda [(1 - \gamma_1^2 - \gamma_2^2)^{1/2} - \gamma_2 \beta_3 (1 - \beta_1^2 - \beta_3^2)^{-1/2}] = 0 ,$$

$$\frac{\partial W_1}{\partial \gamma_1} = - \frac{k_0^2}{I^2} (A - C) \gamma_1 + \lambda [\beta_1 - \beta_3 \gamma_1 (1 - \gamma_1^2 - \gamma_2^2)^{-1/2}] = 0 ,$$

$$\frac{\partial W_1}{\partial \gamma_2} = - \frac{k_0^2}{I^2} (B - C) \gamma_2 + \lambda [(1 - \beta_1^2 - \beta_3^2)^{1/2} -$$

$$- \beta_3 \gamma_2 (1 - \gamma_1^2 - \gamma_2^2)^{-1/2}] = 0$$

which are satisfied by the values

$$R = R_0 , \quad \psi = 0 , \quad \beta_1 = \beta_3 = 0 , \quad \gamma_1 = \gamma_2 = 0 , \quad \lambda = 0 , \quad (4\text{–}55)$$

where R_0 is the root of equation

$$\frac{k_0^2}{I_0^2} MR_0 = \frac{\mu}{R_0^2} \left[M + \frac{3}{2} \frac{A + C - 2B}{R_0^2} \right], \quad I_0 = MR_0^2 + C , \quad (4\text{–}56)$$

which is identical to Eq. (3–25). The solution of (4–55), like the solution of (3–24), describes uniform motion with angular velocity $\omega = k_0/I_0$ of the mass center of the system along a circle with radius R_0, and with its center at point O', situated in the $x_1'\ Ox_2'$ plane; here the x_1 axis is directed along the tangent to the orbit, the x_2 axis is directed along the straight line $O'O$, and the x_3 axis is directed parallel to the x_3' axis.

We shall take this motion as undisturbed and examine its stability. For this we must determine the behavior of the extremum of the function W for the values given by (4–55). For these conditions we have

$$\begin{aligned}
\frac{\partial^2 W}{\partial R^2} &= -M\omega^2 \frac{C - 3MR_0^2}{C + MR_0^2} - 2\mu \frac{M}{R_0^3} + (2B - A - C)\frac{6\mu}{R_0^5},\\
\frac{\partial^2 W}{\partial \psi^2} &= M\omega^2 R_0^2,\\
\frac{\partial^2 W}{\partial \beta_1^2} &= 3\frac{\mu}{R_0^3}(A - B)\\
\frac{\partial^2 W}{\partial \beta_3^2} &= 3\frac{\mu}{R_0^3}(C - B),\\
\frac{\partial^2 W}{\partial \gamma_1^2} &= (C - A)\,\omega^2,\\
\frac{\partial^2 W}{\partial \gamma_2^2} &= (C - B)\,\omega^2,
\end{aligned} \qquad (4\text{–}57)$$

while the remaining second derivatives of W are zero. It is easy to check that for real artificial satellites the conditions for the minimum of the function W reduce to the inequalities

$$C > A > B, \qquad (4\text{–}58)$$

since then $\partial^2 W/\partial R^2 > 0$. According to Theorem I, the inequalities (4–58) are sufficient conditions for the stability of circular motion of an artificial satellite with a cavity completely filled with a fluid, with respect to the quantities

$$R, \psi, \gamma_1, \gamma_2, \gamma_3, \beta_1, \beta_2, \beta_3, \dot{R}, \dot{\psi}, \dot{\varphi}, \omega_1, \omega_2, \omega_3, T_2^{(1)}.$$

Conditions (4–58) are identical with conditions (3–27) for stability with respect to $R, \dot{R}, \psi, \dot{\psi}, \dot{\varphi}, \omega_i, G_{2i}, \beta_i, \gamma_i\ (i = 1, 2, 3), \varrho \int_*^2 v\ d\tau$.

For a viscous fluid the disturbed motion under conditions (4–58) will, in the limit, approach a steady motion consisting of uniform rotation of the system as a single rigid body about the angular momentum vector. If, however, conditions (4–58) are not satisfied and the motion (4–55) is isolated, and in its vicinity the function W can take on negative values, then it is unstable.

[4-6] STABILITY OF ROTATION OF A RIGID BODY CONTAINING A FLUID ABOUT A FIXED AXIS

Let us consider a rigid body, which has one stationary point O and a fluid-filled cavity, in a homogeneous field of gravity forces. The x_3' axis of the fixed coordinate system with its origin in the fixed point of the body is directed vertically upward. The moving axes, which are rigidly fixed to the body, will be drawn so that the x_3 axis passes through the center of gravity of the body.

The potential energy and the moment of inertia of the system about the x_3' axis are, respectively,

$$\begin{aligned} \Pi &= Mg(x_{c1}\gamma_1 + x_{c2}\gamma_2 + x_{c3}\gamma_3), \\ I &= A\gamma_1^2 + B\gamma_2^2 + C\gamma_3^2 - 2D\gamma_2\gamma_3 - 2E\gamma_1\gamma_3 - 2F\gamma_1\gamma_2, \end{aligned} \tag{4–59}$$

where the γ_i are interrelated by

$$\gamma_1^2 + \gamma_2^2 + \gamma_3^2 = 1.$$

Making use of this equality, we shall write the equation of steady motion (4–16) in the explicit form

$$\begin{aligned} \frac{\partial W}{\partial \gamma_1} &= -\omega^2[(A-C)\gamma_1 + D\gamma_1\gamma_2\gamma_3^{-1} - E(\gamma_3 - \gamma_1^2\gamma_3^{-1}) - F\gamma_2] + \\ &\quad + Mg(x_{c1} - x_{c3}\gamma_1\gamma_3^{-1}) = 0, \\ \frac{\partial W}{\partial \gamma_2} &= -\omega^2[(B-C)\gamma_2 - D(\gamma_3 - \gamma_2^2\gamma_3^{-1}) + E\gamma_1\gamma_2\gamma_3^{-1} - F\gamma_1] + \\ &\quad + Mg(x_{c2} - x_{c3}\gamma_2\gamma_3^{-1}) = 0. \end{aligned} \tag{4–60}$$

Of greatest interest is the following solution of these equations:

$$y_1 = y_2 = 0, \quad y_3 = 1, \tag{4–61}$$

which is true for any value of ω if the system's center of gravity is situated on the x_3 axis, which in this case is also the principal axis of inertia:

$$x_{c1} = x_{c2} = 0, \quad D = E = 0. \tag{4–62}$$

If the fluid only partially fills the cavity, the free surface (4–20) in this case has the shape of a paraboloid of revolution

$$\tfrac{1}{2}\omega^2(x_1^2 + x_2^2) - gx_3 = c_0. \tag{4–63}$$

Solution (4–61) corresponds to uniform motion of the system as a single rigid body about the x_3 axis, which coincides with the x_3' axis, with angular velocity ω, and in this case the x_3 axis is the principal central axis of inertia of the system; here the coordinate of the center of gravity of the system is

$$x_{c3}^0 = \frac{1}{M}(M_1 x_{13} + M_2 x_{23}^0).$$

Let us consider the stability of this motion.

We shall first consider the case when the cavity is completely filled with a fluid and find the conditions for the minimum of the function $W(\gamma_1, \gamma_2)$. We direct the x_1 and x_2 axes along the principal axes of inertia of the system for point O. Then for the undisturbed motion we will obtain

$$\frac{\partial^2 W}{\partial \gamma_1^2} = (C - A)\,\omega^2 - Mgx_{c3},$$

$$\frac{\partial^2 W}{\partial \gamma_2^2} = (C - B)\,\omega^2 - Mgx_{c3}, \quad \frac{\partial^2 W}{\partial \gamma_1 \partial \gamma_2} = 0.$$

Consequently the conditions for the minimum reduce to the inequality

$$(C - A)\,\omega^2 - Mgx_{c3} > 0 \quad (A \geqslant B). \tag{4–64}$$

Let us now pass to the case of partial filling. We make the x_i axes coincide with the principal axes of inertia of the rigid body for point O. The function $\Phi(x_1, x_2, x_3, q_j)$ in the given case has the form

$$\begin{aligned}\Phi(x_1, x_2, x_3, \gamma_1, \gamma_2) = {}& \tfrac{1}{2}\omega^2 [x_1^2 + x_2^2 - x_1^2\gamma_1^2 - x_2^2\gamma_2^2 + \\ & + x_3^2(\gamma_1^2 + \gamma_2^2) - 2x_1x_2\gamma_1\gamma_2 - 2x_3(x_1\gamma_1 + x_2\gamma_2)(1 - \gamma_1^2 - \gamma_2^2)^{1/2}] - \\ & - g[x_1\gamma_1 + x_2\gamma_2 + x_3(1 - \gamma_1^2 - \gamma_2^2)^{1/2}].\end{aligned} \tag{4–65}$$

Here

$$\left(\frac{\partial x_3}{\partial \Phi}\right)_0 = -\frac{1}{g}, \quad \mu_0 = -(\omega^2 x_{30} + g)(x_1\gamma_1 + x_2\gamma_2),$$

where

$$x_{30} = \frac{1}{2g}\omega^2(x_1^2 + x_2^2) - \frac{c_0}{g}.$$

Let us first consider the case $\omega = 0$, corresponding to equilibrium of a massive rigid body containing a fluid. The free surface of the fluid in the equilibrium position is here the part Q of the plane $x_3 = \bar{x}_3$ bounded by the cavity's walls.

From condition (4–43) for constancy of the fluid volume, we find

$$\Delta c = -g(x_1^1\gamma_1 + x_2^1\gamma_2),$$

where

$$x_i^1 = \frac{1}{S_Q}\iint_Q x_i\,dx_1dx_2 \quad (i = 1, 2), \quad S_Q = \iint_Q dx_1dx_2.$$

Furthermore, according to Eq. (4–48), we find

$$\Delta_2 W = \Delta_2 \Pi = \frac{\varrho g}{2}\left[S_Q(x_1^1\gamma_1 + x_2^1\gamma_2)^2 - \iint\limits_Q (x_1\gamma_1 + x_2\gamma_2)^2\, dx_1 dx_2\right].$$

Taking into account that the moments of inertia and the inertia product of the area segment Q for the axes parallel to x_1 and x_2 and passing through the center of inertia of Q are, respectively,

$$A_* = \iint\limits_Q x_2^2 dx_1 dx_2 - S_Q x_2^{12}, \quad B_* = \iint\limits_Q x_1^2 dx_1 dx_2 - S_Q x_1^{12},$$

$$F_* = \iint\limits_Q x_1 x_2 dx_1 dx_2 - S_Q x_1^1 x_2^1,$$

we finally obtain

$$\Delta_2 \Pi = -\frac{\varrho g}{2}\left[A_*\gamma_2^2 + B_*\gamma_1^2 + 2F_*\gamma_1\gamma_2\right].$$

We find from Eq. (4–50) that

$$\Pi^{(2)} = -\tfrac{1}{2}g\left\{(Mx_{c3}^0 + A_*\varrho)\,\gamma_2^2 + (Mx_{c3}^0 + B_*\varrho)\,\gamma_1^2 + 2F_*\varrho\gamma_1\gamma_2\right\}.$$

Consequently, the conditions for the minimum of the potential energy reduce to

$$Mx_{c3}^0 + A_*\varrho < 0 \quad (A_* \geqslant B_*), \tag{4–66}$$

which is identical to (3–43).

We now pass to the case $\omega \neq 0$. From now on we assume for simplicity that the region Q is a ring bounded by circles with radii R_1 and $R_2 (R_1 > R_2 \geqslant 0)$ and centers on the x_3 axis. Here, as is easy to check, Eqs. (4–47) are applicable, so that $\Delta c = 0$ and $\Delta I = 0$. Then

$$\varrho \int\limits_{\tau_1} \Phi(x_1, x_2, x_3, \gamma_1, \gamma_2)\, d\tau = \frac{a}{2}(\gamma_1^2 + \gamma_2^2),$$

where

$$a = \pi\varrho g \int\limits_{R_2}^{R_1} \left[\frac{\omega^2}{g^2}\left(\frac{\omega^2}{2}r^2 - c\right) + 1\right]^2 r^3 dr.$$

Thus we find from Eq. (4–49)

$$W^{(2)} = \tfrac{1}{2}\{[(C_0 - A_0)\,\omega^2 - Mgx_{c3}^0 - a]\,\gamma_1^2 + {} \\ + [(C_0 - B_0)\,\omega^2 - Mgx_{c3}^0 - a]\,\gamma_2^2\},$$

where A_0, B_0, and C_0 denote the principal moments of inertia of the system for point O in undisturbed motion. The conditions for the minimum in this case reduce to the single inequality

$$(C_0 - A_0)\,\omega^2 - Mgx_{c3}^0 - a > 0\,, \tag{4–67}$$

if we assume, without loss of generality, that $A_0 \geqslant B_0$.

When ω is large the surface of the paraboloid of revolution (4–63) differs but little from the surface of the circular cylinder

$$r^2 = x_1^2 + x_2^2 = b^2 \tag{4–68}$$

and becomes identical with it in the limit as $2g/\omega^2 \to 0$. Let us consider this limiting case, which holds for a weightless fluid.

Instead of (4–65), we will have in this case

$$\begin{aligned}\Phi(r, \theta, x_3, \gamma_1, \gamma_2) = \tfrac{1}{2}\omega^2 [r^2 - r^2(\cos^2\theta\gamma_1^2 + \sin^2\theta\gamma_2^2) + \\ + x_3^2(\gamma_1^2 + \gamma_2^2) - 2r^2 \sin\theta\cos\theta\gamma_1\gamma_2 - \\ - 2rx_3(\cos\theta\gamma_1 + \sin\theta\gamma_2)(1 - \gamma_1^2 - \gamma_2^2)^{1/2}]\,.\end{aligned} \tag{4–69}$$

Here

$$\left(\frac{\partial\Phi}{\partial r}\right)_0 = \omega^2 b\,, \quad \mu_0 = -\,\omega^2 b x_3(\cos\theta\gamma_1 + \sin\theta\gamma_2)\,,$$

$$\theta = \arctan\frac{x_2}{x_1}\,.$$

We assume that the cylinder (4–68) intersects the surface σ of the cavity along circles with centers at the x_3 axis in points with coordinates $x_3 = h + d$ and $x_3 = h - d$. The condition of constancy of the volume of fluid in the first approximation will take on the form

$$\int\limits_{\tau_1} d\tau = \int\limits_0^{2\pi} d\theta \int\limits_{h-d}^{h+d} dx_3 \int\limits_{\mu_0}^{\mu_1} \left(r\frac{\partial r}{\partial\Phi}\right)_0 d\mu = \frac{4\pi d}{\omega^2}\, b^2 \varrho \left(\Delta c + \frac{\omega^2}{I_0}\,\Delta_2 I\right).$$

Further, we find in the first approximation

$$\Delta_2 I = \varrho \int\limits_{\tau_1} r^2 d\tau = \frac{4\pi d}{\omega^2}\, b^2 \varrho \left(\Delta c + \frac{\omega^2}{I_0}\, b^2 \Delta_2 I\right).$$

It follows from the last two equations that in the first approximation

$$\Delta c = 0\,, \quad \Delta_2 I = 0\,.$$

Finally we find

$$\int\limits_{\tau_1} \Phi(r, \theta, x_3, \gamma_1, \gamma_2)\, d\tau = \frac{3h^2 + d^2}{3}\, \pi d b^2 \omega^2 (\gamma_1^2 + \gamma_2^2)\,.$$

The condition for the minimum of W in this case reduces to the single inequality

$$\left(C_0 - A_0 - 2\pi\varrho b^2 d\,\frac{3h^2 + d^2}{3}\right)\omega^2 - Mgx_{c3}^0 > 0\,. \tag{4-70}$$

Let us now present the solution of the last problem by another method[57]. The configuration f of the fluid in the disturbed motion can be regarded as one formed by imposing a fluid layer of variable thickness χ on the equilibrium configuration f_0 along its free surface S_0. Here we must obviously satisfy the condition

$$\int_{S_0} \chi dS = 0\,.$$

In stable motion of a rigid body, χ is of the order of smallness of γ_1 and γ_2; then, with a precision to within higher-order infinitesimals, we will have

$$\begin{aligned} W - W_0 = \frac{1}{2}\Big\{&[(C_0 - A_0)\,\omega^2 - Mgx_{c3}^0]\,\gamma_1^2 + \\ &+ [(C_0 - B_0)\omega^2 - Mgx_{c3}^0]\gamma_2^2 + 2\varrho\int_{S_0}(\omega^2 x_3 + g)(x_1\gamma_1 + x_2\gamma_2)\chi dS + \\ &+ \varrho\int_{S_0}[\omega^4(x_1^2 + x_2^2) + g^2]^{1/2}\,\chi^2 dS + \frac{\omega^2}{I_0}\left(\varrho\int_{S_0}(x_1^2 + x_2^2)\,\chi dS\right)^2\Big\} + \cdots, \end{aligned} \tag{4-71}$$

where for evaluating integrals taken over the volume τ of the disturbed configuration f we use methods that were developed in the theory of the stability of equilibrium configurations of a rotating fluid[59].

To establish the sufficient condition for stability of rotation of a rigid body containing a fluid, χ can be expanded in a series in terms of the complete system of eigenfunctions of the corresponding boundary-value problem.

When the free surface of the fluid in steady motion is the circular cylinder (4–68) bounded by the planes $x_3 = h \pm d$, χ can be represented in the form

$$\chi = \sum_{k=0}^{\infty}(a_k\cos\theta + b_k\sin\theta)\cos\frac{k\pi}{2d}(x_3 - h + d)\,.$$

Substituting this expression into Eq. (4–71), we obtain

$$\begin{aligned} W - W_0 = \frac{1}{2}\Big\{&[(C_0 - A_0)\,\omega^2 - Mgx_{c3}^0]\,\gamma_1^2 + \\ &+ [(C_0 - B_0)\,\omega^2 - Mgx_{c3}^0]\,\gamma_2^2 + 4\pi\varrho d b^2 h\omega^2(a_0\gamma_1 + b_0\gamma_2) - \\ &- 16\varrho b^2\omega^2\frac{d}{\pi}\sum_{j=0}^{\infty}\frac{a_{2j+1}\gamma_1 + b_{2j+1}\gamma_2}{(2j+1)^2} + \pi\varrho b^2 d\omega^2 \times \\ &\times [2(a_0^2 + b_0^2) + \sum_{k=1}^{\infty}(a_k^2 + b_k^2)]\Big\} + \cdots. \end{aligned} \tag{4-72}$$

The conditions for positive definiteness of the right-hand side of this equality, which is considered a function of $\gamma_1, \gamma_2, a_k, b_k$ $(k=0, 1, 2, \ldots)$, reduce to the inequality (4–70).

Consequently the undisturbed motion of a massive rigid body with a massive fluid is stable with respect to the quantities

$$\omega_i, \gamma_i \quad (i = 1, 2, 3), \quad T_2^{(1)},$$

if condition (4–64) is satisfied when the fluid completely fills the cavity, or if condition (4–67) is satisfied when the cavity is partially filled: For a weightless fluid the condition of stability is the inequality (4–70).

In a viscous fluid, when the above conditions are satisfied, the disturbed motion will tend in the limit to uniform rotation of the entire system as a single rigid body. When these conditions are violated, if W can take on negative values in the vicinity of the isolated steady motion, the undisturbed motion will be unstable.

[4-7] STABILITY OF THE RELATIVE EQUILIBRIUM OF A PHYSICAL PENDULUM CONTAINING A FLUID

Let us examine a physical pendulum containing a fluid, *i.e.*, a massive rigid body with a cavity filled with a massive viscous fluid, which can rotate about a horizontal axis. We assume that the axis of the pendulum can rotate about a vertical axis passing through the point O in which the pendulum axis intersects its orthogonal plane, containing the center of gravity O of the rigid body[53].

We take the fixed point O of the rigid body as the origin of the fixed coordinate system $Ox_1'x_2'x_3'$, with the x_3' axis directed vertically upward and with the x_1' and x_2' axes located in the horizontal plane. We rigidly fix to the rigid body the moving coordinate system $Ox_1x_2x_3$, with the x_1 axis directed along the horizontal axis of rotation of the pendulum, the x_3 axis along the the line OO_1 passing through the center of gravity of the body, and the x_2 axis orthogonal to the first two. In addition, we shall also consider a moving coordinate system $Ox_1\xi_2x_3'$, the ξ_2 axis of which is horizontal and orthogonal to axes x_1 and x_3'. We assume for simplicity that x_1, x_2, and x_3 are the principal axes of inertia of the body for point O. We denote the angle between the x_1' and x_1 axes by ψ, and the angle between the x_3' and x_3 axes by θ (Fig. 16). These two independent parameters completely define the position of the pendulum in the fixed coordinate system $Ox_1'x_2'x_3'$.

We now assume that the pendulum is acted upon by nonconservative forces that give a moment about the x_3' axis such that during the entire motion the angular velocity with which the axis of the pendulum rotates about the vertical remains constant; *i.e.*, $\psi' = \omega$.

The moment of inertia of the system about the x_3' axis is

$$I = B \sin^2\theta + C \cos^2\theta - 2D_2 \sin\theta \cos\theta,$$

where $D_2 = \varrho \int_\tau x_2 x_3 d\tau$ is the centrifugal moment of inertia of the fluid. The potential energy of the gravity forces is

$$\Pi = Mg\,(x_{c2} \sin\theta + x_{c3} \cos\theta).$$

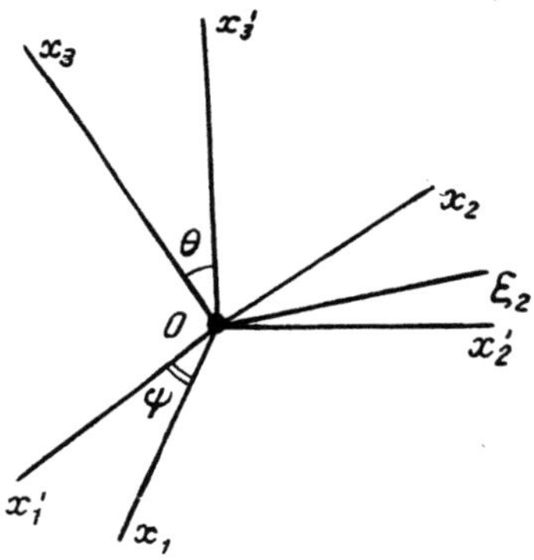

Fig. 16.

Here

$$Mx_{c2} = \varrho \int_\tau x_2 d\tau, \quad Mx_{c3} = M_1 x_{13} + \varrho \int_\tau x_3 d\tau.$$

Thus, in the case under study,

$$W_* = -\tfrac{1}{2}\omega^2 (B \sin^2\theta + C \cos^2\theta - 2D_2 \sin\theta \cos\theta) + + Mg\,(x_{c2} \sin\theta + x_{c3} \cos\theta). \tag{4–73}$$

The equation of relative equilibrium of the pendulum has the form

$$\frac{\partial W_*}{\partial \theta} = [(C - B) \sin\theta \cos\theta + D_2 (\cos^2\theta - \sin^2\theta)]\,\omega^2 + + Mg\,(x_{c2} \cos\theta - x_{c3} \sin\theta) = 0. \tag{4–74}$$

Obviously this equation has two roots for any value of ω, *i.e.*,

$$\theta_1 = 0, \quad \theta_2 = \pi,$$

and, if $B \neq C$ and

$$\omega^2 \geqslant \omega_1^2 \equiv \frac{Mg\,|x_{c3}^0|}{/B - C|},$$

it also has the third root

$$\theta_3 = \arccos \frac{Mgx_{c3}^0}{(C-B)\,\omega^2}.$$

Here

$$\theta_3 \leqslant \frac{\pi}{2}, \quad \text{if} \quad \frac{x_{c3}^0}{B-C} < 0,$$

$$\theta_3 \geqslant \frac{\pi}{2}, \quad \text{if} \quad \frac{x_{c3}^0}{B-C} > 0,$$

if in all of these cases

$$D_2^0 = x_{c2}^0 = 0, \tag{4–75}$$

i.e., if the x_3 axis is then the principal central axis of inertia of the system.

Moving on to the relative equilibrium of a pendulum, we first consider a cavity that is entirely filled with fluid; obviously, we assume that conditions (4–75) have been satisfied. We find

$$\delta \equiv \left(\frac{\partial^2 W}{\partial \theta^2}\right)_0 = (C-B)\,\omega^2 (\cos^2\theta - \sin^2\theta)_0 - Mgx_{c3}\cos\theta_0\,. \tag{4–76}$$

Depending on the magnitude of the difference $B-C$, we shall distinguish between the following three cases.

1) We assume that $B>C$. When $\theta=0$, δ is positive if $x_{c3}<0$, $\omega^2<\omega_1^2$, and negative if $x_{c3}>0$ or if $x_{c3}<0$, $\omega^2>\omega_1^2$. When $\theta=\pi$, δ is positive if $x_{c3}>0$, $\omega^2<\omega_1^2$, and negative if $x_{c3}>0$ or if $x_{c3}>0$, $\omega^2>\omega_1^2$. When $\omega^2>\omega_1^2$, $\theta=\theta_3$, $\delta>0$.

Consequently, when $B>C$, the vertical position of the relative equilibrium of the pendulum for which the x_3 axis coincides with the x_3' axis is stable if the center of gravity is situated below the point of suspension O, and $\omega<\omega_1$, and it is unstable if $\omega>\omega_1$ or if the center of gravity is above the point of suspension. When $\omega>\omega_1$, it is the position of the relative equilibrium of the pendulum that is stable.

2) Let $C>B$. When $\theta=0$, δ is positive if $x_{c3}<0$ or if $x_{c3}>0$, $\omega^2>\omega_1^2$, and negative if $x_{c3}>0$ and $\omega^2<\omega_1^2$. When $\theta=\pi$, δ is positive if $x_{c3}>0$ or if $x_{c3}<0$, $\omega^2>\omega_1^2$, and negative if $x_{c3}<0$ and $\omega^2<\omega_1^2$. When $\omega^2>\omega_1^2$ and $\theta=\theta_3$, δ is negative.

Consequently, when $C>B$, the vertical position of the relative equilibrium of the pendulum is stable if the center of gravity is below the point of suspension; if, however, the center of gravity is above the point of suspension then it is stable when $\omega^2>\omega_1^2$ and unstable when $\omega^2<\omega_1^2$. The position of the relative equilibrium of the pendulum deflected from the vertical is unstable.

Obviously, in both cases a change in stability takes place at the bifurcation points.

3) Let $B=C$. When $\theta=0$, $\delta=-Mgx_{c3}$ is positive if $x_{c3}<0$ and negative if $x_{c3}>0$. When $\theta=\pi$, $\delta=Mgx_{c3}>0$ if $x_{c3}>0$, and $\delta<0$ if $x_{c3}<0$.

Consequently, in the case $B=C$ the vertical equilibrium position of the pendulum is stable if the center of gravity is below the point of suspension and unstable if it is above the point.

We now consider the problem of the stability of the vertical position of the relative equilibrium of a fluid-containing pendulum, with the free surface in the shape of paraboloid of revolution (4–63). We assume for simplicity that the area segment Q is in the shape of a ring bounded by circles of radii $R_1>0$ and $R_2\geqslant 0$ with centers on the x_3 axis.

In the given case we have

$$\begin{aligned}\Phi(x_1, x_2, x_3, \theta) = \tfrac{1}{2}\omega^2 [(x_1^2 + x_3^2)\sin^2\theta &+ \\ + (x_1^2 + x_2^2)\cos^2\theta - 2x_2x_3\sin\theta\cos\theta] &- \\ - g(x_2\sin\theta + x_3\cos\theta)&.\end{aligned} \tag{4–77}$$

Here

$$\left(\frac{\partial x_3}{\partial \Phi}\right)_0 = -\frac{1}{g}, \quad \mu_0 = -(\omega^2 x_{30} + g)\, x_2\theta .$$

It is easy to check that $\Delta c=0$ and, consequently, $\mu_1=0$. Further we determine

$$\Delta_2 W_* = \frac{1}{2}\varrho \iint\limits_Q \left(\frac{\partial x_3}{\partial \Phi}\right)_0 \mu_0^2 dx_1 dx_2 + \cdots = -\frac{a}{2}\theta^2 + \cdots,$$

where

$$a = \pi\varrho g \int\limits_{R_2}^{R_1} \left[\frac{\omega^2}{g^2}\left(\frac{\omega^2}{2}r^2 - c\right) + 1\right]^2 r^3 dr . \tag{4–78}$$

According to Eq. (4–52),

$$W_*^{(2)} = \tfrac{1}{2}[(C_0 - B_0)\,\omega^2 - Mgx_{c3}^0 - a]\,\theta^2 . \tag{4–79}$$

The condition for the minimum of W_* reduces to the single equality

$$(C_0 - B_0)\,\omega^2 - Mgx_{c3}^0 - a > 0 . \tag{4–80}$$

When the above condition is satisfied, the vertical position of the relative equilibrium of a massive pendulum with a cavity partially filled with a fluid is stable.

We note that, by virtue of Theorem V, the steady motions of a massive pendulum that coincide with the stable positions of the relative equilibria of the pendulum will also be stable.

Part Two

Vibrations of a Fluid and of the Body Containing the Fluid

5

Statement of the Problems of the Theory of Vibrations

In this chapter, using Hamilton's principle, we shall formulate in sequence the variational problems of the theory of vibrations of bodies containing an ideal fluid and we shall consider certain elementary problems, on the assumption that the external forces are conservative. It is assumed for simplicity that the vibrating system contains only one "fluid link," *i.e.*, one cavity with an ideal incompressible fluid. All of the results are trivially extended to the case of an arbitrary number of cavities.

[5-1] VIBRATIONS OF A MASSIVE FLUID CONTAINED IN A VESSEL

[5-1.1] Formulation of the Variational Problem

Let us examine the linear vibrations of a massive fluid contained inside a stationary shell Σ (see Fig. 17, which defines the notation). We fix the inertial reference system $Oxyz$ to the gravity force vector

$$\boldsymbol{g} = -g\boldsymbol{z}^0 .$$

It is assumed that the fluid free surface in the equilibrium position coincides with the xOy plane. The volume occupied by the fluid in the equilibrium position is denoted by τ. The kinetic and potential energies of the fluid are expressed by the equations[6]

$$T = \frac{\varrho}{2}\int\limits_{\tau} \boldsymbol{v}^2 d\tau , \quad \Pi = \frac{\varrho g}{2}\int\limits_{S} \zeta^2 dS ,$$

and $z=\zeta(P, t)$ is the equation of the free surface $P \in S$. According to Hamilton's principle for virtual displacements that transfer a system from one fixed state to another during a given time,

$$\delta L = 0 , \tag{5–1}$$

where

$$L = \int_0^{t_1} (T - \Pi)\, dt .$$

Equation (5–1) will be called the *variational equation.*

We shall assume that the fluid bounded within the volume τ is governed by the following conditions:

1) Potential condition:

$$\boldsymbol{v} = \nabla\varphi . \tag{5–2}$$

2) Continuity condition:

$$\operatorname{div} \boldsymbol{v} = 0 . \tag{5–3}$$

3) Shell impermeability condition:

$$\boldsymbol{v} = 0 , \quad P \in \Sigma . \tag{5–4}$$

4) Kinematic condition:

$$\zeta_t = v_z , \quad P \in S . \tag{5–5}$$

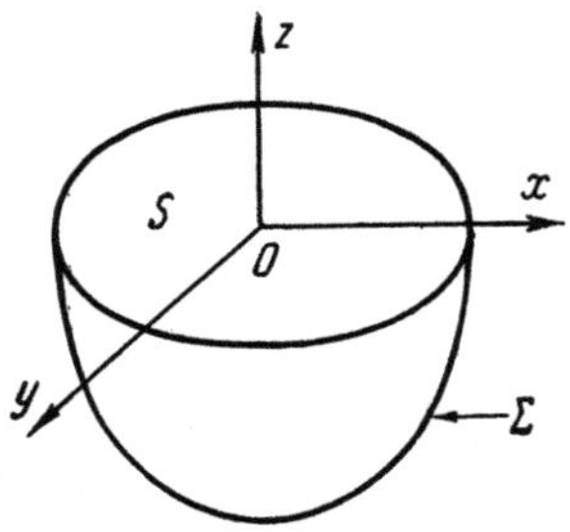

Fig. 17.

Conditions (5–4) and (5–5) follow from the condition of fluid continuity. The potential condition, as will be seen below, can be dropped in the linear theory.

Thus the problem reduces to finding, in the class of functions satisfying conditions (5–2) – (5–5), all of the possible functions $\varphi(P, t)$, $P \in \tau$ and $\zeta(P, t)$, $P \in S$ satisfying (5–1).

Using (5–2) and (5–3), we rewrite (5–1) as

$$\delta L \equiv \varrho \int_0^{t_1} \{ \int_\tau \nabla\varphi \nabla\delta\varphi d\tau - g \int_S \zeta\delta\zeta dS \}\, dt = 0 . \tag{5–6}$$

Using Green's formula and condition (5–4), we obtain

$$\int_\tau \nabla\varphi\nabla\delta\varphi d\tau = \int_S \varphi \frac{\partial\delta\varphi}{\partial n}\, dS .$$

In addition, it follows from (5–5) that

$$\frac{\partial \delta\varphi}{\partial n} = \delta\zeta_t .$$

Hence (5–6) can be rewritten as

$$\varrho \int_0^{t_1} \int_S (\varphi \delta\zeta_t - g\zeta\delta\zeta) \, dS dt = 0 .$$

Integrating by parts and using the conditions at the ends, we obtain

$$-\varrho \int_0^{t_1} \int_S (\varphi_t + g\zeta) \, \delta\zeta dS dt = 0 .$$

From this, by virtue of the arbitrariness of the variation, we find

$$\varphi_t + g\zeta = 0 , \quad P \in S . \tag{5–7}$$

which is a known condition for constancy of pressure at the free surface[6].

Thus the dynamic condition at the free surface was obtained as a result of Hamilton's principle, and this will make it possible to set up other expressions that are more convenient for numerical calculations, for the functional.

Using (5–7), we can represent L in the form

$$L = \frac{\varrho}{2} \int_0^{t_1} \left\{ \int_\tau (\nabla\varphi)^2 \, d\tau - \frac{1}{g} \int_S \varphi_t^2 dS \right\} dt . \tag{5–8}$$

We also introduce Neumann's operator H, which by virtue of the function $f(P)(P \in S, \int_s f dS = 0)$ puts the function $\varphi(Q)$ $(Q) \in \tau$)—harmonic in τ—into correspondence and satisfies the conditions

$$\frac{\partial \varphi}{\partial n} = 0 , \quad P \in \Sigma ,$$

$$\frac{\partial \varphi}{\partial n} = f , \quad P \in S .$$

We shall write this correspondence as

$$\varphi = Hf ,$$

where H is the integral operator

$$Hf \equiv \int_S H(P, Q) \, f(P) \, dP = \varphi(Q) ,$$

the kernel of which is Green's function of Neumann's problem for the region

τ. According to the general theory[65,66], this kernel is symmetrical and, when $P=Q$, has the following property: in the case of the plane problem it is logarithmic while in the case of the three-dimensional problem it is polar. Thus, H is a fully continuous self-adjoint operator.

Using the obvious representation

$$\varphi(Q) = H\frac{\partial\varphi}{\partial z} = H\zeta_t, \tag{5–9}$$

we can eliminate φ from (5–7):

$$g\zeta + H\zeta_{tt} = 0. \tag{5–10}$$

Upon setting $Q\in S$ in (5–10), we will obtain an integro-differential equation. This approach is particularly convenient in the study of general properties, such as properties of the spectrum and the convergence of variational methods.

Using (5–9), the functional L can be represented in the form

$$L = \frac{\varrho}{2}\int\limits_0^{t_1}\int\limits_S (H\zeta_t\cdot\zeta_t - g\zeta^2)\, dSdt. \tag{5–11}$$

[5-1.2] Expression for the Functional in the Case of Free Vibrations

We now consider the problem of natural vibrations of the fluid; hence we assume

$$\begin{aligned}\varphi(P, t) &= \cos\sigma t\Phi(P),\\ \zeta(P, t) &= \sin\sigma t\Psi(P).\end{aligned} \tag{5–12}$$

The number σ is called the natural frequency and is to be determined. We now write an expression for the functional L, taken in the form (5–8) and (5–11) for the case when φ has the form of (5–12). Integrating over t from 0 to $t_1 = 2\pi/\sigma$, and dropping an inconsequential multiplier, we reduce it to the form

$$L = \int\limits_\tau (\nabla\Phi)^2\, d\tau - \lambda\int\limits_S \Phi^2 dS, \tag{5–13}$$

$$L = \lambda\int\limits_S H\Psi\cdot\Psi dS - \int\limits_S \Psi^2 dS, \tag{5–14}$$

where

$$\lambda = \frac{\sigma^2}{g}.$$

Thus the determination of natural vibrations of the fluid reduces to the vibrational problem for the functionals (5–13) and (5–14).

Remark

According to the above, the solution of the extremal problem should be sought in the class of harmonic functions. However, it can be shown that the extremum will be identical with the value obtained if we consider any functions $\varphi \in L_2$ as the allowable functions.* This assertion is a result of the fact that the boundary conditions of our problem are natural boundary conditions [65].

According to the general theory, the smallest of the eigenvalues λ is given by the formula

$$\lambda_1 = \min \frac{\int_\tau (\nabla \Phi)^2 \, d\tau}{\int_S \Phi^2 dS} . \tag{5–15}$$

The second of the eigenvalues, λ_2, is determined as the solution of the variational problem

$$\lambda_2 = \min \frac{\int_\tau (\nabla \Phi)^2 \, d\tau}{\int_S \Phi^2 dS}$$

in the class of functions orthogonal to Φ_1, if Φ_1 is a function solving the variational problem (5–15), *etc.* Orthogonality is used here in the sense of the metrics of L_2, *i.e.*, of functions defined on S.

We now establish the validity of certain relationships that we shall need further on. We assume that Φ_n and ψ_n are functions introduced by Eqs. (5–12) and corresponding to the natural frequency σ_n. Then we find from (5–5), (5–7), (5–9), and (5–10) that

$$\Psi_n = \frac{\sigma_n}{g} \Phi_n , \quad P \in S , \tag{5–16}$$

$$\frac{\partial \Phi_n}{\partial z} = \sigma_n \Psi_n , \quad P \in S , \tag{5–17}$$

$$H \Psi_n = \frac{g}{\sigma_n^2} \Psi_n , \quad P \in S . \tag{5–18}$$

Let Ψ_m and Φ_m be functions corresponding to the natural frequency $\sigma_m \neq \sigma_n$; then, multiplying (5–18) by Ψ_m and integrating with respect to s, we obtain

$$\int_S H\Psi_n \cdot \Psi_m dS = \frac{g}{\sigma_n^2} \int_S \Psi_n \Psi_m dS . \tag{5–19}$$

* We recall that the space L_2 is used to denote a set of functions $f(P)$, $P \in S$ with an integrable square, in which a scalar product is introduced according to the formula

$$(f_1 \cdot f_2) = \int_S f_1 f_2 dt .$$

Similarly,

$$\int_S H\Psi_m \cdot \Psi_n dS = \frac{g}{\sigma_m^2} \int_S \Psi_m \Psi_n dS. \tag{5–20}$$

It immediately follows from the symmetry of Green's function that

$$\int_S H\Psi_m \cdot \Psi_n dS = \int_S H\Psi_n \cdot \Psi_m dS.$$

Now, subtracting (5–20) from (5–19), we obtain

$$\int_S \Psi_n \Psi_m dS = 0. \tag{5–21}$$

Further

$$\int_S H\Psi_n \cdot \Psi_m dS = 0. \tag{5–22}$$

Using Green's formula, we transform

$$\int_\tau \nabla\Phi_n \cdot \nabla\Phi_m dS = \int_{S+\Sigma} \Phi_n \frac{\partial \Phi_m}{\partial n} dS = \int_S \Phi_n \frac{\partial \Phi_m}{\partial z} dS,$$

but $\Phi_n = H\Psi_n, \partial\Phi_m/\partial z = \sigma_m \Psi_m$. Hence, on the basis of (5–21), we have

$$\int_\tau \nabla\Phi_n \nabla\Phi_m d\tau = 0. \tag{5–23}$$

In addition, it immediately follows from (5–16) that

$$\int_S \Phi_n \Phi_m dS = 0. \tag{5–24}$$

[5-1.3] Outline of the Ritz Method

It is expedient to solve the variational problem (5–1) for the functional (5–13) by using the Ritz method. His standard procedure reduces to the following: a system of coordinate functions $\{\chi_n\}$ is introduced and the solution of the problem is written in the form

$$\Phi = \sum_{n=1}^{N} a_n \chi_n.$$

We substitute the above expression into (5–13) and obtain some quadratic form with respect to the unknowns a_n. It follows from the variational equation (5–1) that the coefficients a_n should satisfy the system of algebraic equations

$$\frac{\partial L}{\partial a_n} = 0, \quad n = 1, 2, \ldots, N.$$

Expanded, this system has the form

$$\sum_{m=1}^{N} a_m(\alpha_{nm} - \lambda\beta_{nm}) = 0, \quad n = 1, 2, \ldots, N, \tag{5–25}$$

where

$$\alpha_{nm} = \int_\tau \nabla\chi_n \cdot \nabla\chi_m d\tau; \quad \beta_{nm} = \int_S \chi_n\chi_m dS;$$

and, in addition,

$$\alpha_{nm} = \alpha_{mn}, \quad \beta_{nm} = \beta_{mn}.$$

The system of equations (5–25) is homogeneous. For the existence of nonzero solutions of this system it is necessary and sufficient that the number λ should satisfy the equation

$$|\alpha_{nm} - \lambda\beta_{nm}| = 0, \tag{5–26}$$

which we shall call the *frequency equation.* We shall denote its roots by $\lambda_n(n=1, 2, \ldots, N)$. By virtue of the symmetry of the matrices $\|\alpha_{nm}\|$ and $\|\beta_{nm}\|$, the numbers λ_n are real. If these are found, then the natural frequencies are given by

$$\sigma_n^2 = \lambda_n g. \tag{5–27}$$

Because of their complexity, problems of determining the natural frequencies of a vibrating fluid require the use of electronic computers. Numerical methods should provide for simplicity of standardization of massive calculations, and the Ritz method satisfies this requirement. The main difficulty encountered in practical implementation of this method consists in selection of the coordinate functions. No general recommendations can be given; however, the following remarks should be kept in mind in the process of solving the problem.

I. The value of λ_1 is "only slightly sensitive" to the selection of the functions χ_n. This means that if the function Φ_1, which brings about a minimum of the functional λ_1, is replaced by another function Φ_1^* (such that $\int_\tau \nabla\Phi_1 \nabla\Phi_1^* d\tau \neq 0$) then the value of λ_1 will not change very much.

II. The boundary conditions for the function Φ are classified as natural conditions; hence it cannot be required that the functions χ_n must satisfy rigorously all of the boundary conditions. According to the general theory, the minimizing sequence constructed by the Ritz method in this case also reduces to this solution.

III. Thus the system of coordinate functions $\{\chi_n\}$ can be selected quite roughly. It is only necessary to make sure that this system is complete. Hence it is expedient to take as these functions the eigenfunctions of the problem of vibration of a fluid in some volume that contains the specified volume but has a simpler shape. For example, for a fluid vibrating inside a cone-shaped tank, it is possible to take as the system of coordinate functions the eigenfunctions of the problem of vibration of a fluid inside a cylindrical vessel, the cross section of which is equal to the larger base of the cone.

[5-1.4] The Congruence Theorem

Let us consider the variational problem (5–15) for two volumes τ_1 and τ_2, which have the same free surface S, while the surface Σ_1 embraces the surface Σ_2 (Fig. 18). Thus $\tau_1 > \tau_2$. Then for any function Φ:

$$\int_{\tau_1} (\nabla \Phi)^2 \, d\tau > \int_{\tau_2} (\nabla \Phi)^2 \, d\tau . \tag{5–28}$$

Let $\lambda^{(1)}$, $\lambda^{(2)}$, $\Phi^{(1)}$ and $\Phi^{(2)}$ be a solution of problem (5–15) for the volumes τ_1 and τ_2, *i.e.*,

$$\lambda^{(1)} = \frac{\int_{\tau_1} (\nabla \Phi^{(1)})^2 \, d\tau}{\int_S \Phi^{(1)2} \, dS}, \quad \lambda^{(2)} = \frac{\int_{\tau_2} (\nabla \Phi^{(2)})^2 \, d\tau}{\int_S \Phi^{(2)2} \, dS} . \tag{5–29}$$

Then it is obvious that

$$\lambda^{(2)} \leqslant \frac{\int_{\tau_2} (\nabla \Phi^{(1)})^2 \, d\tau}{\int_S \Phi^{(1)2} \, dS} .$$

Using the first of equations (5–29), we obtain

$$\lambda^{(2)} \leqslant \lambda^{(1)} \frac{\int_{\tau_2} (\nabla \Phi^{(1)})^2 \, d\tau}{\int_{\tau_1} (\nabla \Phi^{(1)})^2 \, d\tau} .$$

But, according to (5–28), the multiplier of $\lambda^{(1)}$ is smaller than unity. Consequently,

$$\lambda^{(2)} < \lambda^{(1)} .$$

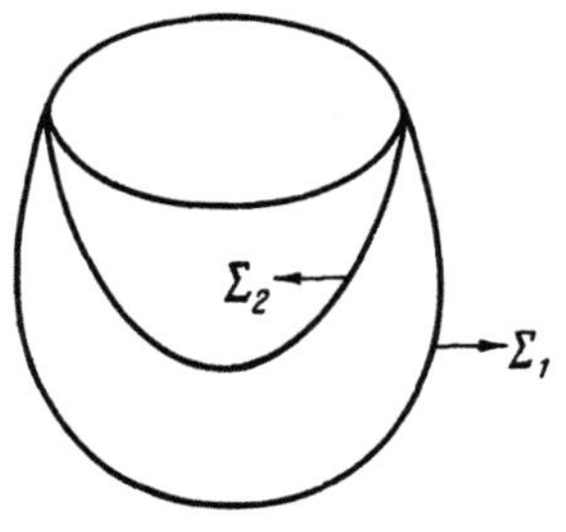

Fig. 18.

Thus, if we have two vessels, with equal free surface areas, but the surface Σ_1 of the first vessel embraces the surface Σ_2 of the second, the natural frequencies will be greater for the vessel with the larger volume.

We shall now use an example to illustrate the possibilities provided by this theorem for the approximate determination of natural frequencies. For simplicity we shall consider a two-dimensional problem.

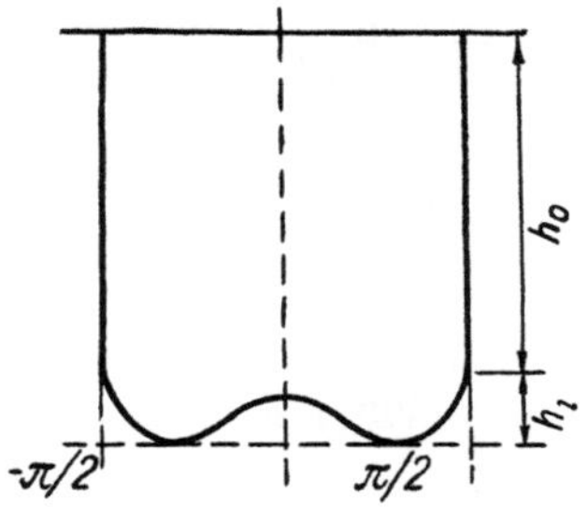

Fig. 19.

Let it be required to estimate the magnitude of the first natural frequency of vibration of a fluid occupying the volume shown in Fig. 19. On the basis of the above considerations, we easily find the following estimate:

$$\tanh(h_0 + h_1) > \lambda_1 > \tanh h_0 .$$

[5-1.5] Approximate Method for Calculating Natural Vibrations

Let us assume that it is required to solve the problem for some region τ close to the region τ^*, for which the problem of the eigenvalues can be effectively solved. This means that, for the latter region, we know a system of eigenfunctions Φ_n^* and eigenvalues λ_n^*. The fact that the regions τ and τ^* are close to one another (S and S^*, respectively) makes it possible to develop effectively the method of the theory of perturbations. An outline of the perturbation method for this problem was first proposed by Petrov[67]. Below, this method is outlined as applied to the case when it is possible to take the functions $\{\Phi_n^*\}$ as functions $\{\chi_n\}$.*

Thus we seek the solution of the problem in the form

$$\Phi = \sum_{n=1}^{N} a_n \Phi_n^* .$$

* More exactly, we are considering the case when the functions $\Phi_n{}^*$ can be extended to the outside of the region τ^* in a manner such that they would form a complete system in region τ. Obviously this will always be true when the region τ^* encloses the region τ.

We obtain the system of equations (5–25), in which

$$\alpha_{nm} = \int_{\tau} \nabla\Phi_n^* \nabla\Phi_m^* d\tau = \int_{\tau^*} \nabla\Phi_n^* \nabla\Phi_m^* d\tau + \int_{\delta\tau} \nabla\Phi_n^* \nabla\Phi_m^* d\tau,$$

$$\beta_{nm} = \int_{S^*} \Phi_n^* \Phi_m^* dS + \int_{\delta S} \Phi_n^* \Phi_m^* dS; \tag{5–30}$$

the meaning of the symbols $\delta\tau$ and δS is obvious. The functions Φ_n^* will be normalized by the condition

$$\int_{S^*} \Phi_n^{*2}\, dS = 1; \tag{5–31}$$

then

$$\int_{\tau^*} (\nabla\Phi_n^*)^2\, d\tau = \lambda_n^*. \tag{5–32}$$

Since the regions $\delta\tau$ and δS are small, we can rewrite Eqs. (5–30) in the form

$$\alpha_{nm} = \lambda_n^* \delta_{nm} + \varepsilon\xi_{nm},$$
$$\beta_{nm} = \delta_{nm} + \varepsilon\eta_{nm},$$

where δ_{nm} is the Kronecker delta: $\delta_{nm}=0$ when $n \neq m$ and $\delta_{nm}=1$ when $n=m$.

The system of equations (5–25) takes the form

$$a_i(\lambda_i^* - \lambda) = \varepsilon A_i, \tag{5–33}$$

where

$$A_i = \sum_{k=1}^{N} a_k(\lambda\eta_{ik} - \xi_{ik}).$$

We shall seek the solution of system (5–33) in the form of the series

$$\left.\begin{aligned} a_i &= \sum_{j=0}^{\infty} a_{ij}\varepsilon^j, \\ \lambda &= \sum_{j=0}^{\infty} \lambda^{(j)}\varepsilon^j. \end{aligned}\right\} \tag{5–34}$$

Substituting the above series into Eqs. (5–33) and equating coefficients for the same powers of ε, we obtain equations for determining the coefficients of the expansion (5–34). Let us assume that we are seeking the solution of the homogeneous system (5–33) for which the roots λ_n of the frequency equation are close to λ_n^*. Then the first of the systems of equations will be

$$\begin{aligned} a_{i0}(\lambda_i^* - \lambda^{(0)}) &= 0, \quad i \neq n, \\ a_{n0}(\lambda_n^* - \lambda^{(0)}) &= 0. \end{aligned} \tag{5–35}$$

We are interested in nontrivial solutions of the system (5–34); hence we should assume

$$a_{i0} = 0, \quad i \neq n, \qquad \lambda^{(0)} = \lambda_n^* . \tag{5–36}$$

The coefficients a_{n0} from system (5–35) cannot be determined, corresponding to the indeterminacy of the amplitude of linear vibrations.

We now consider a system of equations of the first approximation, which are obtained by equating coefficients for terms containing ε,

$$a_{i1}(\lambda_i^* - \lambda^{(0)}) = \sum_{k=1}^{N} a_{k0}(\lambda_n^* \eta_{ik} - \xi_{ik}) + a_{i0}\lambda^{(1)},$$

or, making use of Eqs. (5–36),

$$\begin{aligned} a_{i1}(\lambda_i^* - \lambda_n^*) &= a_{n0}(\lambda_n^* \eta_{in} - \xi_{in}) \quad \text{when} \quad i \neq n, \\ 0 &= a_{n0}(\lambda_n^* \eta_{nn} - \xi_{nn}) + a_{n0}\lambda^{(1)} \quad \text{when} \quad i = n. \end{aligned} \tag{5–37}$$

The first of Eqs. (5–37) yields

$$a_{i1} = \frac{a_{n0}(\lambda_n^* \eta_{in} - \xi_{in})}{\lambda_i^* - \lambda_n^*}. \tag{5–38}$$

In order for the second equation of this system to have a nontrivial solution, it is necessary and sufficient that

$$\lambda^{(1)} = \lambda_n^* \eta_{nn} - \xi_{nn}, \quad etc. \tag{5–39}$$

This method makes it possible to obtain an approximate solution in explicit form for a wide class of cavities. A more detailed presentation with examples showing the effectiveness of this method is given in Ref. [68].

[5-1.6] The Effect of Vorticity on the Vibrations of a Fluid

At the start of this section we noted without proof that the condition requiring that the flow be potential can be dropped in problems of linear vibrations. The meaning of this assertion consists in the following: the natural frequencies, the configurations of the free surface, and the pressure distribution in the linear theory do not depend on the vortex component of the velocity field; they are completely determined by solving the problem of vibration of a fluid with a velocity potential, which was formulated above. We now prove this statement.

We consider a set E of solenoidal vectors $\boldsymbol{v}$, specified in the τ-region, which is bounded by the surfaces Σ and S. The field of $\boldsymbol{v}$ can be represented as the superposition of the potential field $\boldsymbol{u} = \nabla\varphi$ and the vortex field $\boldsymbol{w}$; here the function φ is harmonic in τ, while $\boldsymbol{w}$ satisfies the incompressibility condition $\nabla\cdot\boldsymbol{w}=0$. Thus

$$\boldsymbol{v} = \boldsymbol{u} + \boldsymbol{w}. \tag{5–40}$$

The representation (5–40) can be brought about by innumerable methods. This gives rise to the problem of separating the potential component.

We shall call the separation complete if the vector $\boldsymbol{w}$ is, in one sense or another, orthogonal to the vector $\boldsymbol{u}$. To make this definition precise it is necessary, in addition, to introduce a metric into the set E, thus converting it into a space. It seems most natural to use the energy metric, defining the scalar product by the equality

$$(\boldsymbol{v}_1, \boldsymbol{v}_2)_E = \int_\tau \boldsymbol{v}_1\cdot\boldsymbol{v}_2 d\tau.$$

Thus the completeness of separation in the given case means that

$$(\boldsymbol{u}, \boldsymbol{w})_E = 0.$$

We denote the subspace of potential vectors by E_u and we define the set E_w of the vectors $\boldsymbol{w}\in E$, the normal component of which becomes zero on $\Sigma+S$. It is obvious that E_w belongs to the orthogonal complement to E_u. The proof follows directly from Green's formula, the condition that the field is solenoidal, and the condition that $w_n=0$ on $\Sigma+S$

$$(\boldsymbol{u}, \boldsymbol{w})_E = \int_\tau \nabla\varphi\cdot\boldsymbol{w} d\tau = \int_{\Sigma+S} \varphi w_n dS = 0.$$

This assertion gives a prescription for constructive separation of the potential component of the field. We first consider the Neumann problem

$$\Delta\varphi = 0 \quad \text{inside} \quad \tau,$$
$$\frac{\partial\varphi}{\partial n} = v_n \quad \text{on} \quad \Sigma + S.$$

The solution of this problem makes it possible to match each vector field $\boldsymbol{v}(P)$ to the gradient field $\boldsymbol{u}(P)=\nabla\varphi$. We write this in the form

$$\boldsymbol{u} = \Pi_u \boldsymbol{v},$$

where Π is the operator for the orthogonal projection from E into E_u.

The free surface is determined from a kinematic relationship which, in the linear theory, has the form

$$\frac{\partial\zeta}{\partial t} = v_n.$$

On the surface S, $v_n = v_z$ and, by definition of the field $\boldsymbol{u}$,

$$v_z = \frac{\partial \varphi}{\partial z},$$

i.e.,

$$\frac{\partial \zeta}{\partial t} = \frac{\partial \varphi}{\partial z}.$$

Thus, one of our assertions stating that the configuration of the free boundary is determined by the potential component of the field $\boldsymbol{u}$ follows from the very definition of the field $\boldsymbol{u} = \nabla \varphi$, which we have called the potential component of the field $\boldsymbol{v}$. It only remains to prove that the pressure is also determined only by the $\boldsymbol{u}$ field.

We define the vector $\boldsymbol{w}$ as $\boldsymbol{w} = \boldsymbol{v} - \boldsymbol{u}$. This formula matches each vector field $\boldsymbol{v}$ with some solenoidal field $\boldsymbol{w}$:

$$\boldsymbol{w} = \Pi_w \boldsymbol{v}.$$

On the basis of what has been proven above, the projection operator Π_w is also the orthogonal projection operator.

Let us now write the linearized Euler equation

$$\frac{\partial v}{\partial t} + \nabla \left(\frac{p}{\varrho} - \frac{U}{\varrho} \right) = 0,$$

where U is the force function of the gravitational field.

We note, first, that

$$\nabla \left(\frac{p}{\varrho} - \frac{U}{\varrho} \right)$$

is a potential vector. In order to satisfy ourselves that this is so, it is sufficient to apply the divergence operation to the left-hand side of the Euler equation and to use the condition that the vector $\boldsymbol{v}$ is solenoidal. Applying the operators Π_u and Π_w to both sides of the equation, we obtain

$$\Pi_u \left(\frac{\partial \boldsymbol{v}}{\partial t} + \nabla \frac{p}{\varrho} - \nabla \frac{U}{\varrho} \right) = \frac{\partial \nabla \varphi}{\partial t} + \nabla \frac{p}{\varrho} - \nabla \frac{U}{\varrho} = 0, \tag{5–41}$$

$$\Pi_w \left(\frac{\partial \boldsymbol{v}}{\partial t} + \nabla \frac{p}{\varrho} - \nabla \frac{U}{\varrho} \right) = \frac{\partial \boldsymbol{w}}{\partial t} = 0. \tag{5–42}$$

In fact, since $\nabla(p - U)$ is a potential vector, the field $\nabla(p - U)$ is uniquely determined by its normal components, *i.e.*, $\Pi_u \nabla(p - U) = \nabla(p - U)$.

The first of these equalities yields the Cauchy-Lagrange integral

$$\frac{\partial \varphi}{\partial t} = \frac{U}{\varrho} - \frac{p}{\varrho} + \text{const}. \tag{5–43}$$

Thus the pressure field is determined only by the potential component.

Remark

It is precisely this fact which justifies the introduction of the energy metric and the corresponding definition of the potential component.

It follows from (5–42) that, within the framework of the linear theory, the field $\boldsymbol{w}$ does not change in time; the vorticity is constant not only in the particle but also in the point.

[5-1.7] The Kinematic Condition

We noted at the start of the section that the conditions (4–5) and (5–5) follow from the continuity of the fluid. We now prove this assertion. We set up the continuity condition for the volume T, which is bounded from the top by the surface of the wave at time t_0 and at the bottom by a surface which is the locus of the vertexes of segments with length l. These segments are laid off along the normal to the wave surface and the side surface formed by the segments of the normals to the wave surface. The latter segments pass through points of some closed contour as which is constructed on the surface of the wave (Fig. 20).

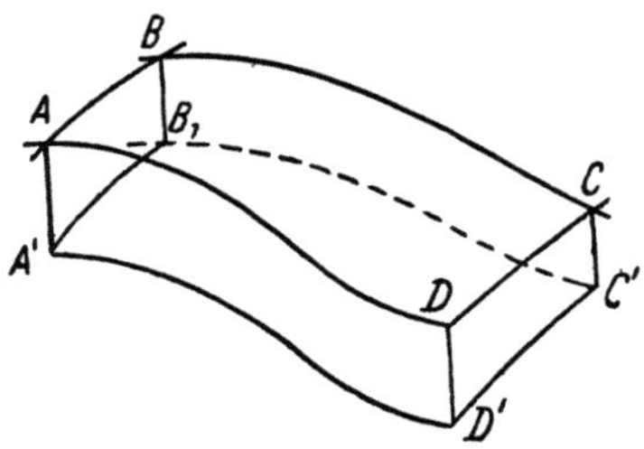

Fig. 20.

Let the equation of the free surface be given in the form

$$F(x, y, z, t) = 0. \tag{5–44}$$

At the time $t = t_0 + \Delta t$, the points on the wave surface $x = x_0 + \Delta x_0$, $y = y_0 + \Delta y_0$, $z = z_0 + \Delta z_0$ will satisfy the equation

$$F(x_0 + \Delta x_0, y_0 + \Delta y_0, \quad z_0 + \Delta z_0, \quad t_0 + \Delta t) = 0, \tag{5–45}$$

where x_0, y_0, z_0 is some point which, at the time $t = t_0$, satisfied (5–44). Then (5–45) can be rewritten, to within first-order infinitesimals, as

$$\left(\frac{\partial F}{\partial x}\right)_0 (x - x_0) + \left(\frac{\partial F}{\partial y}\right)_0 (y - y_0) + \left(\frac{\partial F}{\partial z}\right)_0 (z - z_0) + \left(\frac{\partial F}{\partial t}\right)_0 \Delta t = 0. \tag{5–46}$$

Here, the subscript 0 denotes the points on the wave surface at $t = t_0$. As the points x, y, z, we take the points on the wave surface which at $t_0+\Delta t$ lie on normals at the points $x_0y_0z_0$ to the wave surface at time $t=t_0$. Hence these points, in addition to Eq. (5–46), should also satisfy the equations of the normal

$$z - z_0 = (x - x_0)\frac{\partial F/\partial z}{\partial F/\partial x},$$

$$y - y_0 = (x - x_0)\frac{\partial F/\partial y}{\partial F/\partial x}. \tag{5–47}$$

Equations (5–46) and (5–47) can be used to calculate the quantities $(x-x_0)$, $(y-y_0)$, and $(z-z_0)$:

$$x - x_0 = -\frac{\partial F}{\partial t}\Delta t\cdot\frac{\partial F}{\partial x}\cdot\frac{1}{k^2},$$

$$y - y_0 = -\frac{\partial F}{\partial t}\Delta t\cdot\frac{\partial F}{\partial y}\cdot\frac{1}{k^2},$$

$$z - z_0 = -\frac{\partial F}{\partial t}\Delta t\cdot\frac{\partial F}{\partial z}\cdot\frac{1}{k^2},$$

where

$$k^2 = \left(\frac{\partial F}{\partial x}\right)^2 + \left(\frac{\partial F}{\partial y}\right)^2 + \left(\frac{\partial F}{\partial z}\right)^2,$$

and consequently the length h of the segment of a normal to the wave surface at time t_0, that is bounded between this surface and the surface of the wave at time $t_0+\Delta t$ is given by

$$h = \frac{\partial F}{\partial t}\Delta t\frac{1}{k}.$$

Thus, during the time Δt the volume τ under study has increased by

$$\Delta\tau = \iint_{S_1}\frac{\partial F}{\partial t}\Delta t\frac{1}{k}\,dS,$$

where S_1 is the surface area of figure *ABCD* (see Fig. 20). Since the fluid is incompressible, a change in its mass contained in the volume τ can be compensated for only by an influx of fluid through the surface $S_2(A'B'C'D')$ and through the side surface S_3.

From this (taking into account the direction of the normal) we arrive at the expression

$$\iint_{S_1}\frac{\partial F}{\partial t}\frac{1}{k}\,dS = -\iint_{S_2}v_n dS + \iint_{S_3}v_n dS.$$

We now let l go to zero. Since $\iint_{S_3} v_n dS = O(l)$ and $S_1 = S_2 + O(l)$, making use of the arbitrariness of the surface S_1, we obtain

$$v_n + \frac{1}{k}\frac{\partial F}{\partial t} = 0. \tag{5–48}$$

This expression is precisely the sought kinematic relationship. It can be put into a somewhat different form. Since

$$v_n = v_x\alpha + v_y\beta + v_z\gamma,$$

where α, β, and γ are the direction cosines of the normal:

$$\alpha = \frac{1}{k}\frac{\partial F}{\partial x}, \quad \beta = \frac{1}{k}\frac{\partial F}{\partial y}, \quad \gamma = \frac{1}{k}\frac{\partial F}{\partial z},$$

and v_x, v_y, and v_z are the projections of the velocity vector

$$v_x = \frac{dx}{dt}, \quad v_y = \frac{dy}{dt}, \quad v_z = \frac{dz}{dt},$$

Eq. (5–48) can be rewritten as

$$\frac{dF}{dt} = 0. \tag{5–49}$$

A fluid particle that has risen to the surface at some instant of time remains on it during the entire motion. In the particular case when the equation of the free surface is specified in the form

$$z = \zeta(x, y, t),$$

condition (5–49) is transformed into

$$\frac{d\zeta}{dt} = v_z.$$

Linearizing this condition, we obtain (5–5). Condition (5–4) is derived similarly.

Remark

The formulation of the variational problem is given in Ref. [69]. The relationship between the integral equation (5–10) and the problem of vibration of a fluid in a vessel was utilized in Ref. [70]. The presentation of the variational principle and the statement of the problem were presented following Ref. [71].

[5-2] VIBRATION OF A FLUID IN A FIELD OF VARIABLE-INTENSITY MASS FORCES

[5-2.1] Statement of the Problem

Let us consider the linear vibrations of a fluid in a vessel of finite size under the condition that the mass forces are functions of time. As was shown, it is sufficient to consider only potential motions with potential φ, which is a harmonic function. On Σ this function should satisfy the condition of impermeability of the shell

$$\frac{\partial \varphi}{\partial n} = 0,$$

while at the free surface S it should satisfy the kinematic condition

$$\frac{\partial \varphi}{\partial z} = \frac{\partial \zeta}{\partial t}$$

and the condition of constant pressure obtained from Eq. (5–43):

$$\varrho \frac{\partial \varphi}{\partial t} = U.$$

Restricting ourselves to the simplest case, we shall represent U, the potential of the mass forces, in the form

$$U = \varrho g \left[X(t)\, x + Y(t)\, y - f(t)\, z\right]; \quad f > 0.$$

Hence the condition for constancy of pressure will be

$$\frac{\partial \varphi}{\partial t} + g\ \zeta = g\left[Xx + Yy\right]. \tag{5–50}$$

We shall seek the solution of the problem in the form of the series

$$\varphi = \sum_{n=1}^{\infty} p_n(t)\, \Phi_n,$$

$$\zeta = \sum_{n=1}^{\infty} q_n(t)\, \Psi_n,$$

where Φ_n and Ψ_n are the eigenfunctions of the linear problem [see (5–12)], normalized by the condition

$$(\Psi_n \cdot \Psi_n) = \int_S \Psi_n^2 ds = 1.$$

Using (5–17), we can rewrite the kinematic relationship in the form

$$\sigma_n p_n = q_n' . \tag{5–51}$$

We now define the Fourier coefficients of the functions x and y as

$$x_n = (x \cdot \Psi_n), \quad y_n = (y \cdot \Psi_n).$$

Then Eq. (5–50), making use of (5–16), will take the form

$$p_n' + \sigma_n f q_n = X(t)\, x_n + Y(t)\, y_n, \quad n = 1, 2, \ldots .$$

Eliminating p_n by means of (5–51), we finally obtain

$$q_n'' + \sigma_n^2 f q_n = \sigma_n \{X(t)\, x_n + Y(t)\, y_n\}, \quad n = 1, 2, \ldots . \tag{5–52}$$

Thus, if the problem of free vibrations has been solved, the problem of forced vibrations can be reduced to the study of an infinite system of ordinary differential equations, which breaks up into individual equations.

Consequently, a fluid poured into a bowl has the same properties as an infinite set of pendulums, the natural frequencies of which are σ_n.

[5-2.2] Some Particular Cases

The theory presented here is linear. Hence we cannot consider vibrations at resonance. In fact let, for example, $Y=0$, $f=1$, and $X(t)=\sin\sigma_m t$. Then

$$q_n = \frac{\sigma_n x_n}{\sigma_n^2 - \sigma_m^2} \cos\sigma_m t, \quad n \neq m,$$

$$q_m = \tfrac{1}{2} x_m t \sin\sigma_m t, \quad n = m.$$

These formulas show that waves with increasing amplitude arise on the surface of the fluid if the frequency of the external forces is identical to one of the natural frequencies. Consequently, the theory presented here will become less and less exact with increasing time. This circumstance must be remembered when using the system (5–52).

Let us present another example in which increasing vibrations can arise. We consider a vessel moving vertically with some acceleration. Let us assume that no mass forces, except for gravity, act on the fluid. Then, in describing the motion in a coordinate system moving with a vertical acceleration, we will obtain, according to (5–52), the following system of equations:

$$q_n'' + \sigma_n^2 f(t)\, q_n = 0, \quad n = 1, 2, \ldots . \tag{5–53}$$

Replacing the variable

$$\tau = \int_0^t \sqrt{f(t)}\, dt,$$

we bring (5–53) to the form

$$\frac{d^2 q_n}{d\tau^2} + \frac{1}{2}\frac{d}{d\tau}(\ln f(\tau))\frac{dq_n}{d\tau} + \sigma_n^2 q_n = 0, \quad n = 1, 2, \ldots.$$

We now set up Lyapunov's function for the *n*th "principal vibration"

$$2V_n = \left(\frac{dq_n}{d\tau}\right)^2 + \sigma_n^2 q_n^2.$$

Evaluating dV_n/dt, we come to the conclusion that the energy of each principal vibration will decrease if $df/d\tau > 0$, and increase if $df/d\tau < 0$.

Thus we arrive at the following conclusion: if the intensity of the gravity force increases constantly, all the principal vibrations turn out to be increasing.

The development of standing waves in a vessel can be evaluated quantitatively in many cases, using known formulas of asymptotic representations:

$$q_n \sim \frac{A_n}{\sqrt[4]{f_n(t)}} \cos\left\{\int_0^t \sqrt{\sigma_n^2 f(t)}\, dt + \Theta_n\right\}, \tag{5–54}$$

where A_n and θ_n are arbitrary constants. Equation (5–54) will be more exact the larger the value of σ_n.

[5-2.3] Vibration of Fluid in a Vibrating Vessel

Let us now consider the problem of vibration of fluid in a vessel standing on a support that itself vibrates in the vertical direction. Fixing the coordinate system to the moving support, we again obtain Eqs. (5–53). The function f in this case will be

$$f = 1 + \mu \cos \omega t,$$

where μ is a quantity proportional to the amplitude of vibrations of the support, and ω is the frequency of these vibrations. Thus Eqs. (5–53) in this case will be a system of Mathieu equations, each of which can be studied by known methods. Using the properties of the solutions of these equations for each of the principal vibrations, we know a priori the following. Some instability regions will exist in the plane of the parameters μ and ω and will have increasing vibrations corresponding to them. For the remaining values of the pairs (μ, ω), the vibrations will be damped. But the system (5–53) consists of a denumerable number of equations. Therefore the question arises whether the instability regions of each principal vibration will not cover the entire plane of parameters (μ, ω). A positive answer to this question will mean that, no matter how small "amplitude" of μ, there will always exist at least one principal vibration whose amplitude will increase with time; *i.e.*, the vibration of the fluid as a whole will be unstable.

To resolve this question let us first clarify the structure of the instability regions in the above plane for each of the principal vibrations.

We set $\omega t = \tau$, so that Eqs. (1–53) will take the form

$$\frac{d^2 q_n}{d\tau^2} + \sigma_n^2 \lambda (1 + \mu \cos \tau)\, q_n = 0\,, \tag{5-55}$$

where

$$\lambda = \frac{1}{\omega^2}\,.$$

As is known, the boundaries of the stability regions of Eq. (5–55) will be the curves in the plane (μ, λ) corresponding to periodic solutions of Eq. (5–55) with periods 2π and 4π. The standard method for constructing them is based on the series expansion in terms of the small parameter μ:

$$\lambda = \sum_{s=0}^{\infty} \lambda_s \mu^s\,, \quad q_n = \sum_{s=0}^{\infty} q_{ns} \mu^s\,, \quad n = 1, 2, \ldots,$$

with subsequent determination of periodic solutions (5–55) of the above periods.

We omit here the calculations required for determining the stability boundaries and present only the final result. The equation of the stability boundaries has the form*

$$\omega_{n_i}^{(m)} = \frac{2\sigma_n}{m}\left[1 \pm \frac{\mu^m}{k_{m_i}} + \cdots\right],$$

$$m = 1, 2, \ldots; \quad n = 1, 2, \ldots; \quad i = 1, 2\,.$$

Here n is the number of the principal vibration, and m is the number of the stability region. The numbers k_{m_i} satisfy the condition

$$\lim_{m \to \infty} k_{m_i} = \infty\,.$$

For example, if $m = 1, 2$, then $k_{m_i} = 4, 12$. The form of the plane (ω, μ) is shown in Fig. 21, where the instability regions are shaded.

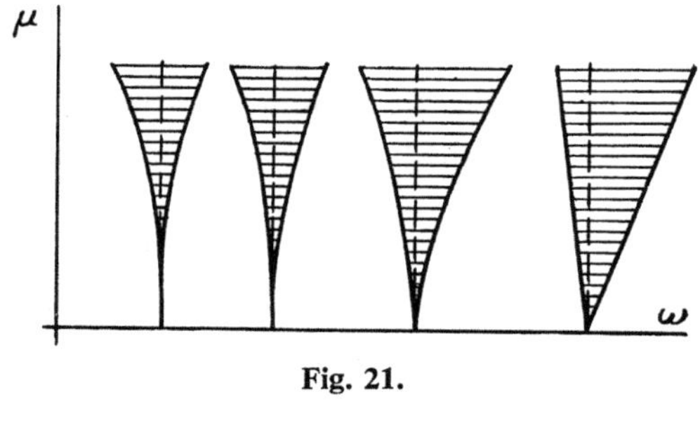

Fig. 21.

* A more detailed presentation of this problem is given in Ref. [73].

If the value of μ is fixed, it is possible to find some limiting frequency $\omega_n^*(\mu)$ for each principal vibration q_n. This frequency has one very important property: if, for the given value of the parameter μ, the frequency with which the intensity of the gravitational field changes is greater than ω_n^*, it will not result in a parametric resonance, *i.e.*, the nth principal vibration will not be excited.

The point $\omega=0$ will be the limiting point of the set $(\omega_n^{(m)})$, of points of intersection of the boundaries of instability regions of the nth principal vibration with the ω axis:

$$\omega_n^{(m)} = \frac{2\sigma_n}{m}.$$

We introduce still another designation. We denote the ensemble of sets $(\omega_n^{(m)})$, for all possible n, by $\Omega(\omega)$.

The question put forward at the beginning of this section is answered by the following theorem: the set $\Omega(\omega)$ is everywhere dense on the half-line $(0\text{-}\infty)$.

Proof

Let l be an arbitrary number. It is required to prove that no matter what the number δ is, the segment $[l-\delta, l+\delta]$ contains at least one point of the set Ω.

The numbers σ_n^2 are proportional to the eigenvalues of Fredholm's integral equation. Hence the natural frequencies can be represented in the form

$$\sigma_n = f(n)\, n^k,$$

where $k>0$ and $f(n)$ does not vanish but has a finite limit f_∞ as $n\to\infty$.

Thus the problem reduces to proving the existence of integral solutions n and m of the inequality

$$l - \delta \leqslant \frac{2n^k}{m} f_\infty + \frac{2n^k}{m}\, \varepsilon(n) \leqslant l + \delta, \tag{5–56}$$

where l and δ are arbitrary positive numbers $(\delta<l/2)$ and $|\varepsilon(n)|\to 0$ when $n\to\infty$.

We now consider the auxiliary inequality

$$\left(m\,\frac{l-\frac{1}{2}\delta}{2f_\infty}\right)^{1/k} \leqslant n \leqslant \left(m\,\frac{l+\frac{1}{2}\delta}{2f_\infty}\right)^{1/k}; \tag{5–57}$$

since

$$\lim_{m\to\infty} m^{1/k}\left\{\left(\frac{l+\frac{1}{2}\delta}{2f_\infty}\right)^{1/k} - \left(\frac{l-\frac{1}{2}\delta}{2f_\infty}\right)^{1/k}\right\} = \infty,$$

it is always possible to find a number S such that, when $m>S$, there will exist at least one integral number n_m satisfying the inequality (5–57). The numbers n_m form a sequence increasing without bound.

We select an $m=S_1>S$, so that

$$l-\frac{\delta}{2}\leqslant\frac{2n_{s_1}^k f_\infty}{S_1}\leqslant l+\frac{\delta}{2}.$$

In addition, it is always possible to specify an S_2 that is sufficiently large that

$$\frac{2n_{S_2}^k\,|\varepsilon(S_2)|}{S_2}<\frac{\delta}{2}.$$

[In fact, the ratio $n_{S_2}^k/S_2$ is contained between the numbers $(l+\frac{1}{2}\delta):2f_\infty$ and $(l-\frac{1}{2}\delta):2f_\infty$.]

Consequently, if we set $m=\max(S_1 S_2)$, and $n=n_m$, then the numbers selected in this manner will satisfy (5–56), which proves the theorem.

Thus, no matter what $\mu>0$ and ω are, it will always be possible to find at least one unstable principal vibration.*

This result is valid only for an ideal fluid. If the fluid is viscous, the motion will be stable for sufficiently small amplitudes of gravitational disturbances μ. This result, within the framework of the phenomenological theory of viscosity, was obtained by Bolotin[74], and in its rigorous statement it was obtained by Krushinskaya[75].

[5-3] THE RITZ METHOD IN THE STOKES-ZHUKOVSKIY PROBLEM

[5-3.1] Formulation of Boundary-Value Problems

It was shown in Sec. 1 of Chapter 2 that the velocity potential of a fluid in absolute motion reduces to determining the function φ, which is represented in the form of (2–9). For convenience in the subsequent presentation, we shall represent it in the form

$$\varphi=\sum_{i=1}^{3}v_{0i}\varphi_i^*+\sum_{i=1}^{3}\omega_i\varphi_{i+s}^*. \tag{5–58}$$

We recall that v_{0i} denotes the projections of the velocities of its mass center, ω_i designates the projections of the angular velocity on the coordinate axes fixed to the body, and φ_i^* is the Stokes-Zhukovskiy potential, which is a

* Moreover, it can be proven that a denumerable set of such vibrations exists.

solution of the following Neumann-Zhukovskiy problems:

$$\frac{\partial \varphi_i^*}{\partial n} = h_i, \quad i = 1, 2, \ldots, 6,$$

where

$$\begin{gathered} h_i = n_i, \quad i = 1, 2, 3, \\ h_4 = n_2 x_2 - n_2 x_3, \quad h_5 = n_1 x_3 - n_3 x_1, \quad h_6 = n_2 x_1 - n_1 x_2, \end{gathered} \tag{5–59}$$

and where n_i (i=1, 2, 3) are the direction cosines of the external normal. The functions φ_i^* (i=1, 2, 3) are immediately determined (see Sec. 1 of Chapter 2):

$$\varphi_i^* = x_i, \quad i = 1, 2, 3.$$

The functions φ_i^* (i=4, 5, 6) must, as a rule, be determined by numerical methods, one of which will be presented later.

[5-3.2] The Ritz Method

In presenting the necessary considerations, we follow, in the main, Ref. [65].

We denote by ϕ_i a function defined in and twice differentiable in τ, such that

$$\frac{\partial \psi_i}{\partial n} = h_i, \quad P \in \Sigma + S, \quad i = 1, 2, \ldots, 6.$$

We set

$$\varphi_i^* = \psi_i + u_i.$$

Then we will have the following boundary-value problem for the function u_i:

$$\begin{gathered} -\Delta u_i = \Delta \psi_i \equiv f, \quad P \in \tau, \\ \frac{\partial u_i}{\partial n} = 0, \quad P \in \Sigma + S. \end{gathered}$$

This problem is equivalent to the variational problem for the functional

$$F(u_i) = -\int_\tau \Delta u \cdot u d\tau - 2\int_\tau u f d\tau.$$

We now transform it to the form

$$F(u_i) = -\int_\tau \Delta\varphi_i^* \cdot \varphi_i^* d\tau + \int_\tau \Delta\varphi_i^* \cdot \psi_i d\tau + \int_\tau \Delta\psi_i \cdot \varphi_i^* d\tau - 2\int_\tau \varphi_i^* \cdot \Delta\psi_i d\tau + C,$$

where

$$C = \int_\tau \Delta\psi_i \cdot \psi_i d\tau.$$

Since C is independent of the sought function, the solution of the variational problem for the functional F reduces to that for the functional F^*:

$$F^* = \int_\tau \{\Delta\varphi_i^* \cdot \varphi_i^* + [\varphi_i^* \Delta\psi_i - \psi_i \Delta\varphi_i^*]\}\, d\tau .$$

Applying Green's formula, we obtain

$$F^* = \int_\tau (\nabla\varphi_i^*)^2\, d\tau - \int_{S+\Sigma} \varphi_i^* h_i dS - \int_{S+\Sigma} [\varphi_i^* h_i - \psi_i h_i]\, dS ,$$

but $\int_{S+\Sigma} \psi_i h_i dS$ is independent of the sought function and can be dropped.

Thus the determination of Stokes-Zhukovskiy potential is equivalent to finding the minimum of the functional

$$F(\varphi) = \int_\tau (\nabla\varphi)^2\, d\tau - 2\int_{\Sigma+S} \varphi h dS , \tag{5-60}$$

where h is a known function. The function φ, which realizes a minimum of the functional (5–60), is sought by the standard Ritz method in the manner pointed out earlier in this chapter. We only note that, in contrast to the problem of determining natural vibrations, the system of equations that we shall obtain here will be inhomogeneous.

[5-3.3] Remark on the Method of Perturbation Theory

Methods of perturbation theory may turn out to be useful in determining the Stokes-Zhukovskiy potentials, as well as in problems of eigenvalues.

To determine the minimum of the functional (5–60), we specify a system of coordinate functions $\{\chi_n\}$ and seek the function φ in the form

$$\varphi = \sum_{n=1}^{N} c_n \chi_n .$$

Substituting this sum into (5–60), we then set up the equations

$$\frac{\partial F}{\partial c_i} \equiv \sum_{j=1}^{N} c_j \alpha_{ij} - 2\beta_i = 0 , \tag{5-61}$$

where

$$\alpha_{ij} = \int_\tau \nabla\chi_i \nabla\chi_j d\tau , \quad \beta_i = \int_{\Sigma+S} h\chi_i dS .$$

Thus the problem reduces to that of solving the linear inhomogeneous system (5–61).

We now assume that we are capable of determining the Stokes-Zhukovskiy potentials for some volume τ^*, for which the system of coordinate functions $\{\chi_n\}$ is known. Then we can use for an approximate solution the approach presented in Sec. 1 for finding eigenvalues.*

Using the known coordinate functions, we now rewrite Eqs. (5–61) in the form

$$\sum_{i=1}^{N} c_i(\alpha_{ij}^* + \varepsilon\tilde{\alpha}_{ij}) = 2\beta_j^* + 2\tilde{\beta}_j\varepsilon, \tag{5–62}$$

where

$$\varepsilon\tilde{\alpha}_{ij} = \int_{\delta\tau} \nabla\chi_i\nabla\chi_j d\tau, \quad \varepsilon\tilde{\beta}_j = \int_{\delta(\Sigma+S)} \chi_j h dS = \int_{\Sigma+S} - \int_{\Sigma^*+S^*} \chi_j h dS.$$

We seek the solution of Eqs. (5–62) in the form

$$c_i = c_i^* + \varepsilon\tilde{c}_i,$$

where c_i^* satisfies the system

$$\sum_{i=1}^{N} c_i^*\alpha_{ij}^* = 2\beta_j^*, \quad j = 1, 2, \ldots, N, \tag{5–63}$$

the solution of which we know.

The unknown $\tilde{c}_i$ are determined from the inhomogeneous system

$$\sum_{i=1}^{N} \tilde{c}_i\alpha_{ij}^* = -\sum_{i=1}^{N} c_i^*\tilde{\alpha}_{ij} + 2\tilde{\beta}_j. \tag{5–63′}$$

The above equations differ from Eqs. (5–63) only in their right-hand sides, and consequently the $\tilde{c}_i$ are determined by the same method as the c_i^*. To determine the right-hand sides of (5–63′), we must evaluate a number of surface integrals.

To solve effectively the problems of the dynamics of a rigid body containing a fluid requires calculating a variety of integrals of the form

$$J = \int_S \varphi_i^* f \, dS$$

and the associated masses

$$m_{ij} = \int_\tau \nabla\varphi_i^*\nabla\varphi_j^* d\tau.$$

Using an approximate solution in the form $\varphi_i^* = \Sigma_{k=1}^N (c_{ik}^* + \tilde{c}_{ik})\chi_k$ (we have set $\varepsilon = 1$), we find

$$m_{ij} = m_{ij}^* + \sum_{k,s=1}^{N} c_{ik}^* c_{js}^* \int_{\delta\tau} \nabla\chi_k\nabla\chi_s d\tau + \sum_{k,s=1}^{N} \tilde{c}_{ik}\tilde{c}_{js} \int_\tau \nabla\chi_k\nabla\chi_s d\tau, \tag{5–64}$$

* We do not repeat here all the limiting considerations that were previously discussed.

where the m_{ij}^* are the associated masses of a known cavity. The use of formulas such as (5–64) often makes it possible to reduce appreciably the number of numerical calculations.

[5-4] THE PENDULUM PROBLEM

[5-4.1] Equations of Motion

We begin the study of the motion of a body containing a fluid with one simple example[51,62]. We consider the oscillations of the pendulum depicted in Fig. 22. The pendulum is an open vessel filled with fluid. Waves may

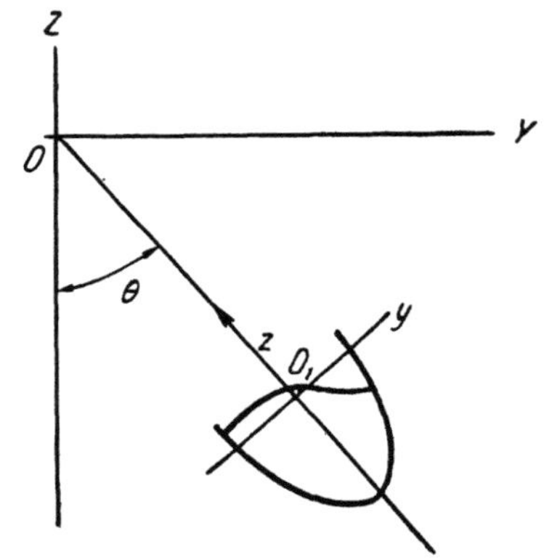

Fig. 22.

arise on the surface of the fluid. The vessel is connected by a rigid weightless rod to an ideal pivot. The stationary coordinate system OZY is placed at the point of suspension. The moving system (Ozy) is fixed to the free surface in its rest position, as is shown in the figure.

We shall consider plane oscillations of the pendulum. The validity of this approach requires some additional assumptions concerning the configuration of the cavity or the properties of the suspension.

We set up an expression for the potential energy of the body-fluid system:

$$T = T_1 + T_2, T_1 = \frac{J_0 \theta'^2}{2}, \quad T_2 = \frac{\varrho}{2} \int_\tau (\nabla \varphi + \boldsymbol{v})^2 \, d\tau,$$

where J_0 is the moment of inertia about hinge O, and $\boldsymbol{v}$ is the velocity vector of the fluid particles, with the assumption that the free surface has been replaced by a rigid partition; the prime denotes differentiation with respect to time;

$$\boldsymbol{v} = \theta' \nabla \varphi^* ,$$

where φ^* is the Stokes-Zhukovskiy potential and is determined solely by the shape of the vessel, while φ is the potential of wave motion in the vessel. Thus

$$T = \frac{J_0 \theta'^2}{2} + \frac{\varrho}{2} \int_\tau (\nabla \varphi)^2 \, d\tau + \theta' \varrho \int_\tau \nabla \varphi \nabla \varphi^* \, d\tau + \varrho \frac{\theta'^2}{2} \int_\tau (\nabla \varphi^*)^2 \, d\tau .$$

The integral in the last term of the right-hand side is a number that does not depend on the motion of the pendulum and is determined solely by the geometry of the cavity. Hence, according to the terminology that has been developed, we shall call the quantity

$$m = \varrho \int_\tau (\nabla \varphi^*)^2 \, d\tau$$

the *associated mass* (in the given case, the associated moment of inertia) of the fluid. Designating

$$J = J_0 + m ,$$

we substitute T in the form

$$T = \frac{J \theta'^2}{2} + \theta' \varrho \int_\tau \nabla \varphi \nabla \varphi^* \, d\tau + \frac{\varrho}{2} \int_\tau (\nabla \varphi)^2 \, d\tau . \tag{5–65}$$

The quantity J is called the moment of inertia of the equivalent rigid body. If the fluid did not have a free surface (*i.e.*, if it were covered), then $\varphi \equiv 0$ and the kinetic energy would have been $J\theta'^2/2$, *i.e.*, the kinetic energy of a rigid body whose moment of inertia is J. We are considering the problem in its linear formulation, and hence we should retain only quadratic terms in the expression for the kinetic energy. Consequently, the integration in (5–65) should be performed over the volume occupied by the fluid in the equilibrium position, and the kinetic energy should be a quadratic functional. This replacement of the region of integration produces errors in third-order infinitesimals.

Similarly, we set up an equation for the potential energy of the system:

$$\Pi = \Pi_1 + \varrho g \int_\tau Z d\tau ,$$

where Π_1 is the potential energy of the vessel:

$$\Pi_1 = \frac{l_0 M_0 g \theta^2}{2} ;$$

M_0 is the mass of the vessel, and l_0 is the distance from its center of gravity to the point O. In this expression the integration should be extended over the entire volume occupied by the fluid, since the replacement of the region occupied by the fluid at the given instant by the region that it occupies in the equilibrium position produces errors in the second- rather than the third-order infinitesimals:

$$\varrho g \int_{\tau} Z d\tau = \varrho g \int_{\tau_0} Z d\tau + \varrho g \int_{\tau_1} Z d\tau .$$

Here τ_0 denotes the volume occupied by the fluid in the equilibrium position, and τ_1 denotes the volume bounded between the free surface $Z = \zeta(P, t)$ and the plane S. The first term is the potential energy that the fluid would have if its free surface were solidly covered:

$$\varrho g \int_{\tau_0} Z d\tau = \varrho g Z^* ;$$

Z^* is the z-coordinate of the center of gravity of the fluid volume. Assuming that the center of gravity of volume τ_0 lies on the straight line OO_1, we have

$$Z^* = - l^* \cos\theta = \frac{l^* \theta^2}{2} + \text{const} + O(\theta^3) .$$

Here l^* is the distance from the pivot O to the center of gravity of the volume τ_0. We transform the second integral by noting that

$$Z = (z - l)\cos\theta + y \sin\theta , \tag{5–66}$$

where l is the length of the segment OO_1 (the distance between the coordinate origins of both systems).
Equation (5–66) follows from Fig. 23:

$$OA = OB - AB ,$$

where $OA = -Z$; $OB = l\cos\theta$; $AB = CD\cos\theta + OD\sin\theta$, $CD = z$, $OD = y$. We now evaluate

$$\varrho g \int_{\tau_1} Z d\tau = \varrho g \int_S dS \int_0^{\zeta} [(z - l)\cos\theta + y \sin\theta]\, dz .$$

Furthermore,

$$\int_S \int_0^{\zeta} l \cos\theta dz dS = l \cos\theta \int_S \zeta dS = 0 ,$$

since the fluid is incompressible. Hence

$$\varrho g \int_{\tau_1} Z d\tau = \frac{\varrho g}{2} \int_S \zeta^2 dS + \varrho g \theta \int_S \zeta y dS + O(\theta^3).$$

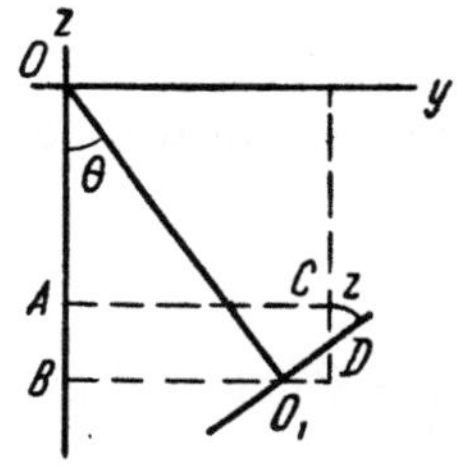

Fig. 23.

Dropping higher-order infinitesimals, we finally obtain

$$\Pi = \frac{k^2\theta^2}{2} + \theta \varrho g \int_S \zeta y dS - \frac{\varrho g}{2} \int_S \zeta^2 dS. \tag{5–67}$$

Here

$$k^2 = l_0 M_0 g + \varrho g \tau_0 l^*.$$

Remark

For our further discussion we shall need still another expression for the kinetic energy. Applying Green's formula to (5–65), and making use of the fact that

$$\frac{\partial \varphi}{\partial n} = \zeta_t, \quad P \in S,$$

we find that

$$T = \frac{J\theta'^2}{2} + \theta' \varrho \int_S \varphi^* \zeta_t dS + \frac{\varrho}{2} \int_S H\zeta_t \cdot \zeta_t dS.$$

Now we can set up the equations of motion. For this we evaluate δL, where

$$L = \int_0^{t_1} (T - \Pi)\, dt,$$

and T and Π are taken in the form (5–65) and (5–67). Using Green's formula and the kinematic relationship, we have

$$\delta L = \int_0^t \{J\theta'\delta\theta' + \delta\theta'\varrho \int_S \varphi^*\zeta_t dS + \theta'\varrho \int_S \varphi^*\delta\zeta_t dS + \varrho \int_S \varphi\delta\zeta_t dS - k^2\theta\delta\theta -$$

$$- \delta\theta\varrho g \int_S \zeta y dS - \theta\varrho g \int_S y\delta\zeta dS - \varrho g \int_S \zeta\delta\zeta dS\} \, dt .$$

Integrating by parts and making use of the fact that the variation becomes zero at the ends, we obtain

$$\delta L = - \int_0^{t_1} \{[J\theta'' + \varrho \int_S \varphi^*\zeta_{tt} dS + k^2\theta + \varrho g \int_S \zeta y dS] \, \delta\theta +$$

$$+ \varrho \int [\varphi^*\theta'' + \varphi_t + gy\theta + g\zeta] \, \delta\zeta dS\} \, dt .$$

Since for virtual displacements $\delta L = 0$, making use of the fact that the variations $\delta\theta$ and $\delta\zeta$ are arbitrary and independent, we arrive at the following system of integro-differential equations:

$$\begin{gathered} J\theta'' + \varrho \int_S \varphi^*\zeta_{tt} dS + k^2\theta + \varrho g \int_S \zeta y dS = 0 , \\ \varrho\theta''\varphi^* + \varrho\varphi_t + \varrho g y\theta + \varrho g\zeta = 0 . \end{gathered} \tag{5–68}$$

This system of equations contains the functions φ and ζ, which are related by the kinematic relationship $\varphi_z = \zeta_t$. This relationship can be used to eliminate ζ. For this it is necessary to integrate the second of the equations with respect to t and to make the replacement $\varphi_z = \zeta_t$. But this process is inconvenient became it artificially increases the order of the system. Okhotsimskiy[77] has suggested that the velocity potential be replaced by a displacement potential, which is defined as

$$Q_t = \varphi .$$

The function Q at the solid wall satisfies the condition

$$\frac{\partial Q}{\partial n} = 0 ,$$

and at the free surface it satisfies the condition

$$\frac{\partial Q}{\partial n} = \zeta .$$

Using the function Q, we can rewrite system (5–68) in the form

$$J\theta'' + \varrho \int_S \varphi^* Q_{ztt} dS + k^2\theta + \varrho g \int_S Q_z y dS = 0, \qquad (5\text{–}68')$$

$$\varrho\theta''\varphi^* + \varrho Q_{tt} + \varrho g y\theta + \varrho g Q_z = 0.$$

For theoretical studies it is more convenient to eliminate the velocity potential φ:

$$\varphi = H\zeta_t.$$

Using the above expression, we have

$$\varrho\theta'' + \varrho \int_S \varphi^* \zeta_{tt} dS + k^2\theta + \varrho g \int_S \zeta y dS = 0, \qquad (5\text{–}69)$$

$$\varrho\theta''\varphi^* + \varrho H\zeta_{tt} + \varrho g y\theta + \varrho g\zeta = 0.$$

[5-4.2] Natural Vibrations of a Pendulum and the Frequency Equation

We now consider the problem of free oscillations. For this we set

$$\theta = X \sin \sigma t,$$
$$\varphi = \Phi \cos \sigma t,$$
$$\zeta = \psi \sin \sigma t.$$

After this substitution, Eqs. (5–69) will become

$$X(k^2 - J\sigma^2) + \varrho \int_S \psi(gy - \varphi^*\sigma^2)\, dS = 0, \qquad (5\text{–}70)$$

$$X\varrho(gy - \varphi^*\sigma^2) + \varrho g\psi - \varrho\sigma^2 H\psi = 0.$$

We denote the resolvent operator $gI - \sigma^2 H$ by R_0, where I is a unit operator. Then

$$\psi = -XR_\sigma(gy - \varphi^*\sigma^2).$$

Substituting into the first of Eqs. (5–70), we obtain the following equation for determining the natural frequencies σ:

$$k^2 - J\sigma^2 - \varrho \int_S R_\sigma(gy - \varphi^*\sigma^2)\cdot(gy - \varphi^*\sigma^2)\, dS = 0.$$

Since R_0 is a meromorphic function of σ^2, the problem reduces to finding zeros of some meromorphic function. Thus a pendulum containing a fluid has a denumerable number of natural frequencies.

If we know how to solve the problem of vibration of the fluid in a vessel, the above process can be implemented effectively and the frequency equation can be written explicitly.

We denote the natural frequencies and the principal modes of vibrations in the vessel normalized by the condition $\int_s \psi_n^2 dS = 1$ by ω_n and ψ_n. These quantities satisfy the expressions

$$H\psi_n = \frac{g}{\omega_n^2}\psi_n .$$

We shall seek the solution of the equation

$$g\psi - \sigma^2 H\psi = f$$

in the form

$$\psi = \sum_{n=1}^{\infty} q_n \psi_n .$$

After obvious calculations, we obtain

$$\psi = \sum_{n=1}^{\infty} \frac{f_n \omega_n^2}{g(\omega_n^2 - \sigma^2)},$$

where

$$f_n = \int_S f \psi_n dS .$$

We note that, if the function f is orthogonal with respect to some eigenfunctions ψ_n, then these functions will not enter the representation for the function ψ.

In our case, $f = X(gy - \varphi^* \sigma^2)$. Hence

$$\psi = -\frac{X}{g} \sum_{n=1}^{\infty} \frac{\omega_n^2}{\omega_n^2 - \sigma^2} \psi_n \int_S (gy - \varphi^* \sigma^2) \psi_n dS ,$$

and consequently the frequency equation can be written as

$$k^2 - J\sigma^2 = \frac{\varrho}{g} \int_{n=1}^{\infty} \frac{\omega_n^2 (a_n + b_n \sigma^2)^2}{\omega_n^2 - \sigma^2}, \tag{5–71}$$

where

$$a_n = \int_S gy\psi_n dS , \quad b_n = -\int_S \varphi^* \psi_n dS .$$

Equation (5–71) is simplest when studied graphically (Fig. 24). The roots that are sought are determined by the point of intersection of the straight line $F_1 = k^2 - J\sigma^2$ with the curve $F_2(\sigma^2)$:

$$F_2(\sigma^2) = \frac{\varrho}{g} \sum_{n=1}^{\infty} \frac{\omega_n^2 (a_n + b_n\sigma^2)^2}{\omega_n^2 - \sigma^2}.$$

The curve $F_2(\sigma^2)$, depicted schematically in Fig. 24, has a denumerable number of poles. Graphical examination of Eq. (5–71) makes it possible to

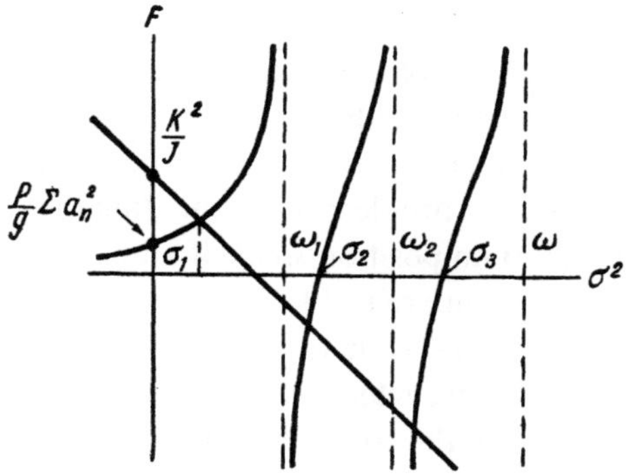

Fig. 24.

discover a number of interesting properties of this motion and, in particular, to construct the asymptotic curve for the natural frequencies as $n \to \infty$.[51,62]

Setting

$$\psi = \sum_{n=1}^{\infty} q_n\psi_n, \quad gy = \sum_{n=1}^{\infty} a_n\psi_n, \quad \text{and} \quad -\varphi^* = \sum_{n=1}^{\infty} b_n\psi_n,$$

we reduce the system of integro-differential equations (5–70) to the infinite system of algebraic equations

$$\begin{gathered} X(k^2 - J\sigma^2) + \varrho \sum_{n=1}^{\infty} q_n(a_n + b_n\sigma^2) = 0, \\ X(a_m + b_m\sigma^2) + gq_m - \frac{g\sigma^2}{\omega_m^2} q_m = 0, \quad m = 1, 2, \ldots. \end{gathered} \tag{5–72}$$

The system (5–72) contains unknowns q_m only with subscripts m for which the eigenfunctions are not orthogonal to the functions φ^* and y. The unknown q_m with indices corresponding to eigenfunctions that are orthogonal to φ^* and y will satisfy the system

$$gq_m = \frac{g\sigma^2}{\omega_m^2} q_m,$$

from which it will follow that $\sigma_m^2=\omega_m^2$.The principal vibrations (we shall call them *even*) cannot be excited by the motion of the rigid body. For translational motion, these are the modes of principal vibration that do not change the position of the center of gravity.

We note still another property of the oscillations of a pendulum. If the pendulum was initially at rest while waves existed on the surface of the fluid, then their energy in the general case is transferred to the pendulum and it begins to oscillate. Even the waves are the special kind of waves whose energy is not transferred to the pendulum. Not only can the oscillations of the body not induce such waves, but the even waves themselves cannot bring about oscillations of the pendulum.

The central problem is that of the stability of the pendulum's equilibrium position. As can be seen from the figure, the presence of the fluid inside the pendulum can only reduce the stability factor, since it moves the roots to the left. This naturally raises the problem of formulating the concept of stability. We shall call the equilibrium position *stable* if all of the principal vibrations of the pendulum-fluid system are stable. This means that all of the roots σ_1^2 of the frequency equation (5–71) lie to the right of the axis $\sigma^2=0$. We note that this concept of stability, which is convenient for the study of linear systems, is not equivalent to the concept of stability according to Lyapunov, which was considered in the first chapters of this book. It is easy to see that stability of the nonlinear system under study does not follow from stability in the sense of boundedness of principal vibrations.

It is quite simple to obtain the necessary conditions for stability of a pendulum in terms of its geometric characteristics.

Since this problem has been the subject of special study for systems of a more general kind, we shall not consider it in more detail.

[5-4.3] The Effect of Dissipative Forces

We now assume that, in addition to the restoring force, the pendulum is also acted upon by the force of viscous friction in the pivot. Then Eqs. (5–69) will take the form

$$\begin{gathered} J\theta'' + \varrho \int\limits_S \varphi^* \zeta_{tt} dS + a\theta' + k^2\theta + \varrho g \int\limits_S \zeta y dS = 0\,, \\ \varrho\theta''\varphi^* + \varrho H\zeta_{tt} + \varrho g\theta + \varrho g\zeta = 0\,. \end{gathered} \tag{5–73}$$

Equations (5–73) differ from Eqs. (5–69) in that the first equation contains the term $a\theta'$, which characterizes the magnitude of the viscous friction ($a>0$).

We note that we are considering only dissipation resulting from external forces. As before, the fluid in the vessel is assumed to be ideal.

Replacing in Eqs. (5–73)

$$\theta = Xe^{i\sigma t}, \quad \zeta = \psi e^{i\sigma t}$$

and repeating the considerations presented above, we obtain the frequency equation in the form

$$k^2 + ia\sigma - J\sigma^2 = \frac{\varrho}{g}\sum_{n=1}^{\infty}\frac{\omega_n^2(a_n + b_n\sigma^2)^2}{\omega_n^2 - \sigma^2}.$$

We rewrite it as

$$\Phi(\sigma^2) \equiv k^2 - J\sigma^2 - \frac{\varrho}{g}\sum_{n=1}^{\infty}\frac{\omega_n^2(a_n + b_n\sigma^2)^2}{\omega_n^2 - \sigma^2} = -\, ia\sigma. \qquad (5\text{–}74)$$

We denote the roots of the equation $\Phi(\sigma^2)=0$ by σ_n^2. All of the σ_n^2 are real.

We note that

$$\left(\frac{d\Phi(\sigma^2)}{d\sigma^2}\right)_{\sigma=\sigma_n} < 0.$$

In fact, the equation $\Phi(\sigma^2)=0$ can be rewritten as $\Phi \equiv F_1 - F_2 = 0$, where

$$F_1 = k^2 - J\sigma^2,$$

$$F_2 = \frac{\varrho}{g}\sum_{n=1}^{\infty}\frac{\omega_n^2(a_n + b_n\sigma_n^2)^2}{\omega_n^2 - \sigma^2}.$$

On the segment $[\omega_n^2, \omega_{n+1}^2]$, the function Φ becomes zero only once; to the left of the point $\sigma^2 = \sigma_n^2$, $F_1 > F_2$, while to the right, $F_1 < F_2$; hence $(d\Phi/d\sigma^2)_{\sigma^2=\sigma_n{}^2} < 0$.

Further, assuming that a is small, we set

$$\sigma = \sigma_n + \delta\sigma.$$

Substituting this expression into Eq. (5–74), and linearizing with respect to $\delta\sigma$, we obtain

$$\delta\sigma = -\frac{ia}{2\left(\dfrac{d\Phi}{d\sigma^2}\right)_{\sigma=\sigma_n}} = +\, i\alpha,$$

where $\alpha > 0$.

Thus in this case we find that the characteristic indicator

$$\lambda = i\sigma = i(\sigma_n + i\alpha) = -\,\alpha + i\sigma_n$$

has a negative real part. Thus all the vibrations that were determined by Eq. (5–71) will become damped by the addition of dissipative forces.

We note, however, that it does not follow from this assertion that all of the waves at the surface of the fluid will be damped with time. In fact, among the modes of vibration of fluid in the vessel, *i.e.*, among the functions ψ_n, there exist functions that are orthogonal to φ^* and y simultaneously. We shall call them "even." The natural frequencies corresponding to them are not given by Eq. (5–71).

Waves described by "even" eigenfunctions cannot be induced by motion of the pendulum. But even if these waves exist on the surface of the fluid, the external dissipative forces cannot damp them. With time, all of the waves, except the "even" waves, will be damped. This is the basic difference between dissipative forces that are external and those produced by the viscosity of the fluid.

Strictly speaking, we have proven our assertion only for the case of sufficiently small values of a. It was shown that under the action of low-intensity friction forces all of the characteristic indicators λ become displaced from the imaginary axis into the left half-plane. But subsequently they will be situated in the left half-plane for any value of $a>0$. In fact, in order to move to the right half-plane when a changes, the characteristic indicators should cross the imaginary axis. At this instant the value of σ should be real. But it is easy to show that Eq. (5–74) does not allow real solutions for any $a \neq 0$.

A theorem analogous to one of Kelvin's theorems is proven just as simply: A position of equilibrium that is unstable for potential forces cannot be stabilized by dissipative forces.

[5-4.4] Stability of the Relative Equilibrium of a Pendulum in an Accelerating Coordinate System

We now consider system (5–73) on the assumption that the intensity of the gravitational field g is a function of time. We also assume that the pendulum is acted upon by dissipative forces, which are also function of time. Setting

$$\zeta(P,t)=\sum_{n=1}^{\infty} q_n(t)\,\psi_n(P),$$

we reduce the integro-differential system of equations (5–73) to an infinite system of differential equations

$$\begin{aligned} J\theta''+\varrho\sum_{n=1}^{\infty} b_n q_n''+a\theta'+k^2\theta+\varrho g\sum_{n=1}^{\infty} a_n q_n=0\,,\\ \varrho b_n\theta''+m_n q_n''+\varrho g\theta+\varrho g q_n=0\,,\quad n=1,2,\dots, \end{aligned} \tag{5–75}$$

where b_n and a_n are the coefficients of the Fourier functions φ^* and y, and m_n is the associated mass of the nth principal vibration. The infinite system (5–75) will always be treated as a system with a finite, but large, number of equations.

The study of Eqs. (5–75) is of interest as a simulation problem in the study of the vibrations of an accelerating missile. We now raise the question of the structure of sufficient stability conditions.

We first consider a somewhat more general problem, which can also be of independent applied interest,

$$\sum_{j=1}^{\infty} (\alpha_{ij} q_j'' + \beta_{ij} q_j) = - \sum_{j=1}^{\infty} c_{ij} q_j', \quad i = 1, 2, \ldots . \tag{5–76}$$

We make the following assumptions with respect to the coefficients of the system (5–76):

a) α_{ij} are constants characterizing the masses of the system elements. They form a symmetrical matrix of positive definite quadratic form.

b) β_{ij} are some functions of time depending only on the one variable $p(t)$:

$$\beta_{ij} = \gamma_{ij} p(t),$$

where γ_{ij} are constants. They form a symmetrical matrix of positive definite quadratic form.

c) c_{ij} are some functions of time, but such that at each instant of time there exists a disturbing function

$$F(t, q_1' \ldots q_n') = 2 \sum_{i,j=1}^{\infty} c_{ij} q_i' q_j' .$$

Equations (5–76) can be treated as Lagrange's equations of some non-steady system performing small vibrations under the action of conservative and dissipative forces, in which the potential energy is a function of only one variable parameter.

We replace, in (5–76), the independent variable by

$$\tau = \int_0^t \sqrt{p(t)}\, dt ,$$

after which we obtain

$$\sum_j \left(\alpha_{ij} \frac{d^2 q_j}{d\tau^2} + \gamma_{ij} q_j \right) = - \sum_i \left(c_{ij} \frac{1}{\sqrt{p(t)}} + \frac{\alpha_{ij} \dfrac{dp}{d\tau}}{2p} \right) \frac{dq_j}{dt} . \tag{5–77}$$

We now examine the Lyapunov function

$$V = \tfrac{1}{2} \sum_{i,j} \alpha_{ij} q_i' q_j' + \tfrac{1}{2} \sum_{i,j} \gamma_{ij} q_i q_j .$$

Because of Eqs. (5–77), its derivative is

$$\frac{dV}{dt} = -\sum_{i,j}\left(c_{ij}\frac{1}{\sqrt{p(t)}} + \frac{\alpha_{ij}\frac{dp}{d\tau}}{2p}\right) \times \frac{dq_i}{d\tau}\frac{dq_j}{d\tau} \equiv -\Phi\left(\frac{dq_i}{d\tau}, \frac{dq_j}{d\tau}\right)$$

from which it follows that positiveness of the quadratic form of Φ is a sufficient condition for stability. Consequently, the sufficient conditions for stability can be written in the form of the following Sylvester's inequalities:

$$\Delta_1 = \frac{c_{11}}{\sqrt{p}} + \alpha_{11}\frac{\frac{dp}{d\tau}}{2p} \geqslant 0,$$

$$\Delta_2 = \begin{vmatrix} \frac{c_{11}}{\sqrt{p}} + \alpha_{11}\frac{\frac{dp}{d\tau}}{2p} & \frac{c_{12}}{\sqrt{p}} + \alpha_{12}\frac{\frac{dp}{d\tau}}{2p} \\ \frac{c_{21}}{\sqrt{p}} + \alpha_{12}\frac{\frac{dp}{d\tau}}{2p} & \frac{c_{22}}{\sqrt{p}} + \alpha_{22}\frac{\frac{dp}{d\tau}}{2p} \end{vmatrix} \geqslant 0$$

etc.

Returning to the variable t, we rewrite these conditions in the form

$$\Delta_1(t) = c_{11} + \alpha_{11}\frac{d\ln\sqrt{p}}{dt} \geqslant 0,$$

$$\Delta_2(t) = \begin{vmatrix} c_{11} + \alpha_{11}\frac{d\ln\sqrt{p}}{dt} & c_{12} + \alpha_{12}\frac{d\ln\sqrt{p}}{dt} \\ c_{21} + \alpha_{21}\frac{d\ln\sqrt{p}}{dt} & c_{22} + \alpha_{22}\frac{d\ln\sqrt{p}}{dt} \end{vmatrix} \geqslant 0 \qquad (5\text{–}78)$$

etc.

Thus, for stability of equilibrium of the pendulum (5–75), it is sufficient for conditions (5–78) to be satisfied at each instant of time.

We now return to our problem of oscillations of a pendulum and we apply the result obtained above. In our case, $c_{11}=a$. All the other $c_{ij}=0$. Hence conditions (5–78) will have the form

$$\Delta_1 = a + J\frac{d\ln\sqrt{g(t)}}{dt} > 0,$$

$$\Delta_2 = \begin{vmatrix} a + J\frac{d\ln\sqrt{g(t)}}{dt} & \varrho b_1\frac{d\ln\sqrt{g(t)}}{dt} \\ \varrho b_1\frac{d\ln\sqrt{g(t)}}{dt} & m_1\frac{d\ln\sqrt{g(t)}}{dt} \end{vmatrix} > 0$$

etc.

[5-5] VIBRATION OF A CONSERVATIVE SYSTEM WITH FLUID LINKS

[5-5.1] Expression for the Energy

We now pass on to the study of the general cases of motion of a body about its equilibrium position. The velocity of the fluid particles with respect to an inertial coordinate system is represented in the form of the sum

$$\boldsymbol{v} = \nabla\varphi + \boldsymbol{v}^*,$$

where $\boldsymbol{v}^*$ is the velocity of the fluid particles on the assumption that the free surface has been replaced by a rigid partition, while φ is the velocity potential for the wave motion of the fluid.

Following Stokes, we set

$$\boldsymbol{v}^* = \sum_{i=1}^{3} v_i \nabla\varphi_i^* + \sum_{i=1}^{3} \omega_i \nabla\varphi_{i+3}^*, \tag{5–79}$$

where v_i is the projection of the velocity of some specified point of the rigid body, ω_i are the components of the instantaneous angular velocity vector, and φ_i^* $(i=1, 2, \ldots, 6)$ are the Stokes-Zhukovskiy potentials satisfying the boundary-value problem (5–59). As to the function φ, it should satisfy the conditions

$$\frac{\partial\varphi}{\partial n} = 0, \quad P \in \Sigma\,; \quad \frac{\partial\varphi}{\partial n} = \zeta_t, \quad P \in S,$$

where $z = \zeta(P, t)$ is the equation of the free surface in the coordinate system rigidly fixed to the rigid body.

Let now the position of the rigid body be determined by the six generalized coordinates $z_1, z_2, \ldots, z_6$. By analogy with (5–79), we write

$$\boldsymbol{v} = \nabla\varphi + \sum_{i=1}^{6} z_i' \nabla\varphi_i^*.$$

Then the expression for the kinetic energy of the system will be

$$T = \frac{1}{2}\sum_{i,j=1}^{6} m_{ij} z_i' z_j' + \frac{\varrho}{2}\int_\tau (\nabla\varphi)^2\, d\tau + \sum_{i=1}^{6} z_i' \varrho \int_\tau \nabla\varphi \nabla\varphi_i^*\, d\tau, \tag{5–80}$$

where

$$m_{ij} = m_{ij}^0 + \varrho \int_\tau \nabla\varphi_i^* \nabla\varphi_j^*\, d\tau,$$

and m_{ij}^0 is the matrix of the coefficients of the quadratic form, which represents the kinetic energy of the rigid body, the quantities $\varrho\int_\tau \nabla\varphi_i^*\nabla\varphi_j^* d\tau$ usually being called the associated masses. Using Zhukovskiy's terminology, we shall call quantities m_{ij} the masses of the equivalent rigid body.

Now we consider a conservative system K with n degrees of freedom. We assume that among the links of this system there exists a body with a cavity containing a fluid, and let z_i ($i=1, 2, \ldots, n$) be the generalized coordinates of this system. Then, without sacrificing generality, we can represent the kinetic energy of this system in the form (5–80), extending the summation from 1 to n.

We now set up the expression for the potential energy Π of the system K. We denote by K_0 a system in which the free surface is covered with a rigid partition. The potential energy of this system is Π_0; for small deviations from the equilibrium position it will represent a quadratic form of the variables z_i:

$$\Pi_0 = \tfrac{1}{2} \sum_{i,j=1}^{n} b_{ij} z_i z_j . \qquad (5\text{–}81)$$

We consider only the case when the equilibrium position of the system K_0 is stable. This means that the form Π_0 is positive definite. This assumption is quite important since, if the system K_0 is unstable, the deviations rapidly cease to be small and the entire theory becomes meaningless.

In the case when the fluid does not fill the entire cavity, the potential energy of the system K is composed of (5–81) and of the potential energy of the vibrating fluid. The latter, in turn, can be represented in the form of two components, one of which is the potential energy of the fluid vibrating in the stationary vessel, $\Pi_1 = \tfrac{1}{2}\varrho g \int_s \zeta^2 dS$. Since the fluid participates in the translational motion of the system, its potential energy should also be a function of the coordinates z_i; consequently, it should contain a term such as $\Pi_2 = \sum_{i=1}^{n} z_i \int_s f_i \zeta dS$, where the functions f_i are determined solely by the geometry of the cavity. Since $\int_s \zeta dS = 0$, without loss of generality, we can assume that $\int_s f dS = 0$.

Thus we will write the potential energy of the system K in the form

$$\Pi = \tfrac{1}{2} \sum_{i,j=1}^{n} b_{ij} z_i z_j + \sum_{i=1}^{n} z_i \int_S f_i \zeta dS + \tfrac{1}{2}\varrho g \int_S \zeta^2 dS .$$

[5-5.2] Canonical Variables; Equations of Motion

To simplify our further study, it is convenient to introduce new variables. Since the system K_0 is conservative and its equilibrium position is stable, this system has the principal coordinates Y_i. This means that there exists a linear transformation $z_i = \sum_{i,j=1}^{n} a_{ij} Y_j$ that simultaneously reduces forms Π_0 and

$T_0 = \frac{1}{2}\sum_{i,j=1}^{n} m_{ij} z_i' z_j'$ to the diagonal form, *i.e.*,

$$T_0 = \tfrac{1}{2}\sum_{i=1}^{n} Y_2'^2, \quad \Pi_0 = \tfrac{1}{2}\sum_{i=1}^{n} \mu_i^2 Y_i^2;$$

μ_k are the natural frequencies of the system when the free surface is covered. In terms of our new variables the expressions for the kinetic and potential energies will be

$$T = T_0 + \sum_{i=1}^{n} Y_i' \varrho \int_\tau \nabla\varphi\nabla\varphi_i^{**} d\tau + \tfrac{1}{2}\varrho \int_\tau (\nabla\varphi)^2 d\tau,$$

$$\Pi = \Pi_0 + \sum_{i=1}^{n} Y_i \int_S v_i \zeta dS + \tfrac{1}{2}\varrho g \int_S \zeta^2 dS. \tag{5–82}$$

Here φ_i^{**} and v_i are linear functions of φ_i^* and f_i. Eliminating φ and using Green's formula and the kinematic relationship, we can represent the first of expressions (5–82) in the form

$$T = T_0 + \sum_{i=1}^{n} Y_i' \varrho \int_S \varphi_i^{**} \zeta_t dS + \tfrac{1}{2}\varrho \int_S H\zeta_t \cdot \zeta_t dS. \tag{5–82'}$$

Using Hamilton's principle and repeating the considerations used in setting up the equations for the pendulum, we obtain the following system of equations for vibrations of the system K:

$$\begin{gathered} Y_i'' + \varrho \int_S \varphi_i^{**} \zeta_{tt} dS + \mu_i^2 Y_i + \int_S v_i \zeta dS = 0, \quad i = 1, 2, \ldots, n, \\ \varrho \sum_{j=1}^{n} \varphi_j^{**} Y_j'' + \varrho H \zeta_{tt} + \sum_{j=1}^{n} v_j Y_j + \varrho g \zeta = 0. \end{gathered} \tag{5–83}$$

A particular case of the above equations was system (5–68), which describes the oscillations of the pendulum. Reference [62] considers a number of specific problems, the solution of which reduces to the study of Eqs. (5–83).

[5-5.3] Free Vibrations

We now consider the problem of free vibrations. We set

$$Y_i = X_i \sin \sigma t, \quad \zeta = \psi \sin \sigma t.$$

Then we set up an expression for the functional L

$$L = \int_0^{t_1} (T - \Pi)\, dt,$$

where T and Π are defined by (5–82) and (5–82′), and we integrate from 0 to $2\pi/\sigma$. Dropping an inconsequential multiplier, we obtain

$$L = \sigma^2 \{\tfrac{1}{2} \sum_{i=1}^{n} X_i^2 + \varrho \sum_{i=1}^{n} X_i \int_S \varphi_i^{**} \psi dS + \tfrac{1}{2}\varrho \int_S H\psi \cdot \psi dS\} - \\ - \{\tfrac{1}{2} \sum_{i=1}^{n} \mu_i^2 X_i^2 + \sum_{i=1}^{n} X_i \int_S v_i \psi dS + \tfrac{1}{2}\varrho g \int_S \psi^2 dS\}. \tag{5–84}$$

Thus the problem of free vibrations of the system K reduces to determining the vector $\boldsymbol{X}$, the function ψ, and the number σ, which cause the variation of the functional (5–84) to vanish.

Let $\{\chi_s\}$ be a system of functions, complete in S, so that we can set

$$\psi = \sum_{n=1}^{N} \alpha_n \chi_n .$$

Setting up the equations for $\partial L/\partial X_i = 0$ and $\partial L/\partial a_s = 0$, we obtain the system of algebraic equations

$$(\sigma^2 - \mu_i^2) X_i + \sum_{s=1}^{N} (\sigma^2 A_{is} - B_{is})\, a_s = 0 ,$$
$$i = 1, 2, \dots, n ,$$
$$\sum_{i=1}^{n} (\sigma^2 A_{is} - B_{is}) X_i + \sum_{i=1}^{N} (\sigma^2 \alpha_{is} - \beta_{is})\, a_i = 0 , \tag{5–85}$$
$$s = 1, 2, \dots, N .$$

Here

$$A_{is} = \varrho \int_S \varphi_i^{**} \chi_s dS , \quad B_{is} = \int_S v_i \chi_s dS ,$$

$$\alpha_{is} = \varrho \int_S H \chi_i \chi_s dS , \quad \beta_{is} = \varrho g \int_S \chi_i \chi_s dS .$$

It is easy to prove that $\alpha_{is} = \alpha_{si}$, for which it is sufficient to apply Green's formula to the integral.

The system (5–85) is a system of homogeneous linear algebraic equations. In order for it to have nontrivial solutions, it is necessary and sufficient that σ^2 be a root of the equation

$$\Delta_N(\sigma^2) = 0 ,$$

where Δ_N is the determinant

$$\left|\begin{array}{ccccc}
\sigma^2-\mu_1^2 & 0 & \ldots & 0 & \sigma^2 A_{11}-B_{11}\ldots\sigma^2 A_{1N}-B_{1N} \\
0 & \sigma^2-\mu_2^2 & \ldots & 0 & \sigma^2 A_{21}-B_{21}\ldots\sigma^2 A_{2N}-B_{2N} \\
\cdots & \cdots & \cdots & \cdots & \cdots \\
\cdots & \cdots & \cdots & \cdots & \cdots \\
0 & 0 & \ldots & \sigma^2-\mu_n^2 & \sigma^2 A_{n1}-B_{n1}\ldots\sigma^2 A_{nN}-B_{nN} \\
\sigma^2 A_{11}-B_{11} & \sigma^2 A_{21}-B_{22} & \ldots & \sigma^2 A_{n1}-B_{n1} & \sigma^2\alpha_{11}-\beta_{11}\ldots\sigma^2\alpha_{1N}-\beta_{1N} \\
\sigma^2 A_{12}-B_{12} & \sigma^2 A_{22}-B_{22} & \ldots & \sigma^2 A_{n2}-B_{n2} & \sigma^2\alpha_{21}-\beta_{21}\ldots\sigma^2\alpha_{2N}-\beta_{2N} \\
\cdots & \cdots & \cdots & \cdots & \cdots \\
\cdots & \cdots & \cdots & \cdots & \cdots
\end{array}\right|.$$

Note that the matrix Δ_N is symmetrical. This means that the roots of the equation $\Delta_N=(\sigma_n^2)$ are real.

We noted in the preceding section that the problem is simplified if we take as χ_n the normalized eigenfunctions of the problem of free vibrations, since in this case

$$H\chi_n = \frac{g}{\omega_n^2}\chi_n,$$

$$(\chi_n, \chi_m) = \begin{cases} 0, & n \neq \mathrm{m}, \\ 1, & n = m, \end{cases}$$

then the form of system (5–85) will be

$$\begin{gathered}
(\sigma^2 - \mu_i^2) X_i + \sum_{s=1}^{N} (\sigma^2 A_{is} - B_{is}) a_s = 0, \\
i = 1, 2, \ldots, n, \\
\sum_{i=1}^{n} (\sigma^2 A_{is} - B_{is}) X_i + (\sigma^2 m_s - \beta_s) a_s = 0, \\
s = 1, 2, \ldots, N,
\end{gathered} \tag{5–86}$$

where $m_s=\varrho g/\omega_s^2$, $\beta_s=\varrho g$, and ω_s are the natural frequencies of vibration of the fluid in the vessel.

Effective calculation of the natural frequencies and modes of vibrations is quite laborious. However, there is a class of problems when the wave motion of the fluid has a "relatively small" effect on the system's vibration. This means that the n natural frequencies of the system K differ little from the numbers $\mu_1,\ldots,\mu_n$. This can occur when the coefficients of the terms responsible for the mutual effect of the vibrations are small. In this case it is possible to apply the procedures of perturbation theory.

Applying the Ritz method to the given problem, it is natural to take the eigenfunctions of the problem of vibrations of a fluid in a stationary vessel as the system of coordinate functions. Hence we shall consider the system (5–86), which we write in the form

$$
\begin{gathered}
(\sigma^2 - \mu_i^2)\, X_i + \varepsilon \sum_{s=1}^{N} (\sigma^2 A_{is} - B_{is})\, a_s = 0\,, \\
i = 1, 2, \ldots, n\,, \\
\varepsilon \sum_{i=1}^{n} (\sigma^2 A_{is} - B_{is})\, X_i + (\sigma^2 \alpha_s - \beta_s)\, a_s = 0\,, \\
s = 1, 2, \ldots, N\,.
\end{gathered} \tag{5–87}
$$

The above equations become Eqs. (5–86) when $\varepsilon = 1$. We shall seek the solution of (5–87) in the form

$$
X_i = \sum_{k=0}^{\infty} X_{ik} \varepsilon^k\,, \quad a_s = \sum_{k=0}^{\infty} a_{sk} \varepsilon^k\,, \quad \sigma^2 = \sum_{k=0}^{\infty} \sigma_k^2 \varepsilon^k\,. \tag{5–88}
$$

For the numbers X_{i0}, a_{s0}, and σ_0^k we will have the following system:

$$
\begin{gathered}
X_{i0}\, (\sigma_0^2 - \mu_i^2) = 0\,, \\
a_{s0}\, (\sigma_0^2 \alpha_s - \beta_s) = 0\,, \\
i = 1, 2, \ldots, n\,; \quad s = 1, 2, \ldots, N\,.
\end{gathered} \tag{5–89}
$$

The above system will have a nontrivial solution if $\sigma_0^2 = \mu_i^2$ or $\sigma_0^2 = \beta_s / \alpha_s$. We now raise the question of determining the frequencies of the system K that are close to the frequencies of system K_0. Hence we should set $\sigma_0 = \mu_i$. For definiteness, we set $\sigma_0 = \mu_1$; then $X_{20} = X_{30} = \cdots = a_{s0} = 0$. Without loss of generality, we can assume $X_{10} = 1$.

We have the following equations for X_i and a_{s1}:

$$
\begin{gathered}
X_{11}\, (\sigma_0^2 - \mu_1^2) = - \sigma_1^2\,, \\
X_{i1}\, (\sigma_0^2 - \mu_i^2) = 0\,, \quad i \neq 1\,, \\
a_{s1}\, (\sigma_0^2 \alpha_s - \beta_s) = - (A_{1s} \mu_1^2 - B_{1s})\,, \\
i = 1, 2, \ldots, n\,; \quad s = 1, 2, \ldots, N\,.
\end{gathered}
$$

The condition for solvability of the first equation of this system is

$$
\sigma_1^2 = 0\,.
$$

We can therefore always assume that $X_{11} = 1$. It is further evident that

$$
X_{i1} = 0 \quad (i \neq 1)\,, \quad a_{s1} = - \frac{A_{1s} \mu_1^2 - B_{1s}}{\mu_1^2 \alpha_s - \beta_s}\,.
$$

The numbers X_{12} and σ_2^2 satisfy the equation

$$
X_{12} (\sigma_0^2 - \mu_1^2) = - \sigma_2^2 - \sum_{s=1}^{N} a_{s1} (A_{1s} \mu_1^2 - B_{1s})\,. \tag{5–90}
$$

For solvability of this equation it is necessary and sufficient to select σ_2^2 in a manner such that the right-hand side of Eq. (5–90) becomes zero:

$$\sigma_2^2 = \sum_{s=1}^{N} \frac{(A_{1s}\mu_1^2 - B_{1s})^2}{\mu_1^2\alpha_s - \beta_s}.$$

Thus we can assume approximately that

$$\sigma^2 = \mu_1^2 + \varepsilon \sum_{s=1}^{N} \frac{(A_{1s}\mu_1^2 - B_{1s})^2}{\mu_1^2\alpha_s - \beta_s}.$$

The above computational scheme is recommended for a large number of engineering calculations.

Remark

It is sometimes expedient in applied problems to replace the integro-differential system of Eqs. (5–83) by an infinite system of ordinary differential equations. For this it is sufficient to set

$$\zeta = \sum_{n=1}^{\infty} q_n(t)\, \chi_n(x, y),$$

where $\{\chi_n\}$ is an orthonormal system of functions, complete on S. If we take as $\{\chi_n\}$ a system of eigenfunctions of the operator H, then the infinite system corresponding to (5–83) will be

$$\begin{gathered} Y_i'' + \sum_{j=1}^{\infty} q_j'' A_{ij} + \mu_i^2 Y_i + \sum_{j=1}^{\infty} q_j B_{ij} = 0, \\ i = 1, 2, \ldots, n, \\ \sum_{i=1}^{n} A_{ij} Y_i'' + m_j q_j'' + \sum_{i=1}^{n} B_{ij} Y_i + \beta_j q_j = 0, \\ j = 1, 2, \ldots. \end{gathered} \qquad (5\text{–}91)$$

The notation used in the above expression is the same as that used in Eqs. (5–86).

[5-5.4] Case of Many Cavities; Existence of Multiple Roots

The considerations that we have presented pertain to the case when the system has only one "fluid link." All of these results can be extended automatically to the general case of r fluid links. In this case the system of integro-differential equations (5–83) will have the form

$$Y_i'' + \sum_{j=1}^{r} \varrho_j \int_{S} \varphi_i^{(j)**} \zeta^{(j)} dS + \mu_i^2 Y_i + \sum_{j=1}^{r} \int_{S_j} \nu_i^{(j)} \zeta^{(j)} dS = 0, \quad i = 1, 2, \ldots, n,$$

$$\varrho_j \sum_{i=1}^{n} \varphi_i^{(j)**} Y_i'' + \varrho_j H_j \zeta_{tt}^{(j)} + \sum_{i=1}^{n} \nu_i^{(j)} Y_i + \varrho_j g \zeta^{(j)} = 0, \quad j = 1, 2, \ldots, r. \tag{5–92}$$

The subscript j here pertains to the jth cavity. Of importance for problems with many cavities is the question of the presence of multiple natural frequencies, since the cavities can be similar. Of course, this problem can also arise in the study of systems with one fluid link, but for systems with many cavities it is much more important. For example, the bunkers of tankerships have practically the same shape and consequently the same eigenvalues. Hence the possibility of internal resonance arises. On the basis of general theorems[31] we can claim that the elementary dividers in this problem will always be mutually simple and no resonance phenomena will take place.

Equations (5–92) can be studied by reducing them to the system of ordinary equations (5–91) and by applying general methods. However, this is frequently not necessary and it is possible, because of the specifics of these equations, to use an elementary method to circumvent the difficulties resulting from the presence of multiple frequencies.

If the system has identical cavities that are equally filled, then its characteristic equation will have multiple roots. To prove this we first reduce our system (5–92) to a system of ordinary differential equations. Since we always have the freedom to alter the sequence in which the principal coordinates are numbered, it is sufficient to consider the system (5–91). The special feature of this system will now be that it contains equations in which the quantities m_j and β_j are identical. We replace Eqs. (5–91) by a finite system and write the corresponding algebraic system

$$X_i(\mu_i^2 - \sigma^2) + \sum_{j=1}^{N} (B_{ij} - \sigma^2 A_{ij})\, a_j = 0, \quad i = 1, 2, \ldots, n,$$

$$\sum_{i=1}^{n} (B_{ij} - \sigma^2 A_{ij})\, X_i + (\beta_j - m_j \sigma^2)\, a_j = 0, \quad j = 1, 2, \ldots, N,$$

or, eliminating a_j,

$$X_i(\mu_i^2 - \sigma^2) + \sum_{j=1}^{N} \sum_{s=1}^{n} \frac{(B_{ij} - \sigma^2 A_{ij})(B_{sj} - \sigma^2 A_{sj})}{\beta_j - m_j \sigma^2} X_s = 0,$$

$$i = 1, 2, \ldots, n.$$

This system can additionally be rewritten in the form

$$X_i(\mu_i^2 - \sigma^2)\, Q + \sum_{j=1}^{N} \sum_{s=1}^{n} (B_{ij} - \sigma^2 A_{ij}) \times (B_{sj} - \sigma^2 A_{sj})\, Q_j X_s = 0, \qquad (5\text{–}93)$$

where

$$Q = \prod_{j=1}^{N} (\beta_j - m_j\sigma^2), \quad Q_j = \prod_{r \neq j}^{N} (\beta_r - m_r\sigma^2).$$

It follows immediately from the form of the system (5–93) that if among the multipliers $(\beta_r - m_r\sigma^2)$ there exist at least two identical ones, then one of the roots of the frequency equation will be $\sigma^2 = \omega_r^2 = \beta_r/m_r$.

Thus we have the following result. If, among the natural frequencies of vibration of the fluid in the cavities, there is at least one multiple frequency (because of the properties of the tank, or because of the presence of similar cavities) with multiplicity k, then this frequency will belong to the spectrum of the body-fluid system and the multiplicity of this frequency will be $k-1$. Thus, among the principal vibrations of a system with fluid links, there exists a principal vibration with this frequency.

This principal frequency can always be separated by means of a linear transformation. For example, among the numbers m_j and β_j let there be two identical $m_{j1} = m_{j2} = m$, and $\beta_{j1} = \beta_{j2} = \beta$. We now write expressions for the kinetic and potential energies:

$$T = \tfrac{1}{2} \sum_{i=1}^{n} Y_i'^2 + \sum_{i=1}^{n} A_{ij_1} Y_i' q_{j_1}' + \sum_{i=1}^{n} A_{ij_2} Y_i' q_{j_2} + \frac{m}{2} (q_{j_1}'^2 + q_{j_2}'^2) + T^*,$$

$$\Pi = \tfrac{1}{2} \sum_{i=1}^{n} \mu_i^2 Y_i^2 + \sum_{i=1}^{n} B_{ij_1} Y_i q_{j_1} + \sum_{i=1}^{n} B_{ij_2} q_{j_2} + \frac{\beta}{2} (q_{j_1}^2 + q_{j_2}^2) + \Pi^*,$$

where T^* and Π^* are the contributions made to the potential and kinetic energies by the remaining generalized coordinates q_j. We can always isolate the principal vibration of interest by a substitution of the kind

$$q_{j_1} = dp_{j_1} + p_{j_2} + \sum Y_i,$$
$$q_{j_2} = p_{j_1} - dp_{j_2} + b \sum Y_i,$$

where the constants b and d are easy to determine. For example, if the system has only one generalized coordinate Y_i,

$$d = \frac{A_{j_2}\beta - mB_{j_2}}{mB_{j_1} - BA_{j_1}},$$

$$b = \frac{A_{j_1}B_{j_2} + mB_{j_2} - A_{j_2}\beta - A_{j_2}B_{j_1}}{mB_{j_1} - \beta A_{j_1}}.$$

After this change of variables, the expressions for T and Π will assume the form

$$T = T^{**} + \frac{m}{2} p_1'^2,$$

$$\Pi = \Pi^{**} + \frac{\beta}{2} p_1^2,$$

where T^{**} and Π^{**} do not depend on p_1. Setting up Lagrange's equations for this system, we find that the variable p_1 satisfies the equation

$$mp_1'' + \beta p_1 = 0. \tag{5–94}$$

The case when $A_{j1} = A_{j2}$ and $B_{j1} = B_{j2}$ is particularly simple, so that

$$p_1 = q_{j_1} - q_{j_2}.$$

In this case it is easy to represent graphically the vibrations corresponding to multiple frequencies. Both vibrations will be waves with the same amplitude, but of opposite phase, which would not change the momentum of the system.

Remark

Multiple roots in the characteristic equation can also arise when the natural frequencies are different. Let us consider one such case.

Writing the system of equations in the form of (5–93), and assuming that $n=1$, it is easy to see that the characteristic equation of this system will have multiple roots if

$$\frac{\beta_k}{m_k} = \frac{B_k}{A_k}. \tag{5–95}$$

In this case it is also easy to isolate the principal vibration, which should have the frequency $\omega_k = \sqrt{\beta_k/m_k}$.

In order to show that this is so, we now write the expressions for the kinetic and potential energies as

$$T = A_k q_k' Y' + \tfrac{1}{2} m_k q_k'^2 + T^*,$$
$$\Pi = B_k q_k Y + \tfrac{1}{2} \beta_k q_k^2 + \Pi^*,$$

where T^* and Π^* are the components of the kinetic and potential energies that do not depend on q_k. We now change variables:

$$q_k = \frac{A_k}{m_k}(p - Y) = \frac{B_k}{\beta_k}(p - Y),$$

after which T and Π will become

$$T = \frac{A_k^2}{2m_k} p'^2 + T^{**},$$

$$\Pi = \frac{B_k^2}{2\beta_k} p^2 + \Pi^{**},$$

where T^{**} and Π^{**} are independent of p.

Setting up Lagrange's equation, we find that the variation in p does not depend on variations in the remaining generalized coordinates of the system and is described by the expression

$$p'' + \Omega^2 p = 0,$$

where

$$\Omega^2 = \frac{B_k^2 m_k}{\beta_k A_k^2}.$$

Using (5–95), we obtain

$$\Omega^2 = \frac{B_k}{A_k} = \frac{\beta_k}{m_k} = \omega_k^2.$$

We have thus isolated the principal vibration with frequency ω_k, which is not excited by the vibration of the remaining system and does not, in turn, affect its vibrations.

[5-6] PROBLEM OF TORSIONAL OSCILLATIONS OF A BEAM WITH A FLUID-CONTAINING CAVITY

[5-6.1] Preliminary Remarks

The problem of oscillations of an elastic body containing a fluid is extremely complex. A relatively simple theory can be developed only for the case of a beam. We shall use this term to denote an elastic body that satisfies the hypothesis of plane sections. We shall consider in detail the problem of torsional oscillations of a beam and then write a general equation for arbitrary flexural-torsional oscillations.

We fix the coordinate system to the neutral axis, as is shown in Fig. 25

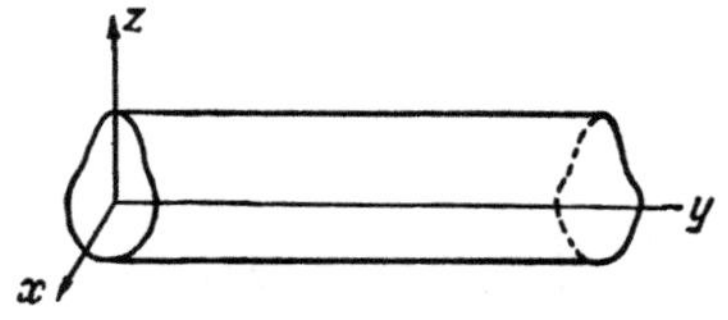

Fig. 25.

(Oy axis). We denote the angle of twist by $\theta(y, t)$. Then, under our assumptions, the kinetic and potential energies of the beam containing no fluid will be

$$T_0 = \frac{1}{2} \int_0^l I(y)\, \theta_t^2 dy\,,$$
$$\Pi_0 = \frac{1}{2} \int_0^l \left[k(y)\, \theta_y^2 + B\theta^2\right] dy\,. \tag{5–96}$$

Here $I(y)$ is the moment of inertia per unit length, $k(y)$ is the torsional rigidity, l is the length of the beam, and B is a characteristic of the moment of the external force. We consider simultaneously the ordinary differential equation

$$\frac{d}{dy}\left[k(y)\frac{du}{dy}\right] - Bu = -\,\mu^2 I(y)\, u\,.$$

The function u is subject to boundary conditions that will make the operators considered self-adjoint. For example, we can assume $u(0)=u'(l)=0$ (the beam is rigidly supported at both ends). The term boundary-value problem for the function u will be used to denote the problem associated with torsional oscillations.

We further denote the eigenfunctions and eigenvalues of this problem by u_n and μ_n. If we assume the normalizing condition in the form

$$\int_0^l I(y)\, u_n^2(y)\, dy = 1\,,$$

then the functions u_n will have the property

$$\int_0^l I(y)\, u_n u_m dy = 0\,, \quad n \neq m\,,$$
$$\int_0^l \left[k(y)\, u_n' u_m' + B u_n u_m\right] dy = \begin{cases} 0\,, & n \neq m\,, \\ \mu_n^2\,, & n = m\,. \end{cases} \tag{5–97}$$

We now return to the problem of oscillation of the beam. We set

$$\theta(y, t) = \sum_{n=1}^{N} \cos \omega t a_n u_n\,,$$

form the expression

$$L \doteq \int_0^{t_1} (T - \Pi)\, dt\,,$$

and integrate from 0 to $2\pi/\omega$. As a result we obtain

$$L = \frac{\pi}{2\omega} \sum_{n=1}^{N} \sum_{m=1}^{N} a_n a_m (\omega^2 \alpha_{nm} - \gamma_{nm}),$$

where

$$\alpha_{nm} = \int_0^l I u_n u_m dy, \quad \gamma_{nm} = \int_0^l [k u_n' u_m' + B u_n u_m]\, dy$$

or, on the basis of (5–97),

$$L = \frac{\pi}{2\omega} \sum_{n=1}^{N} a_n^2 (\omega^2 - \mu_n^2),$$

whence it follows that the natural frequencies of the beam will be equal to the eigenvalues of the associated boundary-value problem.

Remark

The utilization of the eigenfunctions discussed above is particularly convenient if the beam parameters do not change over its span. If they vary slowly, it is expedient to use their asymptotic representations

$$u_n = C \sin\left(\mu_n \int_0^y \sqrt{\frac{I}{k}}\, dy\right) \div \sqrt[4]{Ik}, \tag{5–98}$$

where C is an arbitrary constant selected from the conventional normalization, and

$$\mu_n = \frac{n\pi}{\int_0^l \sqrt{I/k}\, dy}.$$

Equation (5–98) corresponds to the problem of oscillation of an aircraft wing (a beam rigidly supported at the left end $u(0)=0$ and free at the right end $u'(l)=0$).

[5-6.2] Torsional Oscillations of a Beam with a Fluid-Containing Cavity

We now consider the torsional oscillations of a beam containing a fluid, assuming that the cavity is completely filled (see Fig. 26 for the notation). Since the only external force acting on the fluid is, by assumption, gravity, the potential energy of the beam-fluid system can be taken in the form of (5–96). The kinetic energy of the system should be taken in the form

$$T = T_0 + \tfrac{1}{2}\varrho \int_\tau (\nabla\varphi)^2\, d\tau,$$

where φ is the potential of the absolute motion of the fluid.

We set

$$\theta = \cos \omega t \vartheta(y),$$
$$\varphi = -\omega \sin \omega t \Phi(x, y, z)$$

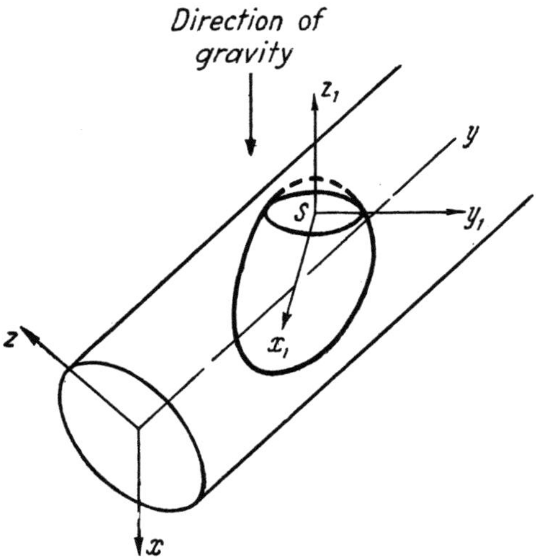

Fig. 26.

and set up an expression for the functional L,

$$L = \frac{\pi}{2\omega} \{\omega^2 [\int_0^l I\vartheta^2 dy + \varrho \int_\tau (\nabla\varphi)^2 d\tau] - \int_0^l [k\vartheta_y^2 + B\vartheta^2] dy\}. \qquad (5\text{–}99)$$

On the surface Σ this potential satisfies the conditions

$$\frac{\partial\varphi}{\partial n} = v_n = (\theta_t \boldsymbol{y}^0 \times \boldsymbol{r}) \cdot \boldsymbol{n}^0.$$

Introducing the notation

$$\boldsymbol{n}^0 = \alpha\boldsymbol{x}^0 + \beta\boldsymbol{y}^0 + \gamma\boldsymbol{z}^0,$$
$$\boldsymbol{r} = x\boldsymbol{x}^0 + y\boldsymbol{y}^0 + z\boldsymbol{z}^0,$$

we obtain

$$\frac{\partial\varphi}{\partial n} = \theta_t f(x, y, z),$$

where

$$f(x, y, z) = z\alpha - x\gamma.$$

The function f is determined solely by the geometry of the cavity. According to the notation used

$$\frac{\partial\Phi}{\partial n} = \vartheta f.$$

Further, to solve the problem by the Ritz method it is necessary to select a system of functions u_n that are complete on $[0, l]$, and set

$$\vartheta = \sum_{i=1}^{N} a_n u_n .$$

(For u_n it is expedient in many cases to select the eigenfunctions of the associated boundary-value problem.)

In addition, we set

$$\Phi = \sum_{i=1}^{N} a_n \Phi_n ,$$

where the Φ_n are functions that are harmonic in τ and are defined as the solutions of the Neumann problems

$$\frac{\partial \Phi_n}{\partial n} = u_n f$$

or

$$\Phi = H u_n(P) f(P), \quad P \in \Sigma .$$

Here $H\psi$ denotes a function harmonic in τ, the normal derivative of which on the surface of the volume τ is ψ. Now the algebraic system defining a_n and ω^2 will be

$$\sum_{m=1}^{N} a_m [\omega^2 (\alpha_{nm} + \beta_{nm}) - \gamma_{nm}] = 0, \quad n = 1, 2, \ldots, N,$$

where α_{nm} and γ_{nm} were defined above, and

$$\beta_{nm} = \varrho \int_{\tau} \nabla \Phi_n \nabla \Phi_m d\tau .$$

In addition, we set up the equation for the oscillations of a beam containing a fluid, using Hamilton's principle. We evaluate

$$\delta L = \int_0^{t_1} \int_0^{l} \{I\theta_t \delta\theta_t - k\theta_y \delta\theta_y - B\theta\delta\theta\} \, dy dt + \varrho \int_0^{t_1} \int_{\tau} \nabla H\theta_t f \cdot \nabla H \delta\theta_t f d\tau dt .$$

Integrating the first term by parts (making use of the fact that the variations become zero at the ends) and applying Green's formula to the second term, we obtain

$$\delta L = - \int_0^{t} \int_0^{l} \{I\theta_{tt} - (k\theta_y)_y + B\theta\} \, \delta\theta dy dt + \varrho \int_0^{t} \int_{\Sigma} H\theta_t f \frac{\partial}{\partial n} H f \delta\theta d\Sigma dt .$$

By definition of the operator H,

$$\frac{\partial}{\partial n} H u = u ;$$

hence the second integral will have the form

$$\varrho \int_0^t \int_\Sigma H\theta_t f \cdot f\,\delta\theta_t d\Sigma dt\,.$$

We denote the contour bounding the plane figure obtained on cutting the cavity by a plane perpendicular to the Oy axis by l_y, so that the second integral can also be represented in the form

$$\varrho \int_{y_1}^{y_2} dy \int_0^t \delta\theta_t \int_{l_y} fH\theta_t f dl dt$$

or, after integration by parts,

$$-\varrho \int_0^t \int_{y_1}^{y_2} \int_{l_y} fHf\theta_{tt} \cdot \delta\theta dl dy dt\,.$$

The meaning of the numbers y_1 and y_2 is evident from the drawing (see Fig. 26). We can always integrate with respect to y from 0 to l since, if $y<y_1$ or $y>y_2$, then $l_y=0$.

If the cavity has plane ends (for example, a cavity in the form of a cylinder), this formula remains valid, since the integral over the ends is zero.

Collecting all of the expressions obtained and making use of the fact that the variations are arbitrary, we arrive at the following expression describing the torsional oscillations of a beam containing a cavity entirely filled with fluid:

$$I\theta_{tt} + \varrho \int_{l_y} fHf\theta_{tt} dl - (k\theta_y)_y + B\theta = 0\,. \tag{5–100}$$

If we set $\varrho \equiv 0$ then the above expression will become the well-known equation for torsional oscillations of a beam.

One the main results of the theory of motion of a rigid body contains a fluid that entirely fills the cavity is that this motion is equivalent to the motion of some other rigid body under the action of the same forces. Equation (5–100) shows that it is not possible to find a beam not containing a fluid whose motion would be equivalent to the motion of a beam containing a fluid. In fact, if in some specified cross section of a beam without a fluid the angle θ and the quantity θ_y are specified, the acceleration of the particles in this section is uniquely determined. If the beam contains a fluid, the acceleration in the section also depends on the angular accelerations of all the other cross sections. This effect is defined by the term

$$\int_{l_y} fHf\theta_{tt} dl_y = \varrho \int_{l_y} f(P_y) \int_\Sigma H(P_y, P)\, f(P)\, \theta_{tt}(P, t)\, dPdP_y\,,$$

where P is a point on the surface Σ of the cavity, and $H(P, Q)$ is Green's function of Neumann's problem for the region τ.

Let us now consider the problem of torsional oscillations of a beam inside which there is a cavity only partially filled with a fluid. A number of new features arise in this case. Equation (5–100) shows that when the cavity is entirely filled with fluid its role reduces to that of modifying the natural frequencies and modes of the principal vibrations. When the mass of the fluid goes to zero, the vibrational modes approach the frequencies and vibrational modes of a beam without a fluid. The situation is entirely different when the fluid has a free surface. In this case new vibrational modes arise due to the vibration of the free surface.

We fix the coordinate system $Ox_1y_1z_1$ to the free surface at rest, as is shown in Fig. 26. The velocity potential φ can be represented as

$$\varphi = \varphi^{(1)} + \varphi^{(2)},$$

where $\varphi^{(1)}$ is the velocity potential for the velocity that the fluid would have if the surface S were not free,

$$\varphi^{(1)} = Hv_n(P), \quad P \in \Sigma + S;$$

i.e., $\varphi^{(1)}$ is the velocity potential considered in the preceding section. Furthermore, $\varphi^{(2)}$ satisfies the conditions

$$\frac{\partial \varphi^{(2)}}{\partial n} = 0, \quad P \in \Sigma, \quad \frac{\partial \varphi^{(2)}}{\partial n} = \zeta_t - v_n, \quad P \in S,$$

where ζ is the amount by which the free surface exceeds the undisturbed level in the absolute coordinate system

$$\varphi^{(2)} = H^*(\zeta_t - v_n) = H^* u_t,$$

where $u = \zeta - \delta_n$, and δ_n is the displacement of the points of the surface resulting from elastic deformations. The expression H^*f defines a function that is harmonic in τ and that has a normal derivative that is zero on the wetted part of the surface and is equal to f on S. Thus the kinetic energy of the fluid can be represented in the form

$$T^* = \tfrac{1}{2}\varrho \int_\tau [\nabla H v_n + \nabla H^* u_t]^2 \, d\tau .$$

We calculate the potential energy of the fluid

$$\Pi = \tfrac{1}{2}\varrho g \int_\tau z_1 dx_1 dy_1 dz_1 = \tfrac{1}{2}\varrho g \Big[\int_{\tau_0} z_1 dx_1 dy_1 dz_1 + \int_{\tau_1} z_1 dx_1 dy_1 dz_1\Big],$$

where τ_0 is the volume occupied by the fluid in the equilibrium position, and τ_1 is the volume bounded by the free surface and the plane S. The last term has the form

$$\tfrac{1}{2}\varrho g \int_S \zeta^2 dS = \tfrac{1}{2}\varrho g \int_S u^2 dS + \varrho g \int_S u\delta_n dS + \tfrac{1}{2}\varrho g \int_S \delta_n^2 dS .$$

We note that the term

$$\tfrac{1}{2}\varrho g \left[\int_S \delta_n^2 dS + \int_{\tau_0} z_1 dx_1 dy_1 dz_1\right]$$

is determined solely by the elastic displacements and has the same structure as the potential energy of the external forces. Hence, without loss of generality, we can include this term in the potential energy of the external forces.

Thus we will write the kinetic and potential energies of the beam-fluid system in the form

$$T = \tfrac{1}{2}\int_0^l I(y)\,\theta_t^2 dy + \tfrac{1}{2}\varrho \int_\tau \{(\nabla H^* u_t)^2 + 2\nabla H\theta_t f\cdot\nabla H^* u_t + (\nabla H\theta_t f)^2\}\, d\tau , \tag{5–101}$$

$$\Pi = \tfrac{1}{2}\int_0^l [k\theta_y^2 + B\theta^2]\, dy + \varrho g \int_S u f\theta dS + \tfrac{1}{2}\varrho g \int_S u^2 dS . \tag{5–102}$$

The integrals that are functions of δ_n in these two equations are written in the $Ox_1y_1z_1$ coordinate system.

Correspondingly, the system of differential equations will be

$$\begin{gathered} I(y)\,\dot{\theta}_{tt} + \varrho \int_{l_y} fH\theta_{tt} f df_y + \varrho \int_{d_y} fH^* u_{tt} dd_y - (k\theta_y)_y + B\theta + \varrho g \int_{d_y} fudd_y = 0 , \\ \varrho H\theta_{tt} f + \varrho H^* u_{tt} + \varrho g f\theta + \varrho g u = 0 . \end{gathered} \tag{5–103}$$

Thus the oscillation of a beam containing a fluid is described by a system of two integro-differential equations. Here d_y denotes the segment that lies at the intersection of the figure S and the plane of the normal cross section, the coordinate of which is y.

Remark

The Ritz method is recommended for calculating principal vibrations and natural frequencies. If the frequencies of the natural vibrations of the fluid in a stationary vessel and of the torsional oscillations of the beam are not close to one another, the spectrum of the combined problem of oscillation of a beam containing a fluid will be close to the superposition of the spectra of both problems. We should select two systems of coordinate functions. In the given case it is expedient to select the system of the eigenfunctions of operator H as one of the systems. The second system should be the system of eigenfunctions of the associated boundary-value problem. To calculate the spectrum of the combined problem, it is expedient to use the methods of the theory of perturbations presented in the preceding section.

[5-7] EQUATIONS OF TORSIONAL-FLEXURAL OSCILLATIONS OF A FLUID-CONTAINING BEAM

We conclude this chapter by writing the general equations for the torsional-flexural oscillations of a beam containing a fluid.

We denote the deflection in the xOy plane by X, and the deflection in the yOz plane by Z. Then the kinetic and potential energies of the beam (without the fluid) are described by the following formulas, which are well known from the elementary course on strength of materials:

$$\begin{aligned} T_0 &= \tfrac{1}{2}\int_0^l \{A_{11}X_t^2 + 2A_{12}X_tZ_t + 2A_{13}X_t\theta_t + A_{22}Z_t^2 + \\ &\qquad + 2A_{23}Z_t\theta_t + A_{33}\theta_t^2\}\, dy\,, \\ \Pi_0 &= \tfrac{1}{2}\int_0^l \{c_1X_{yy}^2 + c_2Z_{yy}^2 + c_3\theta_y^2 + B_{11}X^2 + \\ &\qquad + 2B_{12}XZ + 2B_{13}X\theta + B_{22}Z^2 + 2B_{23}Z\theta + B_{33}\theta^2\}\, dy\,. \end{aligned} \tag{5–104}$$

Here c_i is the torsional and flexural rigidity, the B_{ij} characterize the external conservative forces ($B_{ij}=B_{ji}$), the A_{ij} are the masses per unit length or the moments of inertia per unit length [in the preceding section, $I(y)$], and the A_{ij} (where $i\neq j$) are functions characterizing the fact that the neutral axes and the axes of the center of gravity of the cross sections do not coincide.

If the beam has a cavity containing a fluid with a free surface, the kinetic and potential energies of the system are written in the form

$$\begin{aligned} T &= T_0 + \tfrac{1}{2}\varrho\int_\tau \{\nabla HX_t\gamma_1 + \nabla HZ_t\gamma_2 + \nabla H\theta_t\gamma_3 + \nabla H^*u_t\}^2\, d\tau\,, \\ \Pi &= \Pi_0 + \tfrac{1}{2}\varrho g\int_S u^2 dS + \varrho g\int_S u\,[X\gamma_1 + Z\gamma_2 + \theta\gamma_3]\, dS\,. \end{aligned} \tag{5–105}$$

Here

$$\gamma_1 = \cos(nx)\,, \quad \gamma_2 = \cos(nz)\,, \quad \gamma_3 = z\cos(nx) - x\cos(nz)\,.$$

We recall that $u=\zeta-\delta_n$, where $\delta_n=X\gamma_1+Z\gamma_2+\theta\gamma_3$ is the elastic displacement of points of the plane S. Applying Hamilton's principle, we can now easily set up the general equations for the vibrations of the beam-fluid system:

$$
\left.
\begin{aligned}
&A_{11}X_{tt} + \varrho \int\limits_{l_y} \gamma_1 H \gamma_1 X_{tt} dl_y + A_{12}Z_{tt} + \\
&\quad + \varrho \int\limits_{l_y} \gamma_1 H \gamma_2 Z_{tt} dl_y + A_{13}\theta_{tt} + \varrho \int\limits_{l_y} \gamma_1 H \gamma_3 \theta_{tt} dl_y + \\
&\quad + \varrho \int\limits_{l_y} \gamma_1 H^* u_{tt} dl_y + (c_1 X_{yy})_{yy} + B_{11}X + B_{12}Z + \\
&\quad + B_{13}\theta + \varrho g \int\limits_{d_y} \gamma_1 u dd_y = 0, \\
&A_{12}X_{tt} + \varrho \int\limits_{l_y} \gamma_2 H \gamma_1 X_{tt} dl_y + A_{22}Z_{tt} + \\
&\quad + \varrho \int\limits_{l_y} \gamma_2 H \gamma_2 Z_{tt} dl_y + A_{23}\theta_{tt} + \varrho \int\limits_{l_y} \gamma_2 H \gamma_3 \theta_{tt} dl_y + \\
&\quad + \varrho \int\limits_{l_y} \gamma_2 H^* u_{tt} dl_y + (c_2 Z_{yy})_{yy} + B_{21}X + B_{22}Z + \\
&\quad + B_{23}\theta + \varrho g \int\limits_{d_y} \gamma_2 u dd_y = 0, \\
&A_{13}X_{tt} + \varrho \int\limits_{l_y} \gamma_3 H \gamma_1 X_{tt} dl_y + A_{23}Z_{tt} + \\
&\quad + \varrho \int\limits_{l_y} \gamma_3 H \gamma_2 Z_{tt} dl_y + A_{33}\theta_{tt} + \varrho \int\limits_{l_y} \gamma_3 H \gamma_3 \theta_{tt} dl_y + \\
&\quad + \varrho \int\limits_{l_y} \gamma_3 H^* u_{tt} dl_y - (c_3 \theta_y)_y + B_{13}X + B_{23}Z + B_{33}\theta + \\
&\quad + \varrho g \int\limits_{d_y} \gamma_3 u dd_y = 0, \\
&\varrho H \gamma_1 X_{tt} + \varrho H \gamma_2 Z_{tt} + \varrho H \gamma_3 \theta_{tt} + \varrho H^* u_{tt} + \varrho g \gamma_1 X + \\
&\quad + \varrho g \gamma_2 Z + \varrho g \gamma_3 \theta + \varrho g u = 0.
\end{aligned}
\right\} \tag{5–106}
$$

As a particular case of this system, we have the system (5–103). For this it is sufficient in (5–106) to set $X = Z = 0$. In precisely the same manner we can consider the system (5–83) as a particular case of the above system. For this it is sufficient, for example, to set $X = Z = 0$ and $c_1 = 0$, and to regard θ as an n-dimensional vector.

6

General Properties of Equations of Vibrations of Bodies Containing a Fluid

In this chapter we shall consider the general properties of the equations introduced in the preceding chapter.

The structure of the spectrum, the existence of natural vibrations, and the feasibility of using the Ritz method are clarified for the case of conservative external forces. In Sec. 4 we prove the principal theorem of this theory: that for stability of equilibrium of the body-fluid system it is necessary and sufficient to have stability of some other, appreciably simpler system. The body is considered either perfectly rigid or elastic, but conforming to the hypothesis of plane sections (a beam). In Sec. 5 are considered some general problems of the theory of impact on a body whose cavities contain a fluid with a free surface.

The problem of vibrations of a body floating in a water reservoir of limited dimensions has a good deal in common with the problem of vibrations of a body containing a fluid with a free surface. The last section is devoted to an analysis of this analogy.

[6-1] VIBRATIONS OF A FLUID CONFINED IN A VESSEL

It was shown in Sec. 1 of the preceding chapter that the motion of the fluid is described by the equation

$$g\zeta + H\zeta_{tt} = 0. \tag{5-10}$$

To find the natural frequencies, we set $\zeta = \cos\sigma t\psi(x, y)$, after which we obtain

$$\lambda H\psi = \psi, \quad \lambda = \frac{\sigma^2}{g}. \tag{6-1}$$

It was recommended that the Ritz method be used for finding solutions of the above equation. However, the questions of solvability of this problem, structure of the spectrum, completeness of the system of eigenfunctions, and convergence of this method as $N \to \infty$ (N being the number of coordinate functions) have remained unresolved. These problems are solved in an elementary manner on the basis of the general theorems of functional analysis. For this it is sufficient to show that the operator H has certain properties. We have proved above that H is a bounded, completely continuous self-adjoint operator. This operator is defined on a set of functions that satisfy the condition $\int_S f(P)\,dP = 0$.

We now determine the Hilbert space E of the functions $f(P)$ with the scalar product

$$(f_1, f_2) = \int_S f_1 f_2 dS.$$

Then the operator H is defined on the subspace E, which is orthogonal to unity.

We now show that H is positive on this set. We consider the scalar product

$$(Hf, f) = \int_S \int_S H(P, Q)\, f(P)\, f(Q)\, dP dQ. \tag{6–2}$$

Here $H(P, Q)$ is the Green's function of Neumann's problem for the region τ. By definition, the inner integral

$$u(Q) = \int_S H(P, Q)\, f(P)\, dP$$

is a function that is harmonic in τ and such that

$$\frac{\partial u}{\partial n} = \begin{cases} 0, & Q \in \Sigma, \\ f(Q), & Q \in S. \end{cases}$$

Consequently,

$$(Hf, f) = \int_S u(Q) \frac{\partial u}{\partial n}\, dQ,$$

whence according to Green's formula

$$(Hf, f) = \int_\tau (\nabla u)^2\, d\tau.$$

The above equality proves the claim, since its right-hand side becomes zero if and only if $u \equiv 0$ or, which is the same thing, when $f \equiv 0$.

The mechanical meaning of this assertion is entirely clear. The scalar product (6–2) defines the kinetic energy of the vibrating fluid if the velocity of points of the free surface is given by $f(P) \sin \sigma t$. It is obvious that the kinetic energy of a fluid performing potential motion in a stationary vessel can be zero (for any t) if and only if $f \equiv 0$.

Thus the operator H is bounded in E, completely continuous, self-adjoint and positive. The known properties of these operators (see, for example, Ref. [65]) allow us to formulate the following theorems.

Theorem I

The motion of a fluid about the equilibrium position in a vessel of finite size is accompanied by natural vibrations, *i.e.*, by solutions in the form $\zeta = \cos \sigma t \psi(x, y)$.

Theorem II

The eigenvalues ψ are positive and of finite multiplicity, and they form the sequence

$$\lim_{n \to \infty} \sigma_n = \infty$$

which increases without bounds.

Theorem III

The eigenfunctions ψ_n of the operator H, which describe the principal modes of free vibration, are such that the sequence

$$1, \psi_1, \ldots, \psi_n, \ldots$$

is complete in E.

Theorem IV

The eigenvalues and eigenfunctions can be determined by the Ritz method.

These theorems exhaust the content of all of the theoretical questions that arise in problems of free vibrations of a fluid with infinitesimal amplitudes.

[6-2] EQUATIONS OF VIBRATIONS OF A CONSERVATIVE SYSTEM WITH A FLUID LINK

[6-2.1] Reduction to an Operational Equation

In this section we consider certain general properties of the system of integro-differential equations

$$\begin{aligned} &Y_i'' + \varrho \int_S \varphi_i^{**} \zeta_{tt} dS + u_i^2 Y_i + \int_S v_i \zeta dS = 0\,, \quad i = 1, 2, \ldots, n\,, \\ &\varrho \sum_{j=1}^{n} \varphi_j^{**} Y_j'' + \varrho H \zeta_{tt} + \sum_{j=1}^{n} v_j Y_j + \varrho g \zeta = 0\,, \end{aligned} \tag{5–83}$$

which describes the vibrations of a mechanical system, among the links of which there is a rigid body with cavities partially filled with a fluid. These problems were analyzed in Ref. [78], and here we shall follow the presentation of this paper. First we write Eqs. (5–83) in the operational form. We introduce an n-dimensional space E_0 of vectors Y with components Y_i $(i=1, 2, \ldots, n)$ and a space E of functions $\zeta(P)$ $(P \in S)$ with an integrable square in S. The scalar products are defined as follows:

$$\text{in} \quad E_0\colon \quad (Y, Z)_0 = \sum_{i=1}^{n} Y_i Z_i ,$$

$$\text{in} \quad E\colon (\zeta, \eta) = \int_S \zeta(P)\, \eta(P)\, dP .$$

In addition, we introduce the space $\mathscr{E} = E_0 + E$. The element $x \in \mathscr{E}$ is a vector with components $Y \in E_0$ and $\zeta \in E$, and the scalar product in $\mathscr{E}$ is defined as

$$(x^{(1)}, x^{(2)}) = (Y^{(1)}, Y^{(2)})_0 + (\zeta^{(1)}, \zeta^{(2)}) .$$

We introduce the following operators:

a) Operating in E_0 (unit operator L_{00} and operator M_{00}):

$$L_{00} = \begin{Vmatrix} 1 & 0 & 0 & \ldots & 0 \\ 0 & 1 & 0 & \ldots & 0 \\ 0 & 0 & 1 & \ldots & 0 \\ \cdots & & & & \\ 0 & 0 & 0 & \ldots & 1 \end{Vmatrix}, \quad M_{00} = \begin{Vmatrix} \mu_1 & 0 & 0 & \ldots & 0 \\ 0 & \mu_2 & 0 & \ldots & 0 \\ 0 & 0 & \mu_3 & \ldots & 0 \\ \cdots & & & & \\ 0 & 0 & 0 & \ldots & \mu_n \end{Vmatrix} .$$

b) Operating in E:

$$L_{11} = H , \quad M_{11} = \varrho g .$$

c) Operating from E_0 into E:

$$L_{10} Y = \varrho (\varphi^{**}, Y)_0 , \quad M_{10} Y = (v, Y)_0 .$$

Here φ^{**} and v denote vectors that have projections φ_i^{**} and v_i, respectively $(i=1, 2, \ldots, n)$.

d) Operating from E into E_0:

$$L_{01} \zeta = \gamma^* , \quad M_{01} \zeta = v^* ,$$

where γ^* and v^* are the n-dimensional vectors

$$\gamma^* = \{\varrho \int_S \varphi_1^{**} \zeta\, dS, \varrho \int_S \varphi_2^{**} \zeta\, dS, \ldots, \varrho \int_S \varphi_n^{**} \zeta\, dS\} ,$$

$$v^* = \{\int_S v_1 \zeta\, dS, \int_S v_2 \zeta\, dS, \ldots, \int_S v_n \zeta\, dS\} .$$

Using this notation we can write Eqs. (5–83) in the form

$$L_{00}Y'' + L_{01}\zeta_{tt} + M_{00}Y + M_{01}\zeta = 0\,,$$
$$L_{10}Y'' + L_{11}\zeta_{tt} + M_{10}Y + M_{11}\zeta = 0\,.$$

This system can be written in an even more compact manner by introducing the operators L and M operating in $\mathscr{E}$:

$$L = \begin{Vmatrix} L_{00} & L_{01} \\ L_{10} & L_{11} \end{Vmatrix}, \quad M = \begin{Vmatrix} M_{00} & M_{01} \\ M_{10} & M_{11} \end{Vmatrix}.$$

In this notation, Eqs. (5–83) will be written as

$$Lx'' + Mx = 0\,, \tag{6–3}$$

where the prime denotes differentiation with respect to time, and $x \in \mathscr{E}$.

[6-2.2] Properties of the Spectrum

As in the preceding section we have primary reference here to the study of the structure of the spectrum and the proof of convergence of the Ritz method. It turns out that all of the circumstances of interest follow in an elementary manner from the general theory of linear operators. To prove this it is sufficient to clarify certain properties of the operators L and M.

First, the operators are self-adjoint, *i.e.*,

$$(Lx, \tilde{x}) = (x, L\tilde{x})\,, \quad (Mx, \tilde{x}) = (x, M\tilde{x})\,. \tag{6–4}$$

This property is established by a direct check. We evaluate

$$(Lx, \tilde{x}) = \left(\begin{Vmatrix} L_{00} & L_{01} \\ L_{10} & L_{11} \end{Vmatrix} \cdot \begin{Vmatrix} Y \\ \zeta \end{Vmatrix}, \begin{Vmatrix} \tilde{Y} \\ \tilde{\zeta} \end{Vmatrix} \right) =$$
$$= (L_{00}Y, \tilde{Y}_1)_0 + (L_{01}\zeta, \tilde{Y}_1)_0 + (L_{10}Y, \tilde{\zeta}_1) + (L_{11}\zeta, \tilde{\zeta}_1)\,.$$

But L_{00} is a unit operator and hence $(L_{00}Y, \tilde{Y})_0 = (L_{00}\,\tilde{Y}, Y)$. Further $L_{11} = H$ is a self-adjoint operator, as was proved above, and consequently $(L_{11}\zeta, \xi) = (\zeta, L_{11}\xi)$. Using the definitions of L_{01} and L_{10}, we obtain

$$(L_{01}\zeta, \tilde{Y}_1)_0 = \varrho \int\limits_S \varphi_1^{**}\zeta dS \cdot \tilde{Y}_1 + \cdots + \varrho \int\limits_S \varphi_n^{**}\zeta dS \cdot \tilde{Y}_n =$$
$$= \varrho \int\limits_S (\varphi_1^{**}\tilde{Y}_1 + \cdots + \varphi_n^{**}\tilde{Y}_n)\,\zeta dS = \int\limits_S L_{10}\tilde{Y}\zeta dS = (\zeta, L_{10}\tilde{Y})\,.$$

By analogy

$$(L_{10}Y, \tilde{\zeta}) = (Y, L_{01}\tilde{\zeta})_0 .$$

Collecting the results thus obtained, we arrive at (6–4). The self-adjointness of the operator M is proved similarly.

The assertion we have proven follows, in addition, from the results of Sec. 5 of Chapter 5, where we established that the matrix of the coefficients of the algebraic equations obtained by applying the Ritz method is symmetrical; this property also means that the operators L and M are self-adjoint.

Further, the operator L is fully continuous. In fact, the operator L_{00} is fully continuous, since it operates in a finite-dimensional space. The operator L_{01} is fully continuous, since it operates in finite-dimensional space E_0. The operator L_{10} operates from a finite-dimensional space and reduces to multiplying an element from E_0 by the vector φ^{**}; consequently, it is also fully continuous. Finally, the operator $L_{11}=H$ is fully continuous, which was proven in the preceding section.

Further, the operator L is a positive operator, since the functional (Lx', x') is twice the kinetic energy of system K. Thus L is a fully continuous, self-adjoint, positive operator.

The operator M will always be considered to be positive definite. Since $(Mx, x)=2\Pi$, where 2Π is twice the potential energy of system K, this assumption means that we restrict ourselves to consideration of only the case when the potential energy of system K in the equilibrium position has an isolated minimum. Setting

$$x = qe^{i\sigma t},$$

we reduce the system (6–4) to the form

$$(M - \sigma^2 L)\, q = 0 .$$

Thus the numbers $\lambda = 1/\sigma^2$ are the eigenvalues of the operator $C = M^{-1}L$

$$Cq = \lambda q . \tag{6–5}$$

We introduce the Friedrichs metric into the space E:

$$(x_1, x_2)_F = (Mx_1, x_2).$$

Since the operator L is fully continuous, and M^{-1} is bounded, the operator C is fully continuous in both metrics, while in the Friedrichs metric it will, additionally, be self-adjoint:

$$(Cx_1, x_2)_F = (MM^{-1}Lx_1, x_2) = (Lx_1, x_2) = (x_1, Lx_2) =$$
$$= (x_1, MM^{-1}Lx_2) = (Mx_1, Cx_2) = (x_1, Cx_2)_F .$$

In addition $(Cx_1, x_1)_F = (Lx_1, x_1) > 0$ if $x_1 \not\equiv 0$. It follows, on the basis of general theorems of analysis, that[65]

a) in the space E the operator C has a complete system of eigenelements, which are orthogonal in the Friedrichs metric;

b) all of the $\sigma_n^2 > 0$ and $\sigma_n \to \infty$ as $n \to \infty$, and each σ_n has corresponding to it a finite number of eigenelements;

c) the eigenelements q_n and the eigenvalues σ_n can be determined by the Ritz method, which is convergent.

We can formulate the ensemble of results obtained above in the form of the following theorem.

Theorem V

If the system K consists of a finite number of conservative links and contains a finite number of cavities partially filled with a fluid, and if in the equilibrium position the potential energy of the system has an isolated minimum, then

a) the motion of this system about the equilibrium position is accompanied by principal vibrations, *i.e.*, the system (5–83) has solutions of the form

$$Y_i = X_i e^{i\sigma t}, \quad \zeta = z e^{i\sigma t};$$

b) the frequencies of these vibrations are real numbers and $\sigma_n \to \infty$ as $n \to \infty$; *i.e.*, the equilibrium position is stable;

c) any unconstrained motion of the system K can be represented as a superposition of vibrations; *i.e.*, the system of principal vibrations is complete;

d) the free vibrations and frequencies can be found by the Ritz method.

In formulating this theorem we have used the concept of stability. We impart to this concept the following meaning: the equilibrium position is regarded as stable if any principal vibration is bounded. Thus assertion (c) is an analog of Lagrange's theorem: for stability of the equilibrium position of a system K it is sufficient that the potential energy of the system have a minimum in the equilibrium position. In this case Lagrange's theorem also has an inverse.

Theorem VI

If the potential energy is not a minimum in the equilibrium position, then among the numbers σ^2 there exists at least one that is negative or zero.

The proof of this theorem was suggested in Ref. [78], and is based on the results obtained in Ref. [79] on the theory of Hermitian operators in spaces with indefinite metrics.

[6-3] GENERAL EQUATIONS FOR THE OSCILLATIONS OF A BEAM WITH A FLUID-CONTAINING CAVITY

[6-3.1] Reduction to an Operational Equation

We now pass on to the study of the general properties of the system (5–106). In a manner similar to that of the preceding section, we should write (5–106) in the operational form. We introduce the following functional spaces:

a) the spaces E_1, E_2, and E_3 of functions $u_1(y)$, $u_2(y)$, and $u_3(y)$, which are defined and summed on the segment $[0, l]$ with the scalar product

$$(u_i, v_i)_i = \int_0^l u_i v_i dy, \quad u_i, v_i \in E_i, \quad i = 1, 2, 3;$$

b) the space E_4 of the functions $u_4(P)$, $P \in S$ with a summable square on S and with the scalar product

$$(u_4, v_4)_4 = \int_S u_4 v_4 dS;$$

c) the direct sum of spaces $\mathscr{E} = \sum_{i=1}^{4} E_i$ with the scalar product

$$(x, y) = \sum_{i=1}^{4} (u_i v_i)_i .$$

Here the element x is a vector with components u_1, u_2, u_3, and u_4. The functions u_i should satisfy some additional boundary conditions, which are determined by the kind of beam supports. We shall not spell out these conditions. They will be assumed to be homogeneous and such as to ensure self-adjointness of the operators describing the elastic oscillations of the beam without the fluid.

We further introduce the operators L_{ij} and M_{ij}, operating from E_j into E_i:

$$L_{ij} u_j = A_{ij} u_j + \varrho \int_{l_y} \gamma_i H \gamma_j u_j dl, \quad i, j = 1, 2, 3,$$

$$L_{4j} u_j = \varrho H \gamma_j u_j, \quad j = 1, 2, 3,$$

$$L_{j4} u_4 = \varrho \int_{l_y} \gamma_j H^* u_4 dl, \quad j = 1, 2, 3,$$

$$L_{44} u_4 = \varrho H^* u_4,$$

$$M_{ii} u_i = (c_i u_{iyy})_{yy} + B_{ii} u_i, \quad i = 1, 2,$$

$$M_{33} u_3 = -(c_3 u_{3y})_y + B_{33} u_3,$$

$$M_{44} u_4 = \varrho g u_4,$$

$$M_{ij} u_j = B_{ij}, \quad i, j = 1, 2, 3,$$

$$M_{4j}u_j = \varrho g \gamma_j u_j\,, \quad j = 1, 2, 3\,,$$

$$M_{j4}u_4 = \varrho g \int\limits_{d_y} \gamma_j u_4 dl\,, \quad j = 1, 2, 3\,.$$

Since H is an integral operator, the kernel of which is Green's function, in order for the operators L_{ij} and M_{ij} to operate from E_j into E_i, it is necessary to impose still more limitations on the functions A_{ij}, B_{ij}, and C_i. We shall not consider these problems and we shall assume once and for all that these functions satisfy all of the required conditions.

Using the notation introduced earlier, it is possible to write (5–106) in the form

$$\sum_{j=1}^{4} (L_{ij}u_{jtt} + M_{ij}u_j) = 0\,, \quad i = 1, 2, 3, 4\,. \tag{6–6}$$

If, in addition, we introduce the operator matrices

$$L = \|L_{ij}\|\,, \quad M = \|M_{ij}\|\,,$$

we can write (6–6) as

$$Lx_{tt} + Mx = 0\,. \tag{6–7}$$

[6-3.2] Properties of the Operators L and M

We now clarify some properties of the operator L. It is, first of all, evident that this operator is bounded. Further, L is a self-adjoint operator; *i.e.*, it satisfies the condition

$$(Lx, y) = (x, Ly)\,. \tag{6–8}$$

We write the above expression in more detail as

$$\sum_{i,j=1}^{4} (L_{ij}u_j, v_i)_i = \sum_{i,j=1}^{4} (u_i, L_{ij}v_j)_i\,. \tag{6–9}$$

The validity of the above equality can be established by a direct check. We evaluate, for example,

$$(L_{41}u_1, v_4)_4 = \varrho \int\limits_S H\gamma_1 u_1 \cdot v_4 dS = \varrho \int\limits_{\Sigma+S} H\gamma_1 u_1 \cdot \frac{\partial}{\partial n} H^* v_4 dS =$$

$$= \varrho \int\limits_\tau \nabla H\gamma_1 u_1 \cdot \nabla H^* v_4 d\tau = \varrho \int\limits_{\Sigma+S} H^* v_4 \frac{\partial}{\partial n} H\gamma_1 v_1 dS =$$

$$= \varrho \int\limits_{\Sigma+S} H^* v_4 \cdot \gamma_1 u_1 dS = \varrho \int\limits_0^l u_1 \{\int\limits_{l_y} \gamma_1 H^* v_4 dl\}\, dy = (u_1, L_{14}v_4)_1\,.$$

The remaining relationships that validate (6–9) are established similarly.

Unlike the operator L considered in the preceding section [see (6–3)], the operator L introduced in this section is not fully continuous. This difference is because the space E_0 of the preceding section was finite-dimensional. In the given case the role of E_0 is played by the infinite-dimensional space $E_1+E_2+E_3$.

The operator M is unbounded and, in addition, self-adjoint. The latter property is an elementary result of the symmetry of the matrix $\|B_{ij}\|$ and of the self-adjointness of the boundary conditions for u_i.

We evaluate additionally the scalar product

$$(Mx, x) = \int_0^l \{c_1 u_{1yy}^2 + c_2 u_{2yy}^2 + c_3 u_{3y}^2 + 2 \sum_{i,j=1}^{3} B_{ij} u_i u_j + \\ + \varrho g \int_S \{u_4^2 + 2u_4(u_1\gamma_1 + u_2\gamma_2 + u_3\gamma_3)\}\, dS\}\, dl\,. \tag{6–10}$$

Comparing (6–10) and (5–105), we can convince ourselves that, within the accuracy of the expressions,

$$(Mx, x) = 2\Pi\,.$$

We shall consider only cases where Π is minimal in the equilibrium position. Hence the operator M will be regarded as positive definite.

Thus the difference between Eq. (6–7) and Eq. (6–3) is that the operator L is not fully continuous. This circumstance makes it impossible to utilize directly the general theorems of the theory of linear operators.

Remark

We note that if the operator M^{-1} had been fully continuous, the above difficulty would be inconsequential, since the operator $M^{-1}L$, because L is bounded, would have been continuous.

[6-3.3] An Auxiliary Transformation

Instead of Eqs. (5–106) we now set up a system of equivalent equations. For this, using the general rules, we calculate the variation of the functional $\int_0^t (T-\Pi)\, dt$ and in the expression obtained we replace the variable

$$u = v - \gamma_1 X - \gamma_2 Z - \gamma_3 \theta\,. \tag{6–11}$$

Then, again using general rules, we set up equations for the variables X, Z, θ, and v. Skipping cumbersome calculations (for more details see Ref. [69]), we arrive at the system of integro-differential equations

$$
\begin{aligned}
&A_{11}X_{tt} + A_{12}Z_{tt} + A_{13}\theta_{tt} + (c_1 X_{yy})_{yy} + B_{11}^* X + B_{12}^* Z + B_{13}^* \theta + \\
&\qquad + \varrho \int\limits_{l_y - d_y} \gamma_1 [H_1 X_{tt}\gamma_1 + H_1 Z_{tt}\gamma_2 + H_1 \theta_{tt}\gamma_3 + H^* v_{tt}]\, ds = 0, \\
&A_{21}X_{tt} + A_{22}Z_{tt} + A_{23}\theta_{tt} + (c_2 Z_{yy})_{yy} + B_{21}^* X + B_{22}^* Z + B_{23}^* \theta + \\
&\qquad = \varrho \int\limits_{l_y - d_y} \gamma_2 [H_1 X_{tt}\gamma_1 + H_1 Z_{tt}\gamma_2 + H_1 \theta_{tt}\gamma_3 + H^* v_{tt}]\, ds = 0, \\
&A_{31}X_{tt} + A_{32}Z_{tt} + A_{33}\theta_{tt} - (c_3 \theta_y)_y + B_{31}^* X + B_{32}^* Z + B_{33}^* \theta + \\
&\qquad + \varrho \int\limits_{l_y - d_y} \gamma_3 [H_1 X_{tt}\gamma_1 + H_1 Z_{tt}\gamma_2 + H_1 \theta_{tt}\gamma_3 + H^* v_{tt}]\, ds = 0, \\
&\qquad \varrho [H_1 X_{tt}\gamma_1 + H_1 Z_{tt}\gamma_2 + H_1 \theta_{tt}\gamma_3 + H^* v_{tt} + gv] = 0.
\end{aligned}
\tag{6–12}
$$

In this equation, $H_1 = H - H^*$, $B_{ij}^* = B_{ij} - \varrho_{ij}$, $\varrho_{ij} = \varrho_{ji}$ are some functions of the variable y that depend solely on the configuration of the volume τ (the expressions for ϱ_{ij} will be written in more detail in the following section).

We recall the notation introduced in the last section of the preceding chapter: H is an operator that matches to values $f(P)$ $(P \in \Sigma + S)$ a function φ, which is harmonic in τ and such that $\partial\varphi/\partial n = \partial Hf/\partial n = f$; H^* is an operator which matches the functions $f(P)$ $(P \in S)$ with the function φ, harmonic in τ and such that $\partial\varphi/\partial n = 0$ when $p \in \Sigma$, and $\partial\varphi/\partial n = f$ when $P \in S$; l_y is a contour formed by the intersection of the plane normal to the Oy axis with the surface $\Sigma + S$; d_y is a segment formed by the intersection of a plane parallel to the Oy axis and the figure S.

The main difference between the transformed system (6–12) and (5–106) is that its first three equations do not contain terms in which v enters (these equations contain only derivatives of the function v with respect to the parameter t).

We now introduce the space $E_0' = E_1 + E_2 + E_3$, and let $\omega \in E_0'$. Here ω is a vector with projections X, Z, and θ. We can now write (6–12) in the form

$$
\begin{aligned}
L'\omega_{tt} + Dv_{tt} + M'\omega &= 0, \\
\varrho H_1 \Gamma' \omega_{tt} + \varrho H^* v_{tt} + \varrho g v &= 0.
\end{aligned}
\tag{6–13}
$$

Here v is an element from E_0, and

$$
L' = \begin{Vmatrix} L_{11}' & L_{12}' & L_{13}' \\ L_{21}' & L_{22}' & L_{23}' \\ L_{31}' & L_{32}' & L_{33}' \end{Vmatrix}.
\tag{6–14}
$$

The operators L_{ij}' differ from the operators L_{ij} only in that the operator H

in the integrand is replaced by the operator H_1, and the integration is extended over $l_y - d_y$;

$$M' = \begin{Vmatrix} M'_{11} & M'_{12} & M'_{13} \\ M'_{21} & M'_{22} & M'_{23} \\ M'_{31} & M'_{32} & M'_{33} \end{Vmatrix}. \tag{6–15}$$

The operators M'_{ij} differ from the operators M_{ij} in that multiplication by the functions B_{ij} has been replaced by multiplication by B^*_{ij}. Further,

$$Dv = \varrho \int_{l_y - d_y} \Gamma H^* v dl,$$

where Γ is a column matrix

$$\Gamma = \begin{Vmatrix} \gamma_1 \\ \gamma_2 \\ \gamma_3 \end{Vmatrix},$$

and Γ' is a row matrix

$$\Gamma' = \|\gamma_1, \gamma_2, \gamma_3\|.$$

If we introduce the operators Λ and N in the space $\mathscr{E}' = E'_0 + E_4$:

$$\Lambda = \begin{Vmatrix} L' & D \\ \varrho H_1 \Gamma' & \varrho H^* \end{Vmatrix}, \quad N = \begin{Vmatrix} M' \\ \varrho g \end{Vmatrix},$$

then (6–13) can be written in the form

$$\Lambda z_{tt} + Nz = 0. \tag{6–16}$$

It is not difficult to establish that Λ, like L, is a self-adjoint operator. In order to convince ourselves of this, it is sufficient to check the validity of the expressions

$$(\Lambda z, z_1) = (Lx, x_1), \quad x \in \mathscr{E},$$
$$(z, \Lambda z_1) = (x, Lx_1), \quad z \in \mathscr{E}'.$$

For this, in turn, it is sufficient to write out explicitly the corresponding scalar products and in the expressions obtained to replace

$$H = H_1 + H^*,$$
$$v = u + \gamma_1 X + \gamma_2 Z + \gamma_3 \theta.$$

As to the operator M', like operator M, it is unbounded and positive definite, but unlike the latter, its inverse M'^{-1} is fully continuous. This can be checked easily by considering the equation

$$M'\omega = f. \tag{6–17}$$

In the particular case of torsional oscillations, (6–17) has the form

$$-(c_3\theta_y)_y + B_{33}^*\theta = f\,, \tag{6–18}$$

where θ satisfies the boundary conditions ensuring that M is self-adjoint.

Using Green's function of the boundary-value problem for (6–18), we obtain

$$\theta(P) = \int_0^l G(P, Q)\, f(Q)\, dQ \equiv Gf\,.$$

Since Green's function of the differential expression in the left-hand side of (6–18) is continuous, the kernel of the integral operator G is integrable with the square. Consequently[65], G is a fully continuous operator.

We convince ourselves similarly that in general the operator $G = M'^{-1}$ is fully continuous.

Thus the operator M is broken up into the operator M' operating in E_0', and an operator for multiplying by a constant; here the breaking up was performed in a manner such that the operator M'^{-1} is fully continuous. This is the purpose of the above change in variables.

[6-3.4] Structure of the Spectrum

We set in (6–13)

$$\omega = W \cos\sigma t\,, \quad v = V \cos\sigma t\,. \tag{6–19}$$

Then it will take on the form

$$\begin{aligned} L'W + DV &= \frac{1}{\sigma^2} M'W\,, \\ \varrho H_1 \Gamma' W + \varrho H^* V &= \frac{1}{\sigma^2} \varrho g V\,. \end{aligned} \tag{6–20}$$

We replace the variables

$$\varphi = \sqrt{M'}W\,, \quad \psi = \sqrt{\varrho g}V$$

and rewrite (6–19):

$$M'^{-\frac{1}{2}} L' M'^{-\frac{1}{2}} \varphi + M'^{-\frac{1}{2}} D \frac{\psi}{\sqrt{\varrho g}} = \frac{1}{\sigma^2}\varphi\,,$$

$$\frac{\varrho}{\sqrt{\varrho g}} H_1 \Gamma' M'^{-\frac{1}{2}} \varphi + \frac{\varrho}{\sqrt{\varrho g}} H^* \frac{\psi}{\sqrt{\varrho g}} = \frac{1}{\sigma^2}\psi\,.$$

Since the operator L' is bounded, $M'^{-\frac{1}{2}} L' M'^{-\frac{1}{2}}$ is fully continuous (since M'^{-1} is fully continuous). The operators H_1 and H^* are fully continuous, as well as integral with nonessential singularity; hence the operator D is also fully continuous. Therefore the operator

$$R = \begin{Vmatrix} M'^{-\frac{1}{2}} L' M'^{-\frac{1}{2}} & M'^{-\frac{1}{2}} D \dfrac{1}{\sqrt{\varrho g}} \\ \dfrac{1}{\sqrt{\varrho g}} H_1 \Gamma' M'^{-\frac{1}{2}} & \dfrac{1}{\sqrt{\varrho g}} H^* \dfrac{1}{\sqrt{\varrho g}} \end{Vmatrix}$$

is fully continuous.

The symmetry of the operator R is established by a simple check. The problem is reduced to the problem of the eigenvalues

$$Rf = \lambda f\,, \quad \lambda = \frac{1}{\sigma^2}$$

for a fully continuous self-adjoint operator. This reduction terminates the study, since, on the basis of known theorems, the eigenvalues λ_n of the operator R form a denumerable sequence of positive numbers such that $\lambda_n \to_{n\to\infty} 0$; the eigenelements are orthogonal, have finite multiplicity and can be found by the Ritz method. We can formulate the results thus obtained in the form of the following theorem.

Theorem VII

If the form Π for the beam-fluid system is positive definite, the system has principal vibrations of the form of (6–19), where σ_n are positive numbers (natural frequencies), forming a denumerable sequence such that $\sigma_n \to_{n\to\infty} \infty$. Each number σ_n has corresponding to it a finite number of modes of principal vibrations, which can be found by the Ritz method.

For the problem at hand this theorem includes, as a particular case, Lagrange's theorem concerning the minimum of the potential energy, since it means that for the vibrations to be bounded it is sufficient for the potential energy to have a minimum in the equilibrium position.

In systems that have, in addition to the fluid link, a finite number of degrees of freedom, it is possible to prove the converse theorem. This theorem has not been proven for the general case of a beam.

The system of principal vibrations is complete in $\mathscr{E}$ in the sense of the Friedrichs norm. We perform the replacement (6–19) in (6–7) and rewrite it in the form

$$M^{-1} L x = \lambda x\,.$$

We introduce the Friedrichs norm

$$(x, y)_F = (Mx, y)\,.$$

To prove completeness it is sufficient to show that $h \equiv 0$ follows immediately from the condition that

$$M^{-1} L h = 0\,.$$

We consider

$$(M^{-1} L h, h)_F = (Lh, h)\,.$$

The scalar product in the right-hand side is the kinetic energy. Hence, if $h \not\equiv 0$, this expression cannot be zero.

[6-4] STUDY OF THE POSITIVE DEFINITENESS OF OPERATOR *M*

[6-4.1] Transformation of the Expression for the Potential Energy

The considerations employed in analyzing the structure of the spectrum of our problem have depended to a great extent on the assumption that the form Π is positive definite. In this section we shall consider certain conditions, the satisfaction of which guarantees the positive definiteness of the potential energy.

We first recall some definitions. The self-adjoint operator M (or, which is the same thing, the form Π) is called positive if, for any $x \in E$

$$2\Pi = (Mx, x) \geqslant 0,$$

where $(Mx, x) = 0$ only if $x = 0$. The mechanical meaning of this is as follows: transition from the state of equilibrium to any other state involves expenditure of work; or a system cannot be taken out of the equilibrium position without the expenditure of work.

In this case, if x is a finite-dimensional vector, the positiveness of the operator M (when the other conditions of Theorem I of Sec. 3 have been satisfied) is sufficient for the boundedness of the principal vibrations. In the infinite-dimensional case we use for this purpose a more rigorous requirement that the operator M be positive definite. The symmetrical operator M is called positive definite if there exists a constant $C > 0$ such that for any $x \in E$

$$2\Pi = (Mx, x) \geqslant C(x, x).$$

From the positive definiteness of operator M follows its positiveness, the converse not being true.

We now write an expression for the functional Π:

$$\Pi = \tfrac{1}{2} \int_0^l \{c_1 X_{yy}^2 + c_2 Z_{yy}^2 + c_3 \theta_y^2 + B_{11} X^2 + 2B_{12} XZ + + 2B_{13} X\theta + B_{33}\theta^2 + 2B_{23} Z\theta\}\, dy + \tfrac{1}{2}\varrho g \int_S u^2 dS + + \varrho g \int_S u [X\gamma_1 + Z\gamma_2 + \theta\gamma_3]\, dS. \tag{6–21}$$

In this expression we now change variables according to (6–11). This transformation was utilized in the preceding section. Let us now show the procedure in more detail. In terms of the new variables, Π has the form

$$\Pi = \Pi^* + \tfrac{1}{2}\varrho g \int\limits_S v^2 dS,$$

where

$$\begin{aligned}\Pi^* = \tfrac{1}{2}\int\limits_0^l \{&c_1 X_{yy}^2 + c_2 Z_{yy}^2 + c_3\theta_y^2 + B_{11}X^2 - 2B_{12}XZ + \\ &+ 2B_{13}X\theta + B_{22}Z^2 + 2B_{23}Z\theta + B_{33}\theta^2\}\, dy - \varrho g \int\limits_S \gamma_1\gamma_2 XZ dS - \\ &- \varrho g \int\limits_S \gamma_1\gamma_3 X\theta ds - \varrho g \int\limits_S \gamma_2\gamma_3 Z\theta dS - \tfrac{1}{2}\varrho g \int\limits_S \gamma_1^2 X^2 dS - \\ &- \tfrac{1}{2}\varrho g \int\limits_S \gamma_2^2 Z^2 dS - \varrho g \int\limits_S \gamma_3^2 dS.\end{aligned} \tag{6–22}$$

This expression can be written in a more graphic form, which clarifies the meaning of a number of its components.

Since $\gamma_1 = \cos(nX)$, and $\gamma_2 = \cos(nZ)$, where these quantities are normal on S (the normal to S is everywhere vertical), and $\gamma_3 = Z\gamma_1 - X\gamma_2$, we have

$$\varrho g \int\limits_S \gamma_1\gamma_2 XZ dS = \int\limits_0^l \varrho_{12}(y)\, XZ dy, \quad \varrho_{12} = \varrho g \gamma_1\gamma_2 d_y,$$

$$\varrho g \int\limits_S \gamma_1^2 X^2 dS = \int\limits_0^l \varrho_{11}(y)\, X^2 dy, \quad \varrho_{11} = \varrho g \gamma_1^2 d_y,$$

$$\varrho g \int\limits_S \gamma_2^2 Z dS = \int\limits_0^l \varrho_{22}(y)\, Z^2 dy, \quad \varrho_{22} = \varrho g \gamma_2^2 d_y,$$

$$\varrho g \int\limits_S \gamma_2\gamma_3 Z\theta dS = \int\limits_0^l \varrho_{23}(y)\, Z\theta dy, \quad \varrho_{23} = \varrho g (\gamma_1\gamma_2 J_z - \gamma_2^2 J_x),$$

$$\varrho g \int\limits_S \gamma_1\gamma_3 X\theta dS = \int\limits_0^l \varrho_{13}(y)\, X\theta dy, \quad \varrho_{13} = \varrho g (\gamma_1^2 J_s - \gamma_1\gamma_2 J_x),$$

$$\varrho g \int\limits_S \gamma_3^2\theta^2 dS = \int\limits_0^l \varrho_{33}(y)\, \theta^2 dy, \quad \varrho_{33} = \varrho g (\gamma_2^2 J_{xx} - 2\gamma_1\gamma_2 J_{zx} + \gamma_1^2 J_{zz}).$$

Here J_x and J_z denote the static moments of the segment $d_y (J_x = \int_{d_y} x\,dl, \ldots)$, and J_{xx}, J_{zx}, and J_{zz} are its moments of inertia. Thus the numbers $\varrho_{ij} = \varrho_{ji}$ are determined solely by the geometric parameters of the cavity.

Introducing the notation

$$B^*_{ij} = B_{ij} - \varrho_{ij},$$

we rewrite the functional Π^* in the form

$$\Pi^* = \tfrac{1}{2} \int_0^l \{c_1 X^2_{yy} + c_2 Z^2_{yy} + c_3 \theta^2_y + B^*_{11} X^2 + 2B^*_{12} XZ + \\ + 2B^*_{13} X\theta + B^*_{22} Z^2 + 2B^*_{23} Z\theta + B^*_{33} \theta^2\}\, dy. \tag{6–23}$$

[6-4.2] Fundamental Theorems

Thus, by transformation (6–11), the functional Π has been represented in the form of a sum of two functionals, one of which, $\int_s v^2 dS$, is always positive definite, while the other does not depend on v. Hence we have the following theorem.

Theorem VIII

In order for the beam-fluid system to be statically stable, *i.e.*, for the functional Π to be positive definite, it is necessary and sufficient that the functional Π^* be positive definite. Since the functional Π^* is defined in a finite-dimensional space, it suffices to require that it be positive.

The mechanical meaning of this theorem is simple. For stability of a beam containing a fluid it is necessary and sufficient that some beam, which has the same elastic properties but is acted upon by other elastic forces, be stable.

We established above that there does not exist a beam without fluid that would be dynamically equivalent to a beam with a fluid. However, for stability of a beam with a fluid it is necessary and sufficient for some equivalent beam without a fluid to be stable.

The theorem we have proved, includes the following corollary as a special case.

Theorem IX

For the stability of the equilibrium position of a rigid body containing a cavity that is partially filled with a fluid, it is necessary and sufficient to have stability of equilibrium of some other rigid body.

We now consider the expression for Π given by Eq. (5–82). The transformation (6–11) in the given case will be

$$\zeta = v - \frac{1}{\varrho g} \sum_{j=1}^{n} v_j Y_j .$$

Then the formula for the potential energy will take the form

$$\Pi = \Pi_1 + \Pi^* ,$$

where

$$\Pi_1 = \tfrac{1}{2}\varrho g \int_S v^2 dS ,$$

$$\Pi^* = \tfrac{1}{2} \sum_{i,j=1}^{n} (\delta_{ij}\mu_i^2 - a_{ij}) \, Y_i Y_j ,$$

δ_{ij} is Kronecker's delta, and $a_{ij} = 1/\varrho g \int_S v_i v_j dS$.

Thus, for the positive definiteness of a form of an infinite number of variables, it is necessary and sufficient to have positive definiteness of the form of Π^*, which is a form of a finite number of variables. But since for the stability of the equilibrium position of the body-fluid system it is necessary and sufficient to have positive definiteness of form Π, we arrive at the following result. For stability of the equilibrium position of the system it is necessary and sufficient to have positive definiteness the form Π^*, which is a form of a finite number of variables. The number of these variables is equal to the number of degrees of freedom of a system without the fluid. If the system contains only one rigid body, the form Π^* depends on not more than six variables and it describes the potential energy of some rigid body, which we have the right to call equivalent with respect to stability.

[6-4.3] Stability of a Pendulum

As an example we now consider the problem of stability of the equilibrium position of a pendulum containing a fluid with a free surface. This problem was examined in Sec. 4 of Chapter 5, where we obtained the following expression for the potential energy:

$$\Pi = \frac{k^2\theta^2}{2} + \theta \varrho g \int_S \zeta y dS + \frac{\varrho g}{2} \int_S \zeta^2 dS , \qquad (5\text{–}67)$$

upon which we now perform transformation (6–11), which here will be

$$\zeta = v - y\theta .$$

The expression for Π now has the form

$$\Pi = \tfrac{1}{2}\varrho g \int_S \zeta^2 dS + \Pi^*,$$

where

$$\Pi^* = \tfrac{1}{2}\theta^2 [k^2 - \varrho g J_S],$$

and J_S is the moment of inertia of the free surface with respect to axis Ox, drawn through the center of inertia of the free surface

$$J_S = \int_S y^2 dS.$$

The expression for Π^* we have obtained makes it possible to find immediately the condition for stability

$$k^2 > \varrho g J_s. \tag{6–24}$$

In other words, for the stability of a pendulum containing a fluid it is necessary and sufficient to have stability of some pendulum not containing a fluid. Its reduced length is determined from the condition

$$l_2 = \frac{Ml_1 - \varrho J_s}{M}, \tag{6–24'}$$

where l_1 is the reduced length of a pendulum the free surface of which is covered with a partition, and M is the mass of the entire system. Thus the presence of the fluid can only detract from the stability of the pendulum.

Condition (6–24) shows that the stability (all other characteristics being those specified) is determined by the area of the free surface. This is well known to specialists in the theory of ships. Equation (6–24) is derived from this theory and is called the formula for the modified metacentric height when the ship is carrying fluid cargo. A similar condition, as we have seen in Chapter 3, also applies for stability according to Lyapunov. Thus, in the particular case under study, the same condition (6–24) ensures stability of the equilibrium position according to Lyapunov as well as the boundedness of the principal oscillations.

[6-4.4] Some Sufficient Conditions for the Stability of a Beam Containing a Fluid

The study of the sign-definiteness of the functional Π^* is generally not trivial, and criteria that would be not only sufficient but also necessary do not exist. Hence we shall point out here only several conditions that are sufficient for static stability.

The functional (6–23) can be written in the form

$$\Pi^* = \Pi_1^* + \Pi_2^*,$$

where

$$\Pi_1^* = \tfrac{1}{2}\int_0^l \{c_1 X_{yy}^2 + c_2 Z_{yy}^2 + c_3 \theta_y^2\}\, dy,$$

$$\Pi_2^* = \tfrac{1}{2}\int_0^l \{B_{11}^* X^2 + 2B_{12}^* XZ + 2B_{13}^* X\theta + B_{22}^* Z^2 + 2B_{23}^* Z\theta + B_{33}^* \theta^2\}\, dy.$$

If the rigidity coefficients are positive and do not vanish anywhere (as is typical for many applied problems), it is sufficient for positive definiteness of Π^* to have positive definiteness of Π_2^*. For this, in turn, it is sufficient to satisfy Sylvester's inequalities

$$\begin{aligned} \Delta_1 &= B_{11}^* > 0, \\ \Delta_2 &= \begin{vmatrix} B_{11}^* & B_{12}^* \\ B_{21}^* & B_{22}^* \end{vmatrix} > 0, \\ \Delta_3 &= \begin{vmatrix} B_{11}^* & B_{12}^* & B_{13}^* \\ B_{21}^* & B_{22}^* & B_{23}^* \\ B_{31}^* & B_{32}^* & B_{33}^* \end{vmatrix} > 0. \end{aligned} \tag{6–25}$$

The above conditions, which can be made graphic, impose limitations on the geometric characteristics. In the case of torsional oscillations of a beam, Eq. (6–25) has the form

$$B_{33}^* > 0$$

or, expanded,

$$B_{33} - \varrho g(\gamma_2^2 J_{xx} - 2\gamma_1\gamma_2 J_{zx} + \gamma_1^2 J_{zz}) > 0. \tag{6–26}$$

If the axis of the beam is horizontal, then $\gamma_1 = 0$, $\gamma_2 = 1$, and (6–26) simplifies to

$$B_{33} - \varrho g J_{xx} > 0; \tag{6–27}$$

i.e., we have arrived at a condition that is equivalent to (6–24). For a pendulum this condition was also a necessary, but for a beam it is only sufficient.

If the axis of the beam is vertical, then $\gamma_1 = \gamma_2 = 0$, and Eq. (6–26) becomes

$$B_{33} > 0;$$

i.e., for stability of the beam it is sufficient that the external forces have the character of a restoring force.

Let us now consider the flexural oscillations of a beam in two mutually perpendicular directions. Then Π_2^* will be

$$\Pi_2^* = \tfrac{1}{2}\int_0^l [B_{11}^* X^2 + 2B_{12}^* XZ + B_{22}^* Z^2]\, dy\,.$$

Conditions (6–25) yield the inequalities

$$B_{11}^* > 0\,, \quad B_{11}^* B_{22}^* - B_{12}^{*2} > 0\,. \tag{6–28}$$

If we return to our old notation, the second of the above inequalities will yield

$$B_{22} > \varrho_{22} + \frac{(B_{12} - \varrho g \gamma_1 \gamma_2)^2}{B_{11} - \varrho_{11}}\,. \tag{6–29}$$

It follows from (6–28), in addition, that

$$B_{11} > \varrho_{11}\,, \quad B_{22} > \varrho_{22}\,,$$

therefore Eq. (6–29) defines the region of stability depicted in Fig. 27. For static stability it is sufficient that the "rigidities" B_{11} and B_{22} define a point on the plane that would lie above the hyperbola

$$B_{22} = \varrho_{22} + \frac{(B_{11} - \varrho g \gamma_1 \gamma_2)^2}{B_{11} - \varrho_{11}}.$$

For comparison we have drawn a hyperbola (broken curve in Fig. 27) bounding the stability region for the case when the fluid inside the beam does not have a free surface.

The conditions we have considered are a natural analog of the stability conditions for systems with a finite number of degrees of freedom, for which they were both necessary and sufficient. In the case of a beam, however, they are not necessary conditions. In fact, they require, for example, that the external forces must have the character of restoring forces and should compensate for the destabilizing effect of the fluid. At the same time it appears that, because of its rigidity, the beam can also remain stable when the external forces are "tipping." Conducting such a study in the general case is difficult. Here, for the special case of torsional oscillations of a cantilever beam we shall point out one of the possible ways for constructing criteria which do not assume that the external forces are restoring forces.

For torsional oscillations

$$\Pi^* = \tfrac{1}{2}\int_0^l c_3 \theta_y^2\, dy + \tfrac{1}{2}\int_0^l B_{33}^* \theta^2\, dy\,. \tag{6–30}$$

We assume that there is a $\delta > 0$ such that

$$B_{33}^* \geqslant -\delta, \quad y \in [0, l].$$

from which we obtain the estimate

$$\Pi^* \geqslant \tfrac{1}{2} \int_0^l c_3 \theta_y^2 dy - \frac{\delta}{2} \|\theta\|^2. \tag{6–31}$$

where

$$\|\theta\|^2 = \int_0^l \theta^2 dy.$$

Since

$$\theta = \int_0^l \theta_y dy = \int_0^l \frac{1}{\sqrt{c_3}} \theta_y \cdot \sqrt{c_3} \theta_y dy,$$

we obtain from Schwarz's inequality the estimate

$$\theta^2 \leqslant \int_0^l \frac{dy}{c_3} \int_0^l c_3 \theta_y^2 dy.$$

Integrating both parts of the above expression, we have

$$\int_0^l c_3 \theta_y^2 dy \geqslant \frac{\|\theta\|^2}{l \int_0^l (dy/c_3)}.$$

Hence Eq. (6–31) should be replaced by the estimate

$$\Pi^* \geqslant \tfrac{1}{2} \|\theta\|^2 \left(\frac{1}{l \int_0^l c_3^{-1} dy} - \delta \right).$$

whence it follows that for positive definiteness of the functional Π^* it is sufficient that the function c_3 satisfy the inequality

$$\left(\int_0^l \frac{dy}{c_3(y)} \right)^{-1} > l\delta.$$

This condition imposes a limitation on rigidity from below. It will be satisfied, for example, if the function c_3 is bounded from below by some constant η_3, which depends on the external forces and on the geometry of the cavity ($c_3 > \eta_3$).

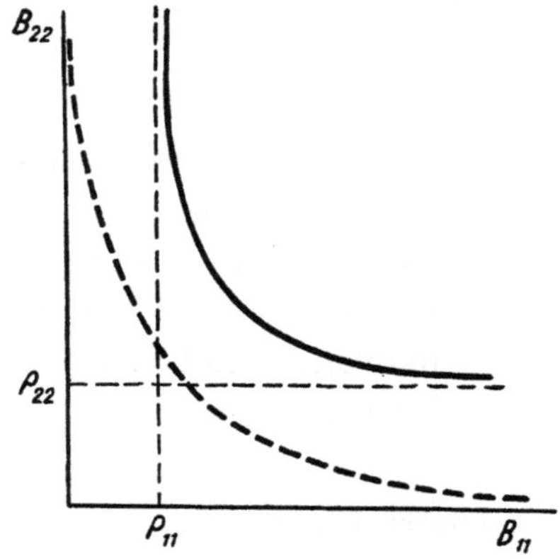

Fig. 27.

[6-5] SOME PROBLEMS OF THE THEORY OF FORCES SUDDENLY APPLIED TO A BODY CONTAINING A FLUID

[6-5.1] Equations of the Theory of a Suddenly Applied Force

The motion of the body-fluid system satisfies the equations of linear and angular momenta

$$\frac{d\boldsymbol{Q}}{dt} = \boldsymbol{F} + \boldsymbol{R},$$
$$\frac{d\boldsymbol{G}}{dt} = \boldsymbol{L}_F + \boldsymbol{L}_R. \tag{6–32}$$

Here $\boldsymbol{F}$ and $\boldsymbol{L}_F$ denote the principal vector and principal moment of the external nonimpulsive forces; $\boldsymbol{R}$ and $\boldsymbol{L}_R$ are the principal vector and principal moment of the impulsive forces. The quantities

$$\lim_{t_1 \to 0} \int_0^{t_1} \boldsymbol{R} dt = \underset{\cdot}{\boldsymbol{R}},$$

$$\lim_{t_1 \to 0} \int_0^{t} \boldsymbol{L}_R dt = \underset{\cdot}{\boldsymbol{L}}$$

will be called the principal impulsive vector and the principal impulsive moment or, simply, the impulse and the moment.

Integrating Eq. (6–32) with respect to t from 0 to t_1, and taking the limit as $t_1 \to 0$, we obtain

$$\Delta \boldsymbol{Q} = \underset{\cdot}{\boldsymbol{R}}, \qquad \Delta \boldsymbol{G} = \underset{\cdot}{\boldsymbol{L}}, \tag{6–33}$$

where $\Delta \boldsymbol{Q}$ and $\Delta \boldsymbol{G}$ denote the increments in the linear and angular momenta as a result of application of the impulsive forces.

The problem of the present section consists in examining some qualitative features of forces suddenly applied to a body containing a fluid. This examination will be performed with reference to Eqs. (6–33), which we shall call the principal equations of the impact theory.

[6-5.2] Equations of the Impact Theory for a Rigid Body Containing Fluid Masses with a Free Surface

In order to write the equations of the impact theory for the case under study, we must first of all write out explicitly the expressions for the vectors $\boldsymbol{Q}$ and $\boldsymbol{C}$. We introduce the coordinate system $\theta x_1 y_1 z_1$, which performs translational motion and the origin of which is located in the mass center of the system K_0. We recall that we used the designation K_0 for the body-fluid system on the assumption that the free surface was covered with a rigid partition. As in the preceding sections, the velocity potential in the coordinate system selected is written in the form of the sum

$$\varphi^* + \varphi,$$

where φ^* is the velocity potential for the velocities that the fluid would have if the free surface were rigidly covered, while φ is the potential of the wave motions of the fluid.

We further denote the mass of the entire body-fluid system by M, and the velocity of the mass center of the system K_0 by $\boldsymbol{v}_0$. Then

$$\boldsymbol{Q} = M\boldsymbol{v}_0 + m_1\boldsymbol{v}_1 + \varrho \int_\tau \nabla\varphi^* d\tau + \varrho \int_\tau \nabla\varphi d\tau. \tag{6–34}$$

Here $\boldsymbol{v}_1$ denotes the velocity of the mass center of the rigid body with respect to point O, and m_1 denotes the mass of the rigid body; $\varrho \int_\tau \nabla\varphi^* d\tau$ is the momentum that the fluid would have if the surface S were fixed:

$$\varrho \int_\tau \nabla\varphi^* d\tau = m_2 \boldsymbol{v}_2,$$

where m_2 is the mass of the fluid.

Consequently,

$$m_1\boldsymbol{v}_1 + m_2\boldsymbol{v}_2 = M\boldsymbol{v}^*,$$

where $\boldsymbol{v}^*$ is the velocity of the mass center of system K_0 about point O, *i.e.*,

$v^* = 0$. Thus

$$Q = M\boldsymbol{v}_0 + \varrho \int_\tau \nabla\varphi d\tau . \tag{6–35}$$

We restrict ourselves to the consideration of the case when the body is at rest at the moment of impact and the free surface of the fluid is horizontal. We fix the coordinate system $Oxyz$ in this plane in a manner such that the Oz axis is always vertical and the coordinate system can perform only translational motion. For example, we assume that the axes of the system $Oxyz$ are parallel to the axes of the system $Ox_1y_1z_1$. Noting that

$$\int_\tau \varphi_x d\tau = \int_S x\varphi_z dS ,$$

$$\int_\tau \varphi_y d\tau = \int_S y\varphi_z dS ,$$

$$\int_\tau \varphi_z d\tau = \int_S z\varphi_z dS = 0 ,$$

we rewrite (6–35) in the scalar form of projections on the Ox, Oy, and Oz axes:

$$Q_x = Mv_x + \varrho \int_S x\varphi_z dS ,$$

$$Q_y = Mv_y + \varrho \int_S y\varphi_z dS ,$$

$$Q_z = Mv_z .$$

According to this we rewrite the first of equations (6–33) in the form

$$M\Delta v_x + \varrho \int_S x\Delta\varphi_z dS = \underset{\cdot}{R}x ,$$

$$M\Delta v_y + \varrho \int_S y\Delta\varphi_z dS = \underset{\cdot}{R}y ,$$

$$M\Delta v_z = \underset{\cdot}{R}_z , \tag{6–36}$$

where $\Delta v_x, \ldots, \Delta\varphi$ are the velocity and velocity potential increments acquired on impact.

We now set up an expression for the kinetic moment about some point

$$\boldsymbol{G} = \boldsymbol{G}_1 + \varrho \int_\tau (\boldsymbol{r} \times \nabla\varphi^*)\, d\tau + \varrho \int_\tau (\boldsymbol{r} \times \nabla\varphi)\, d\tau . \tag{6–37}$$

The first term is the angular momentum of the rigid body

$$\boldsymbol{G}_1 = I_0 \boldsymbol{\theta}' , \tag{6–38}$$

where I_0 is the inertia tensor of the rigid body, and $\boldsymbol{\theta}'$ is the vector of the instantaneous angular velocity of the rigid body. We shall denote its projections by θ_i'. If small motions are considered, θ_i coincides with Euler's angles to within second-order infinitesimals.

We now consider the second term

$$\boldsymbol{G}_2 = \varrho \int_\tau (\boldsymbol{r} \times \nabla\varphi^*) \, d\tau .$$

It was shown in Sec. 1 of Chapter 2 that the angular momentum is a linear vector, *i.e.*, a function of the instantaneous angular velocity. Using the notation of this section, Eq. (2–18) has the form

$$\boldsymbol{G}_2 = \boldsymbol{I}^* \boldsymbol{\theta}' .$$

Here $\boldsymbol{I}^*$ is a tensor, the component of which is determined by Eqs. (6–13) and which depends solely on the geometry of the cavity. The vector $\boldsymbol{G}_2$ can be combined with the vector $\boldsymbol{G}_1$:

$$\boldsymbol{G}_1 + \boldsymbol{G}_2 = \boldsymbol{I}\boldsymbol{\theta}' , \tag{6–39}$$

where $\boldsymbol{I} = \boldsymbol{I}^* + I_0$ is the inertia tensor of the equivalent rigid body. We shall write its matrix as $\boldsymbol{I} = \{I_{ij}\}$. The last component of the kinetic moment is written similarly:

$$\boldsymbol{G}_3 = \int_\tau (\boldsymbol{r} \times \nabla\varphi) \, d\tau .$$

For example,

$$G_{3x} = \varrho \int_{S+\Sigma} \varphi \frac{\partial \varphi_4^*}{\partial n} \, dS .$$

Further, applying Green's formula and utilizing the fact that $\sum \partial\varphi/\partial n = 0$, we obtain

$$G_{3x} = \varrho \int_S \varphi_4^* \varphi_z dS .$$

Similarly

$$\begin{aligned} G_{3y} &= \varrho \int_S \varphi_5^* \varphi_z dS , \\ G_{3z} &= \varrho \int_S \varphi_6^* \varphi_z dS . \end{aligned} \tag{6–40}$$

Collecting the results (6–37) – (6–40) obtained above, we can now write the second of Eqs. (6–33) as projections on the Ox, Oy, and Oz axes:

$$J_{11}\Delta\theta_1' + J_{12}\Delta\theta_2' + J_{13}\Delta\theta_3' + \varrho \int_S \varphi_4^* \Delta\varphi_z dS = \underset{\cdot}{L}_x,$$

$$J_{21}\Delta\theta_1' + J_{22}\Delta\theta_2' + J_{23}\Delta\theta_3' + \varrho \int_S \varphi_5^* \Delta\varphi_z dS = \underset{\cdot}{L}_y, \qquad (6\text{–}41)$$

$$J_{31}\Delta\theta_1' + J_{32}\Delta\theta_2' + J_{33}\Delta\theta_3' + \varrho \int_S \varphi_6^* \Delta\varphi_z dS = \underset{\cdot}{L}_z.$$

In order to close the system of Eqs. (6–36) and (6–41), we should also write out the conditions defining the harmonic function φ.

The function $\varphi^* + \varphi$ satisfies the condition of constancy of pressure on S. In order to form this, we write the Cauchy-Lagrange integral

$$\frac{\partial(\varphi + \varphi^*)}{\partial t} + U + \frac{p}{\varrho} = \text{const}.$$

Here U is the potential of the inertia forces and of the gravity forces

$$U = v_{0x}'x + v_{0y}'y + (v_{0z}' + g)\zeta.$$

Using the representation for φ^*, we can now write the pressure constancy condition in the form

$$\frac{\partial\varphi}{\partial t} = \sum_{i=1}^{3} \theta''\varphi_{i+3}^* + v_{0x}'x + v_{0y}'y + (v_{0z}' + g)\zeta = 0. \qquad (6\text{–}42)$$

Integrating the above equation with respect to t from 0 to t_1, letting t_1 go to zero, and making use of the fact that when $t=0$ the free surface was horizontal, *i.e.*, $z=\zeta=0$, we obtain

$$\Delta\varphi = \sum_{i=1}^{3} \Delta\theta_i'\varphi_{i+3}^* + \Delta v_{0x}x + \Delta v_{0y}y = 0. \qquad (6\text{–}43)$$

This equation determines the boundary values of the function Δ in terms of the increments $\Delta\theta_i'$, Δv_{0x}, and Δv_{0y}. The harmonic function $\Delta\varphi$ can thus be uniquely defined by the values of $\Delta\theta_i'$, Δv_{0x}, and Δv_{0y}:

$$\Delta\varphi = -\left[\sum_{i=1}^{3} \Delta\theta_i' A\varphi_{i+3}^* + \Delta v_{0x}Ax + \Delta v_{0y}Ay\right]. \qquad (6\text{–}44)$$

Here A is some linear operator that matches the function f with the function φ, which is harmonic in τ, the normal derivative of which on Σ is 0, and which on S takes on the value $(\varphi)_s = f$. If the solution of the problem of free vibrations is known, it is easy to construct operator A effectively.

For this we set

$$\varphi = \sum_{n=1}^{\infty} a_n \varphi_n ,$$

where the φ_n are the potentials of the principal vibrations. According to what was proven above, they form on S (when $z=0$) a complete and orthogonal system of functions, which can be regarded as normed. Then, introducing the notation $f_n = (\varphi_n, f)$, we find that $a_n = f_n$.

Utilizing (6–44), we can now eliminate φ from (6–36) and (6–41). Denoting

$$\Delta\varphi_z = -\sum_{i=1}^{3} \Delta\theta_i' A_z \varphi_{i+3}^* + \Delta v_{0x} A_z x + \Delta v_{0y} A_z y ,$$

we arrive at the following system of equations for the increments Δv_{0x}, Δv_{0y}, Δv_{0z}, and $\Delta\theta_i'$:

$$\begin{aligned} &\sum_{j=1}^{3} \alpha_{ij} \Delta x_j' + \sum_{j=1}^{3} \alpha_{ij+3} \Delta\theta_j' = \underset{\cdot}{R}_i , \qquad i = 1, 2, 3 , \\ &\sum_{j=1}^{3} \alpha_{ij} \Delta x_j' + \sum_{j=1}^{3} \alpha_{ij+3} \Delta\theta_j' = \underset{\cdot}{L}_{i-3} , \quad i = 4, 5, 6 . \end{aligned} \tag{6–45}$$

Here $x_1' = v_{0x}$, $x_2' = v_{0y}$, $x_3' = v_{0z}$, and the coefficients α_{ij} are defined by

$$\begin{aligned} &\alpha_{11} = M - \varrho \int_S x A_z x dS; \quad \alpha_{12} = -\varrho \int_S x A_z y dS; \quad \alpha_{13} = 0; \\ &\alpha_{14} = -\varrho \int_S x A_z \varphi_4^* dS; \ldots \\ &\alpha_{22} = M - \varrho \int_S y A_z y dS; \quad \alpha_{21} = -\varrho \int_S y A_z x dS; \quad \alpha_{23} = 0; \\ &\alpha_{24} = -\varrho \int_S y A_z \varphi_4^* dS; \ldots \\ &\alpha_{33} = M; \quad \alpha_{3i} = 0, \quad \text{if} \quad i \neq 3, \\ &\alpha_{44} = J_1 - \varrho \int_S \varphi_4^* A_z \varphi_4^* dS; \quad \alpha_{41} = -\varrho \int_S \varphi_4^* A_z x dS; \\ &\alpha_{42} = -\varrho \int_S \varphi_4^* A_z y dS; \quad \alpha_{43} = 0; \quad \alpha_{45} = -\varrho \int_S \varphi_4^* A_z \varphi_5^* dS; \quad \text{etc.} \end{aligned} \tag{6–46}$$

The system (6–45) is a set of algebraic equations describing the distribution of angular and linear velocities acquired by a solid body containing in its cavities a fluid with free surfaces, as a result of the application of a sudden impulse to the body. To set up this system we must know the Stokes-Zhukovskiy potentials. In addition, we should have at our disposal an effective formulation for the operator A. For this it is sufficient, for example, to have the complete solution of the problem of natural vibrations of the fluid in the cavity.

Preliminary analysis of the system (6–45) shows that not all of the directions are equally favored. The anisotropy is in the direction Oz, which is the direction of gravity. The presence of this peculiar direction, which is a consequence of the existence of the free surface, results in the appearance of a number of characteristic features that account for the difference in behavior on suddenly striking a rigid body containing a fluid with a free surface, from that of a rigid body containing fluid-filled cavities.

[6-5.3] The Central Impact Theorem

We shall use the term *central impact* to denote an impact in which the line of action of the impact impulse passes through the mass center of the system.

If the body is rigid, as a result of the central impact it will have imparted to it only a translational motion, which is collinear with the impulse. If the body contains a cavity filled with a fluid, it reacts to the impact in precisely the same manner as a rigid body; *i.e.*, it acquires only translational motion that is collinear with the impulse, no matter what the kind of impulse as long as its line of action passes through the mass center of the system. When the fluid has a free surface it turns out that the body acquires a translational motion that is collinear with the impulse only under certain exceptional conditions.

Thus, we shall assume that the body with the fluid has applied to it an impact impulse $\underset{\cdot}{L}$, the line of action of which passes through the mass center of the system; consequently, the moment of the impact impulse relative to the mass center, $\underset{\cdot}{\boldsymbol{R}}$ is zero. We denote the direction cosines of the vector $\underset{\cdot}{\boldsymbol{R}}$ by α, β, and γ. We assume further that the body has acquired only a translational displacement which is collinear with R, *i.e.*, that the vector $\Delta\boldsymbol{\theta}'=0$, while the vector $\Delta\boldsymbol{x}'(\Delta x_1', \Delta x_2', \Delta x_3')$ is collinear with $\underset{\cdot}{\boldsymbol{R}}$, *i.e.*,

$$\Delta x_1' = \alpha\Delta v, \quad \Delta x_2' = \beta\Delta v, \quad \Delta x_3' = \gamma\Delta v,$$

where Δv is the absolute magnitude of the velocity increment: $\Delta v = |\Delta\boldsymbol{x}'|$. Then the first three equations of the system (6–45) will take the form

$$\left.\begin{aligned} \Delta v(\alpha_{11}\alpha + \alpha_{12}\beta) &= \underset{\cdot}{R}\alpha, \\ \Delta v(\alpha_{21}\alpha + \alpha_{22}\beta) &= \underset{\cdot}{R}\beta, \\ \Delta v\alpha_{33}\gamma &= \underset{\cdot}{R}\gamma. \end{aligned}\right\} \tag{6–47}$$

If the body had been rigid, then $\alpha_{12}=\alpha_{21}=0$, while $\alpha_{11}=\alpha_{22}=\alpha_{33}=M$, and the system (6–47) would have been identically satisfied with respect to α, β, and γ. This also means that, no matter what the direction of the central impact, it results only in a translational displacement. The quantity Δv would then be defined quite simply as $\Delta v = \underset{\cdot}{R}/M$. In our case, the system (6–46) can be regarded as a system of equations for Δv, α, β, and γ. We consider, in sequence, two cases: $\gamma=0$ and $\gamma \neq 0$.

Let $\gamma \neq 0$. Then (since $\alpha_{33}=M$)

$$\Delta v = \frac{\underset{\cdot}{R}}{|M|}.$$

Substituting this quantity into the first two of Eqs. (6–47), and making use of the notation (6–46), we find that α and β satisfy the homogeneous system

$$\begin{aligned} \alpha \int_S x A_z x dS + \beta \int_S x A_z y dS &= 0, \\ \alpha \int_S y A_z x dS + \beta \int_S y A_z y dS &= 0. \end{aligned} \tag{6–48}$$

The determinant of this system is

$$\Delta = \int_S x A_z x dS \cdot \int_S y A_z y dS - \int_S x A_z y dS \cdot \int_S y A_z x dS.$$

We set $\varphi_1 = Ax$. This function is harmonic in τ, and its normal derivative on Σ is zero and satisfies the condition $(\varphi_1)_S = x$ on S.

Similarly, $\varphi_2 = Ay$. Then we can rewrite Δ as

$$\Delta = \int_S \varphi_1 \frac{\partial \varphi_1}{\partial z} dS \cdot \int_S \varphi_2 \frac{\partial \varphi_2}{\partial z} dS - \int_S \varphi_1 \frac{\partial \varphi_2}{\partial z} dS \cdot \int_S \varphi_2 \frac{\partial \varphi_1}{\partial z} dS$$

or, applying Green's formula,

$$\Delta = \int_\tau (\nabla \varphi_1)^2 \, d\tau \cdot \int_\tau (\nabla \varphi_2)^2 \, d\tau - \Big(\int_\tau \nabla \varphi_1 \nabla \varphi_2 d\tau\Big)^2.$$

By virtue of Schwarz's inequality, we immediately obtain

$$\Delta \geqslant 0.$$

The equal sign can appear if and only if φ_1 is proportional to φ_2. But this cannot occur because the functions x and y are independent. Thus, $\Delta > 0$. Consequently, the system (6–48) has only the trivial solution

$$\alpha = \beta = 0.$$

In this case $\gamma = 1$.

Thus, if $\gamma \neq 0$, the only direction of central impact resulting in a translational displacement is the direction along the vertical.

It is not too difficult to check that the solution obtained ($\alpha = \beta = 0$, $\gamma = 1$, $\Delta\Theta_i' = 0$) is satisfied by the remainder of Eqs. (6–45).

Now let $\gamma = 0$. Then the last of Eqs. (6–47) is satisfied identically. The first two equations of this system and the normalizing condition

$$\alpha^2 + \beta^2 = 1$$

allow us to find α, β, and Δv.

Let us now consider the three last equations of the system (6–45). Since $\underset{\cdot}{L} = 0$ and $\boldsymbol{\theta}' = 0$, these equations yield

$$\begin{aligned} \alpha_{41}\alpha + \alpha_{42}\beta &= 0, \\ \alpha_{51}\alpha + \alpha_{52}\beta &= 0, \\ \alpha_{61}\alpha + \alpha_{62}\beta &= 0. \end{aligned} \tag{6–49}$$

These equations, for fixed α and β, can be satisfied only when the coeffients α_{ij} satisfy some specified conditions. Using the notation (6–46), we can represent these conditions in the form

$$\begin{aligned} \int_\tau \nabla\varphi_4^*\nabla\varphi_1^*\,d\tau\cdot\int_\tau \nabla\varphi_5^*\nabla\varphi_2^*\,d\tau &= \int_\tau \nabla\varphi_4^*\nabla\varphi_2^*\,d\tau\cdot\int_\tau \nabla\varphi_5^*\nabla\varphi_1^*\,d\tau, \\ \int_\tau \nabla\varphi_5^*\nabla\varphi_1^*\,d\tau\cdot\int_\tau \nabla\varphi_6^*\nabla\varphi_2^*\,d\tau &= \int_\tau \nabla\varphi_5^*\nabla\varphi_2^*\,d\tau\cdot\int_\tau \nabla\varphi_6^*\nabla\varphi_1^*\,d\tau, \\ \int_\tau \nabla\varphi_6^*\nabla\varphi_1^*\,d\tau\cdot\int_\tau \nabla\varphi_4^*\nabla\varphi_2^*\,d\tau &= \int_\tau \nabla\varphi_6^*\nabla\varphi_2^*\,d\tau\cdot\int_\tau \nabla\varphi_4^*\nabla\varphi_1^*\,d\tau. \end{aligned} \tag{6–50}$$

Since the Stokes-Zhukovskiy potentials are determined solely by the configuration of the cavity, Eqs. (6–50) are conditions imposed on the geometry of the cavity.

Thus, we arrive at the following result: If the shape of the cavity is such that conditions (6–50) are not satisfied, the only direction for the central impact that results in translational displacement that is collinear with the impulse will be the vertical. In the exceptional case when the shape of the cavity does satisfy conditions (6–50), in addition to the vertical direction there exists one additional direction for a horizontal central impact that results in translational motion that is collinear to the impulse.

[6-5.4] Cone of Characteristic Directions

We have thus proven that in the general case a central impact applied to a body with a fluid with a free surface produces not only translational but also rotational motion. It is therefore natural to raise the following question: is it at all possible to strike a body with a fluid in such a manner that it acquires translational motion only and, if so, how should the body be struck?

As a result of the impact, let the body perform translational motion only. We denote the direction cosines of this displacement by α, β, and γ. Then Eqs. (6–45) can be written

$$\begin{aligned} \Delta v(\alpha_{11}\alpha + \alpha_{12}\beta) &= \underset{\cdot}{R}_1\,, \\ \Delta v(\alpha_{21}\alpha + \alpha_{22}\beta) &= \underset{\cdot}{R}_2\,, \\ \Delta v\alpha_{33}\gamma &= \underset{\cdot}{R}_3\,, \end{aligned} \tag{6–51}$$

$$\begin{aligned} \Delta v(\alpha_{41}\alpha + \alpha_{42}\beta) &= y_0\underset{\cdot}{R}_3 - z_0\underset{\cdot}{R}_2\,, \\ \Delta v(\alpha_{51}\alpha + \alpha_{52}\beta) &= z_0\underset{\cdot}{R}_1 - x_0\underset{\cdot}{R}_3\,, \\ \Delta v(\alpha_{61}\alpha + \alpha_{62}\beta) &= x_0\underset{\cdot}{R}_2 - y_0\underset{\cdot}{R}_3\,, \end{aligned} \tag{6–52}$$

where x_0, y_0, and z_0 are the coordinates of the point of application of the impulse. Eliminating $\underset{\cdot}{\boldsymbol{R}}_i$ from (6–52) and dividing through by Δv, we have

$$\begin{aligned} y_0\alpha_{33}\gamma - z_0(\alpha_{21}\alpha + \alpha_{22}\beta) &= \alpha_{41}\alpha + \alpha_{42}\beta\,, \\ x_0\alpha_{33}\gamma + z_0(\alpha_{11}\alpha + \alpha_{12}\beta) &= \alpha_{51}\alpha + \alpha_{52}\beta\,, \\ x_0(\alpha_{21}\alpha + \alpha_{22}\beta) - y_0(\alpha_{11}\alpha + \alpha_{12}\beta) &+ \alpha_{61}\alpha + \alpha_{62}\beta\,. \end{aligned} \tag{6–53}$$

The above system can be regarded as a system of equations for determining x_0, y_0, and z_0. It is easy to check that its determinant is equal to zero for any α, β, and γ.

Thus, for solutions of the system (6–53) to exist, it is necessary and sufficient that the rank of the expanded matrix of the system (6–53) be equal to the rank of the matrix of the homogeneous problem. This condition imposes certain limitations on α, β, and γ; not every direction can serve as the direction for translational motion. These exclusive directions along which translational motion of a body with a fluid is possible after impact will be called *characteristic directions.*

The characteristic directions form a ruled surface, *i.e.*, a characteristic cone, the equation of which we now proceed to find.

The condition for solutions of the system (6–53) to exist can be written

$$\alpha_{41}\alpha + \alpha_{42}\beta + (\alpha_{51}\alpha + \alpha_{52}\beta)\frac{\alpha_{21}\alpha + \alpha_{22}\beta}{\alpha_{11}\alpha + \alpha_{12}\beta} + (\alpha_{61}\alpha + \alpha_{62}\beta)\frac{\alpha_{33}\gamma}{\alpha_{11}\alpha + \alpha_{12}\beta} = 0\,. \tag{6–54}$$

Using the normalization condition

$$\gamma^2 = 1 - \alpha^2 - \beta^2$$

and solving (6–54) for β, we have

$$\bar{\beta} = f(\alpha). \tag{6–55}$$

The equations of the characteristic directions have the form

$$\frac{x}{\alpha} = \frac{y}{\beta} = \frac{z}{\gamma}$$

or, making use of normalization and of Eq. (6–55),

$$\frac{x}{y} = \frac{\alpha}{f(\alpha)}, \quad \frac{z}{x} = \frac{\sqrt{1 - \alpha^2 - f^2(\alpha)}}{\alpha}.$$

Eliminating α from these equations, we obtain

$$\frac{z}{x} = \Phi\left(\frac{x}{y}\right). \tag{6–56}$$

This is the equation sought for the characteristic cone.

We now fix the three numbers α, β, and γ, which belong to the characteristic cone (6–56), *i.e.*, which determine the possible translational motion of the body with a fluid. Let it now be required to determine the characteristic impulse, *i.e.*, the impulse resulting in translational motion of the body along a selected direction. This problem can be resolved by Eqs. (6–53), since α, β, and γ were selected in a manner such that the system (6–53) can be solved for x_0, y_0, and z_0.

From this system, we find

$$\begin{aligned} z_0 &= \frac{\alpha_{33}\gamma}{\alpha_{21}\alpha + \alpha_{22}\beta}\, y_0 - \frac{\alpha_{41}\alpha + \alpha_{42}\beta}{\alpha_{21}\alpha + \alpha_{22}\beta}, \\ z_0 &= \frac{\alpha_{33}\gamma}{\alpha_{11}\alpha + \alpha_{22}\beta}\, x_0 + \frac{\alpha_{51}\alpha + \alpha_{52}\beta}{\alpha_{11}\alpha + \alpha_{22}\beta}. \end{aligned} \tag{6–57}$$

The intersection of these planes determines the line of action of the sought impulse.

We denote the direction cosines of the line of action of the impulse $\underset{\cdot}{\boldsymbol{R}}$ by α_1, β_1, and γ_1. Then the third of Eqs. (6–51) will yield the absolute value of the velocity increment

$$\Delta v = \frac{\underset{\cdot}{\boldsymbol{R}}}{M} \frac{\gamma_1}{\gamma}. \tag{6–58}$$

As can be seen from (6–57), the characteristic impulse will, in general, not be the central impulse. It will be central only in the case when the shape of the cavity satisfies conditions (6–52) or when its line of action is a vertical straight line.

The associated mass M^* of the body-fluid system, as shown by Eq. (2–58), is also not equal to the sum of the masses of the body and the fluid

$$M^* = M\frac{\gamma}{\gamma_1}.$$

Using Eqs. (6–45) we can study a large number of additional problems, for example, the problem of the impact center if the body has two stationary points, or of the impact surface if the body has one stationary point.

[6-5.5] Cauchy's Problem

The impact theory resolves the problem of the velocity distribution of a system after an impact impulse has been applied to it. If the magnitudes of the vectors $\underset{\bullet}{\boldsymbol{R}}$ and $\underset{\bullet}{\boldsymbol{L}}$ have been specified, Eqs. (6–45) can be used to determine $\Delta\boldsymbol{v}_0$ and $\Delta\boldsymbol{\theta}'$, which are the velocity vector of the mass center and the vector of the instantaneous angular velocity. Thus all the quantities, with the exception of $\Delta\varphi$, will be determined by the condition (6–43). We now arrive at the mixed problem of the theory of harmonic functions: to determine a harmonic function $\Delta\varphi$ that takes on specified values on S, and the normal derivative of which becomes zero on Σ.

It was shown above that this problem can be solved if we know the solution of the problem of free vibrations of the fluid inside the cavity.

Having found the function $\Delta\varphi$, we can calculate $\Delta\varphi_z$ and, using the kinematic relationship, determine the initial rate of deformation of the free surface

$$\left(\frac{\partial\zeta}{\partial t}\right)_{t=0} = (\Delta\varphi_z)_{z=0}\,.$$

Thus the solution of the problem of impact on a body with a fluid makes it possible to state Cauchy's problem for a body-fluid system: to define the motion of a body with a fluid if the initial velocity of the mass center of the system, the angular velocity of the rigid body, and the rate of deformation of the free surface are given; the free surface proper in this case is plane (it is undisturbed).

This problem is easily reduced to an infinite system of ordinary differential equations for $q_n(t)$, the initial conditions for the functions $q_n(t)$ in the expansion

$$\zeta(t,P) = \sum q_n \psi_n(xy),$$

where ψ_n are the eigenfunctions of operator H, defined by the formulas

$$q_n(0) = 0,$$
$$q_n'(0) = \int\limits_S (\Delta\varphi_z)_{z=0}\psi_n(x,y)\,dS\,.$$

[6-6] CONCERNING THE OSCILLATIONS OF BODIES FLOATING IN A RESERVOIR OF LIMITED DIMENSIONS

The problem of small oscillations of a body (or a system of bodies) floating in a reservoir of limited dimensions (Fig. 28) is fully analogous to the problems considered above of motion of a rigid body containing fluid masses with free surfaces. From the mathematical point of view, these problems are simply equivalent since they are described by the same system of operational equations[92].

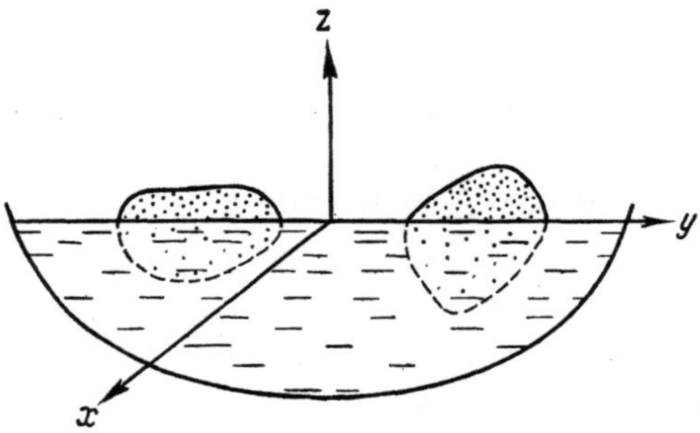

Fig. 28.

For example, the oscillation of one rigid body floating in a vessel of limited dimensions can be described by six time functions Y_i (the generalized coordinates of the rigid body) and the function $\zeta(P, t)$ [free surface], which satisfy the system of integro-differential equations

$$\begin{gathered} Y_i'' + \int\limits_S v_i(x, y)\, \zeta_{tt}(x, y, t)\, dS + \mu_i^2 Y_i = L_i\,, \\ \varrho H \zeta_{tt} + \sum_{j=1}^{6} Y_j'' v_j + \varrho g \zeta = 0\,, \quad i = 1, 2, \ldots, 6\,. \end{gathered} \tag{6–59}$$

In this system, H is the Neumann operator for the region τ, bounded by the walls of the reservoir (vessel), the plane of the free surface at rest, and the boundary of the floating body in the equilibrium position; $\boldsymbol{L}$ is the tensor of the external forces. We note that the region τ is simply connected (see Fig. 28). We will omit the derivation of this system. A detailed presentation of these problems will be found in Ref. [92].

The system (6–59) can be regarded as a particular case of system (5–83), the properties of which were examined in detail in Sec. 2 of this chapter. Hence the principal results formulated there are valid for the system at hand. Thus:

1. For free oscillations of a rigid body floating in a limited volume of fluid, there exist normal oscillations about the equilibrium position ($L \equiv 0$); *i.e.*, the

system (6–59) has solutions in the form

$$Y_{jn} = p_{jn}e^{i\omega_n t}, \quad \zeta_n = z_n e^{i\omega_n t}, \quad j = 1, 2, \ldots, 6, \quad n = 1, 2, \ldots. \tag{6–60}$$

2. The equilibrium position of a rigid body floating in a limited volume of fluid will be stable if and only if the potential energy of the body-fluid system in the equilibrium position is minimum. We use the term stability in the sense that all of the natural frequencies ω_n in (6–60) are real.

3. The set $\{\omega_n\}$ has a single limiting point at infinity, *i.e.*, $\lim_{n\to\infty}\omega_n = \infty$.

4. Any solution of the system (6–59) can be obtained by superposition of solutions such as (6–60).

Let us in addition consider the stability conditions. It is easy to see that the potential energy of the system in this case has the form

$$\Pi = \tfrac{1}{2}\{\sum_{i=1}^{6} \mu_i^2 Y_i^2 + \varrho g \int_S \zeta^2 dS\}.$$

Thus, for positive definiteness of the form Π, it is necessary and sufficient that all the $\mu_i^2 > 0$. This condition has a simple mechanical meaning. For stability of a floating rigid body it is necessary and sufficient that the total moment of the external and hydrostatic forces be restoring.

We note that the result we have formulated points to a qualitative difference in the character of oscillations of a rigid body in an infinite fluid and in a fluid of limited extent. In the boundless fluid the oscillatory motion is damped, since the oscillations of the rigid body bring about vibrations of the fluid, *i.e.*, they induce waves on the surface of the fluid which, receding to infinity, carry away with them the energy of the oscillating body. If the fluid volume is limited, the waves generated by the oscillations of the body are reflected from the walls and again interact with the floating body. This gives rise to undamped vibrations of the rigid body and the fluid.

The system of Eqs. (6–59) can be used for describing various phenomena that arise in engineering. For example, this system describes the rocking of a vessel in a canal lock.

The system (6–59) can be used to study a number of general properties of the phenomenon of impacts on bodies floating in a fluid. In particular, it is easy to establish the following fact: if two bodies are floating in a limited volume of fluid and if one body is struck in a particular manner, the moments of inertia of the second body can be selected in a manner such that the first body will remain stationary after impact, while the second acquires some motion.

7

Fluid Surface Phenomena and Their Effect on the Motion of a Body Containing a Fluid

In the preceding chapters of Part Two we have considered problems in which it was assumed that the fluid is acted upon only by mass forces, *i.e.*, gravitational forces or inertial forces. The free surface of the fluid was considered as free of any stresses. For an extensive group of problems this model is completely satisfactory. But actually each real fluid is also subjected to surface tension forces and, when the mass forces are weak (for example, under conditions close to weightlessness), it is precisely the surface forces that determine the character of the fluid motion. Consequently, study of the dynamics of a body with a fluid under conditions of small overloads requires the consideration of more complex models that take into account the action of capillary forces. This chapter is devoted to the study of these models. We note that the study of the motion of a fluid subjected to surface tension forces is also of intrinsic interest.

[7-1] STATEMENT OF PROBLEMS OF THE THEORY OF VIBRATIONS

[7-1.1] Formulation of the Variational Problem

We shall consider the motion of a fluid inside some stationary shell $\Sigma = \Sigma_1 + \Sigma_2$. The fluid volume τ fills only a part of the vessel and the remaining part is filled with a gas. This gas will be regarded as being at rest and its mass will be disregarded. The mass force field is assumed to be homogeneous. We fix the stationary coordinate system $Oxyz$ to the vector $\boldsymbol{g}^0$, which is a unit vector of the mass force intensity, in such a manner that the Oz axis is directed along the negative direction of the vector $\boldsymbol{g}^0$, as shown in Fig. 29.

We assume, for definiteness, that the xOy plane coincides with the free surface of the fluid in the equilibrium position, where the fluid is subjected to mass forces only. In this case the free surface is a plane. The projection of the free surface S on this plane will be denoted by S^0.

We denote the surface tension coefficient at Σ_2, the rigid body–gas interface, by α_2; the surface tension coefficient at Σ, the rigid body–fluid interface, by α_1; and the surface tension coefficient at S, the fluid-gas interface, by α.

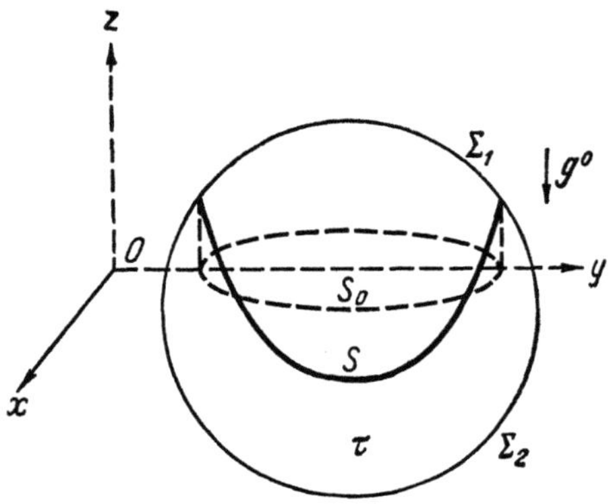

Fig. 29.

Chapter 1 presented the formulation of Hamilton's principle for the general problem of motion of a body with cavities containing a fluid subjected to surface tension forces. Hence we shall restrict ourselves here only to formulating the final result for the particular case considered in this chapter.*

The problem of determining the motion of a massive fluid in a stationary vessel, acted upon by surface tension in addition to gravity reduces to defining the velocity field $\boldsymbol{v}(x, y, z, t)$, the pressure distribution $p(x, y, z, t)$, and the configuration of the free surface, which is specified by the equations

$$\Phi(x, y, z, t) = 0, \tag{7–1}$$

which satisfy the Euler equation

$$\frac{d\boldsymbol{v}}{dt} = \boldsymbol{F} - \frac{1}{\varrho}\nabla p, \tag{7–2}$$

the continuity equation

$$\nabla \cdot \boldsymbol{v} = 0, \tag{7–3}$$

and the following boundary conditions: the condition of impermeability of the shell of the vessel Σ:

$$(v_n)_\Sigma = 0; \tag{7–4}$$

* See also the appendix to this chapter, which presents the derivation of the equations and boundary conditions for the particular case of motion of a fluid in a stationary vessel in more detail than in Chapter 1.

the dynamic condition

$$p - \alpha K = C, \tag{7–5}$$

where C is some constant; and the kinematic condition

$$\frac{d\Phi}{dt} = 0. \tag{7–6}$$

The dynamic condition (7–5) is frequently called the condition of thermodynamic equilibrium of the free surface.

Conditions (7–5) and (7–6) are satisfied at points of the free surface, *i.e.*, at points satisfying Eq. (7–1). We recall that the expression for K in Eq. (7–5) denotes the average curvature

$$K = \left(\frac{1}{R_1} + \frac{1}{R_2}\right). \tag{7–7}$$

Here the principal radius of curvature is regarded as positive if the center of curvature lies on the same side as the fluid. The expression for the average curvature contains second derivates of the configuration of the boundary. Consequently, the boundary-value problem for a fluid subjected to capillary forces contains, in the dynamic boundary condition, second-order derivatives. These boundary conditions (in contrast to the case when capillary forces do not act) are insufficient for isolating a unique solution in the steady-state case. For this we must have yet another condition [see (1–31)], *i.e.*, the contact angle θ at the contour Γ is equal to some fixed quantity, which is determined solely by the physical properties of the fluid, gas, and shell:

$$\cos\theta = \frac{\alpha_2 - \alpha_1}{\alpha}. \tag{7–8}$$

In the study of nonsteady-state problems these conditions must be supplemented by Cauchy's conditions.

[7-1.2] Linearization

To formulate the problem of small vibrations of a fluid subjected to surface tension forces, we must linearize Eqs. (7–1)–(7–8), assuming that the velocities of fluid particles are low. For this it is first necessary to be able to solve the static problem, *i.e.*, the problem of determining the equilibrium configuration. This problem does not arise in the theory of vibrations of a massive fluid that is not subjected to surface tension forces, since under these conditions the free surface is a plane.

Setting $\boldsymbol{v} \equiv 0$ and assuming that the functions are not time dependent, we find that the solution of the static problem reduces to finding a function Φ_0

that defines the configuration of the free surface (S^0) when it is at rest

$$\Phi_0(x, y, z) = 0 \tag{7-9}$$

and the pressure distribution $p(x, y, z)$ that satisfies the expressions

$$\begin{aligned} \boldsymbol{F} - \frac{1}{\varrho} \nabla p_0 &= 0, \\ p_0 - \alpha K_0 &= 0. \end{aligned} \tag{7-10}$$

Here K_0 denotes the average curvature of the surface Φ_0.

Denoting the potential energy of the unit volume $F = -\nabla \Pi_1$ by Π_1, we can rewrite the first of Eqs. (7–10) in the form

$$p_0 = -\varrho \Pi_1 + C, \tag{7-11}$$

where C is an arbitrary constant which, without loss of generality, can be assumed to be zero. The function Φ_0 should, in addition, satisfy condition (7–8). We now assume that the static problem has been solved, *i.e.*, that the functions Φ_0 and p_0 are known. We now linearize problem (7–1)–(7–8) and set

$$p = p_0 + p'.$$

Regarding p' and v as first-order infinitesimals, we rewrite (7–2), dropping second- and higher-order infinitesimals,

$$\frac{\partial \boldsymbol{v}}{\partial t} = -\nabla \left(\frac{p'}{\varrho} \right). \tag{7-12}$$

The equation of the free surface (7–1) is now written in the form

$$\Phi \equiv \Phi_0(x, y, z) + N\psi(x, y, z, t) = 0, \tag{7-13}$$

where N is a normalizing multiplier. The function ψ is regarded as a first-order infinitesimal. Hence in the linear problem the normalizing multiplier is calculated on the basis of values of Φ in the state of rest:

$$N = |\nabla \Phi_0|. \tag{7-14}$$

We now consider the kinematic condition (7–6). It is easy to see that this condition can be rewritten in the form

$$\frac{\partial \Phi}{\partial t} + \nabla \Phi \boldsymbol{v} = 0. \tag{7-15}$$

But $\nabla \Phi = \boldsymbol{n}^0 |\nabla \Phi|$, where $\boldsymbol{n}^0$ is the unit vector of the normal to surface S.

Consequently, (7–15) can additionally be rewritten in the form

$$\frac{\partial \Phi}{\partial t} + |\nabla \Phi| v_n = 0. \tag{7-16}$$

Now substituting (7–13) into the above expression, we obtain in the linear case a new form for the kinematic relationship:

$$\frac{\partial \psi}{\partial t} + v_n = 0. \tag{7–17}$$

Condition (7–17) should be satisfied at the surface S^0. Since, by virtue of incompressibility, $\int_{s^0} v_n dS = 0$, we have $\int_{s^0} \psi dS = \text{const}$. However, it is not difficult to show that this constant is equal to zero. For this we introduce in the vicinity of S^0 the curvilinear system of coordinates ξ, η, and δ:

$$\begin{aligned} x &= X(\xi, \eta) + \delta(\boldsymbol{n}^0, \boldsymbol{x}^0) = X(\xi, \eta) + \frac{\delta}{N}\left(\frac{\partial \Phi_0}{\partial x}\right), \\ y &= Y(\xi, \eta) + \delta(\boldsymbol{n}^0, \boldsymbol{y}^0) = Y(\xi, \eta) + \frac{\delta}{N}\left(\frac{\partial \Phi_0}{\partial y}\right), \\ z &= Z(\xi, \eta) + \delta(\boldsymbol{n}^0, \boldsymbol{z}^0) = Z(\xi, \eta) + \frac{\delta}{N}\left(\frac{\partial \Phi_0}{\partial z}\right). \end{aligned} \tag{7–18}$$

Here δ is the length of the perpendicular dropped from the point (x, y, z) to the surface S^0. The values of the derivatives and of the normalizing multiplier N are taken on the surface S^0.

We now refine the structure of the coordinate grid (ξ, η) on the surface S^0. Consider any point A on the line of intersection Γ_0 of surfaces S^0 and Σ. We construct a line $\eta = \text{const}$ that will, in the region Γ_0, coincide with the line along which the surface S is cut by the plane D, which is drawn through the unit vectors of the normals $\boldsymbol{n}_{S^0}$ and $\boldsymbol{n}_\Sigma$ to the surfaces S^0 and Σ in point A (Fig. 30). The line $\xi = 0$ is made to coincide with Γ_0. Then the coordinate grid (ξ, η) will be orthogonal in the vicinity of Γ_0.

In this coordinate system the equation of the surface S^0 will be $\delta = 0$. Consequently,

$$\Phi_0(\xi, \eta) = \Phi_0[X(\xi, \eta), \quad Y(\xi, \eta), \quad Z(\xi, \eta)] = 0.$$

Utilizing this condition, we perform the substitution of (7–18) in Eq. (7–13). Then, retaining first-order infinitesimals, we reduce the latter equation to the form

$$\Phi \equiv \frac{\delta}{N}\left[\left(\frac{\partial \Phi}{\partial x}\right)^2 + \left(\frac{\partial \Phi}{\partial y}\right)^2 + \left(\frac{\partial \Phi}{\partial z}\right)^2\right] + N\psi(\xi, \eta, t) = 0.$$

Here

$$\psi(\xi, \eta, t) = \psi(X(\xi, \eta), \quad Y(\xi, \eta), \quad Z(\xi, \eta)).$$

Utilizing the definition of N, we obtain the equation of the free surface in the form

$$\Phi \equiv N\tilde{\Phi} = 0$$

or

$$\tilde{\Phi} \equiv \delta + \psi(\xi, \eta, t) = 0, \tag{7–19}$$

where

$$|\nabla\tilde{\Phi}| = 1 + O(\psi^2).$$

Since at any given time instant, by virtue of incompressibility, $\int_{^0S} \delta dS = 0$, we immediately have from (7–19) the desired result

$$\int_{^0S} \psi(\xi, \eta, t)\, dS = 0. \tag{7–20}$$

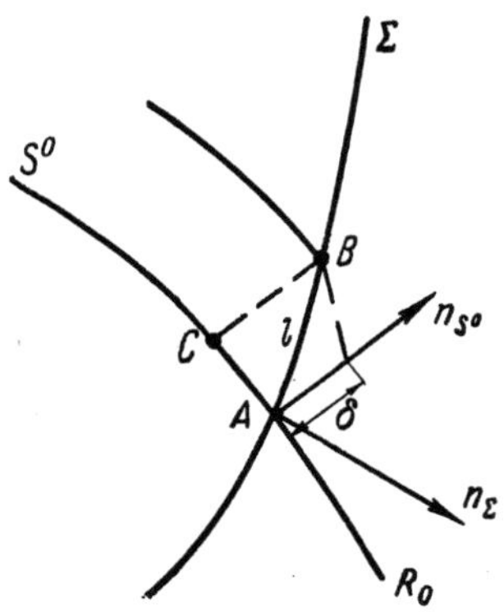

Fig. 30.

The average curvature K can be regarded as some operator that operates in the space of functions describing the free surface. It matches to any function ψ a function $K(x, y, z)$, which is defined in each point of this surafce

$$K = K\psi.$$

We shall assume that this operator is differentiable, and we denote its linear part by L.* Then operator $K\psi$ can be represented in the form

$$K\psi = K_0 + L\psi + O(\psi^2),$$

where K_0 is the value of the average curvature of surface S^0. We rewrite Eq. (7–5) in the form

$$p_0 + p' - \alpha[K_0 + L\psi + O(\psi^2)] = \text{const}.$$

This condition should be satisfied at the surface S. Making use of (7–10) and (7–11), from which it follows that

$$(\Pi_1)_{S_0} + \alpha K_0 = \text{const},$$

* More precisely, L is the Fréchet derivative of operator K at the point $\psi = 0$.

we transform the dynamic relationship

$$-\varrho\Delta\Pi_1 + p' - \alpha\big(L\psi + O(\psi^2)\big) = 0.$$

Here $\Delta\Pi_1 = (\Pi_1)_S - (\Pi_1)_{S^0}$. Dropping second-order infinitesimals, we finally obtain

$$-\varrho\Delta\Pi_1 + p' - \alpha L\psi = 0. \tag{7–21}$$

The above condition should be satisfied along the surface S^0.

Let us now consider condition (7–8). It is more convenient to write it in the form

$$(\boldsymbol{n}_S \cdot \boldsymbol{n}_\Sigma) = \gamma = \cos\theta. \tag{7–22}$$

Here $\boldsymbol{n}_S$ and $\boldsymbol{n}_\Sigma$ are the unit vectors of the external normals to the surfaces S and Σ, respectively. The quantity $\boldsymbol{n}_S = \nabla\tilde{\Phi}/|\nabla\tilde{\Phi}|$. The function $\tilde{\Phi}(\xi, \eta, \delta) = \delta + \psi(\xi, \eta)$ is defined at any point $P(\xi, \eta, \delta)$ if the function $\psi(\xi, \eta)$, which is regarded as a first-order infinitesimal, is determined. Consequently, the vector $\boldsymbol{n}_S$ is defined at any point $P(\xi, \eta, \delta)$. This makes it possible to evaluate the scalar product (7–22) at point B on the basis of values of the vectors $\boldsymbol{n}_S$ and $\boldsymbol{n}_\Sigma$ and their derivatives at the nearby point A (see Fig. 30). At any point of $A \in \Gamma_0$, which is the line of intersection between the surfaces S^0 and Σ, the following condition is satisfied:

$$(\boldsymbol{n}_{S^0} \cdot \boldsymbol{n}_\Sigma) = \gamma. \tag{7–23}$$

We draw plane D through point A in such a manner that the vectors $\boldsymbol{n}_{S^0}$ and $\boldsymbol{n}_\Sigma$ lie in it. Then the plane D is perpendicular to the line Γ_0, *i.e.*, to the vector $\boldsymbol{\eta}^0$ in point A. In Fig. 30 the line AB designates the line of intersection of plane D and surface Σ. Points A and B lie on the lines of intersection of the surfaces S^0 and Σ^0 with S and Σ.

We now transform (7–22)

$$(\boldsymbol{n}_S, \boldsymbol{n}_\Sigma)_B = (\boldsymbol{n}_S, \boldsymbol{n}_\Sigma)_A + l_{AB}\left\{\left(\frac{\partial \boldsymbol{n}_S}{\partial l_{AB}} \cdot \boldsymbol{n}_\Sigma\right) + \left(\frac{\partial \boldsymbol{n}_\Sigma}{\partial l_{AB}} \cdot \boldsymbol{n}_S\right)\right\}_A. \tag{7–24}$$

We evaluate the first term on the right-hand side of the above equation:

$$(\boldsymbol{n}_S \cdot \boldsymbol{n}_\Sigma)_A = (\boldsymbol{F}_{S^0} \cdot \boldsymbol{n}_\Sigma) + \frac{\partial\psi}{\partial\xi}(\xi^0 \cdot \boldsymbol{n}_\Sigma) + \frac{\partial\psi}{\partial\eta}(\boldsymbol{\eta}^0 \cdot \boldsymbol{n}_\Sigma) + \cdots.$$

In fact,

$$\boldsymbol{n}_S = \frac{\boldsymbol{\delta}^0 + \dfrac{\partial\psi}{\partial\xi}\xi^0 + \dfrac{\partial\psi}{\partial\eta}\boldsymbol{\eta}^0}{\left(1 + \left(\dfrac{\partial\psi}{\partial\xi}\right)^2 + \left(\dfrac{\partial\psi}{\partial\eta}\right)^2\right)^{1/2}} = \boldsymbol{\delta}^0 + \frac{\partial\psi}{\partial\xi}\xi^0 + \frac{\partial\omega}{\partial\eta}\boldsymbol{\eta}^0 + O(\psi^2).$$

But the direction of the line δ on S^0 coincides with the direction of the normal to this surface; *i.e.*, $\boldsymbol{n}_{S^0} = \boldsymbol{\delta}^0$.

We note further that

$$(\xi^0 \boldsymbol{n}_\Sigma) = \sin\theta\,, \quad (\boldsymbol{\eta}^0 \boldsymbol{n}_\Sigma) = 0\,.$$

Hence, finally dropping second-order infinitesimals, we obtain

$$(\boldsymbol{n}_S \cdot \boldsymbol{n}_\Sigma)_A = \cos\theta + \frac{\partial\psi}{\partial\xi}\sin\theta\,.$$

Using the representation (7–24), we rewrite (7–22), retaining in it only first-order infinitesimals:

$$\frac{\partial\psi}{\partial\xi}\sin\theta + l_{AB}\left\{\left(\frac{\partial \boldsymbol{n}_S}{\partial l_{AB}}\cdot \boldsymbol{n}_\Sigma\right) + \left(\frac{\partial \boldsymbol{n}_\Sigma}{\partial l_{AB}}\cdot \boldsymbol{n}_S\right)\right\} = 0\,. \tag{7–25}$$

We note that in the above expression the quantities contained in the braces can be evaluated without taking into account first-order infinitesimals, since the braces are multiplied by the factor l_{AB}, which is a first-order infinitesimal.

The vector n_Σ coincides with the normal to the plane curve AB only at point A; however, on the assumption of continuous differentiability of the surface Σ, it differs from it by first-order infinitesimals. Hence both of these vectors can be assumed to be identical over the entire arc AB. On this basis

$$\frac{\partial \boldsymbol{n}_\Sigma}{\partial l_{AB}} = -\,K_\Sigma \boldsymbol{l}^0\,,$$

where K_Σ is the curvature of the normal section of the surface Σ at point A in a plane perpendicular to Γ_0, and $\boldsymbol{l}^0$ is unit vector of the tangent to line AB. In addition, the vector $\boldsymbol{n}_{S0}$ differs from the vector $\boldsymbol{n}_S$ by a first-order infinitesimal. Thus,

$$\left(\frac{\partial \boldsymbol{n}_\Sigma}{\partial l_{AB}}\cdot \boldsymbol{n}_S\right) = -\,K_\Sigma(\boldsymbol{l}^0\cdot \boldsymbol{n}_{S0}) = -\,K_\Sigma\sin\theta\,.$$

We further note that (see Fig. 30)

$$\frac{\partial \boldsymbol{n}_S}{\partial l_{AB}} = \frac{\partial \boldsymbol{n}_S}{\partial l_{AB}}\cos\theta = -\,\frac{\partial \boldsymbol{n}_S}{\partial\xi}\cos\theta = -\,\frac{\partial \boldsymbol{n}_{S0}}{\partial\xi}\cos\theta + O(\psi)\,.$$

Repeating the considerations presented above, we obtain

$$\left(\frac{\partial \boldsymbol{n}_S}{\partial l_{AB}}\cdot \boldsymbol{n}_\Sigma\right) = -\cos\theta\left(\frac{\partial \boldsymbol{n}_{S0}}{\partial\xi}\cdot \boldsymbol{n}_\Sigma\right) = \cos\theta K_S(\xi^0 \boldsymbol{n}_\Sigma) = \cos\theta\sin\theta K_S\,.$$

Here K_S denotes the curvature of the normal section of the surface S at point A in a plane perpendicular to Γ_0. We note further (see Fig. 30) that $l_{AB} = \delta/\sin\theta = -\psi/\sin\theta$. Collecting all the results obtained up to now, we bring condition (7–22) to the form

$$\frac{\partial \psi}{\partial n_\Gamma} + \omega\psi = 0, \tag{7-26}$$

where

$$\omega = \frac{K_\Sigma - K_S \cos\theta}{\sin\theta},$$

and $\boldsymbol{n}_\Gamma$ is a unit direction vector characterizing the intersection of the plane D by the plane tangent to S^0 at point A.

Condition (7–26) should be satisfied on the curve Γ_0. In contrast to the nonlinear problem, the boundary condition in the linear problem is homogeneous; here the coefficient ω is determined by a physical property of the material (angle θ), the vessel, and the fluid, and by the shape of the vessel (K_S and K_Σ).

Thus, in the linear formulation the problem reduces to determining the vector function $\boldsymbol{v}(x, y, z, t)$, the function $p(x, y, z, t)$, *i.e.*, the pressure, and the function $\psi(x, y, z, t)$ which defines the free surface. These functions must be satisfied by the equations and conditions (7–12), (7–20), (7–21) and (7–26).

It was established in Sec. 1 of Chapter 5 that, within the framework of the linear theory, it is sufficient to consider only potential flows, since the natural frequencies, the configuration of the free surface, and the pressure distribution do not depend on the vortex component of the velocity field. The same result is also valid in the linear theory of the motion of a fluid subjected to surface tension forces. In fact, as was shown in Chapter 1, the pressure in the linear theory is determined only by the potential component of the velocity field; here this fact follows solely from the structure of the linearized Euler equation and does not in any way depend on the character of the boundary conditions. Consequently, the dynamic relationship (7–21) will contain the function p', which is defined by the Cauchy-Lagrange integral. Using the linearized equation (7–12), we can write the Cauchy-Lagrange integral for the disturbances as

$$\frac{\partial \varphi}{\partial t} + \frac{p'}{\varrho} = 0. \tag{7-27}$$

By definition of the potential component at the boundary of the region τ, $v_n = \partial\varphi/\partial n$; hence the function ψ, according to Eq. (7–17), is also solely determined by the potential component.

Thus, in the linear formulation it is sufficient to study only those problems in which the flow of the fluid is potential. Consequently, the problem of the preceding subsection can be simplified even further. It reduces to finding a function $\varphi(x, y, z, t)$ that is harmonic in τ and a function $\psi(x, y, z, t)$ satisfying the following conditions on S:

$$\varrho \frac{\partial \varphi}{\partial t} + \varrho \Delta \Pi_1 - \alpha L \psi = 0, \quad \frac{\partial \psi}{\partial t} + \frac{\partial \varphi}{\partial n} = 0. \tag{7-28}$$

In addition, the function φ should satisfy the condition of impermeability of the shell (surface Σ)

$$\frac{\partial \varphi}{\partial n} = 0, \tag{7-29}$$

while the function ψ should satisfy condition (7–26) on Γ_0. Since for small ψ, Π_1 on the surface S can be written as $\varrho \Delta \Pi_1 = \kappa(x, {}^5, z)\, \psi$, where κ is some known function, the first of conditions (7–28) will be written in the form

$$\varrho \frac{\partial \varphi}{\partial t} + \kappa \psi + \alpha L \psi = 0. \tag{7-30}$$

[7-2] PROBLEM OF THE EQUILIBRIUM CONFIGURATION OF THE FREE SURFACE

[7-2.1] Formulation of the Static Problem

Conditions (7–28) are satisfied on the surface S^0; consequently, the first problem that arises in the theory is that of defining the surface S^0, *i.e.*, the equilibrium configuration of the free surface. We shall assume that the fluid is at rest and shall consider the boundary-value problem defined by Eqs. (7–8), (7–10), and (7–11).

Equation (7–11) shows that along surface S^0

$$P = -\varrho \Pi_1 + \text{const}.$$

Consequently, the second of Eqs. (7–10) takes on the form

$$\varrho \Pi_1 + \alpha K = \text{const}. \tag{7-31}$$

Conditions (7–4) and (7–6) will then be satisfied automatically.

Thus the static problem reduces to determining a single unknown function $\Phi(x, y, z)$, the zero equality of which denotes a free surface. This function should satisfy Eq. (7–31) and boundary condition (7–8). From the incompressibility condition there also follows the isoperimetric condition, *i.e.*, the volume bounded by the surface $\Phi = 0$ is specified

$$\tau = \text{const}. \tag{7-32}$$

Equations (7–31) and (7–32) are, in substance, linear equations. The operator

$\tau(\Phi)$ is an integral operator. The operator K is a differential operator and contains second-order partial derivatives. In the general case of a non-homogeneous gravitational field, or in the case where the mutual attraction of masses is taken into account, the potential energy will also be some integral operator, defined on the set of function Φ. Thus the static problem reduces to a complicated boundary-value problem for a nonlinear integro-differential equation. In the general case it is impossible to obtain a solution to this problem in closed form. Hence it is important to study possible simplifications and to isolate integrable cases. One of the simplifications involves the assumption that the function $z=\zeta(x, y)$ is a single-valued function of the point (x, y). In this case, $\Pi_1 = \varrho g \zeta$ and Eq. (7–31) takes on the form

$$\zeta + \frac{\alpha}{\varrho g} K\zeta = \text{const}. \tag{7–33}$$

Here $K=K\zeta$ is the average curvature, given by

$$K\zeta = \pm \frac{\zeta_{xx}(1+\zeta_y^2) + 2\zeta_x\zeta_y + \zeta_{yy}(1+\zeta_x^2)}{(1+\zeta_x^2+\zeta_y^2)^{1/2}}. \tag{7–34}$$

The plus sign is used when the fluid and the center of curvature are located on the same side of the surface $z=\zeta$; in the opposite case the minus sign is used.

We denote the characteristic linear dimension by l, whereupon we can rewrite Eq. (7–33) in the dimensionless form

$$B\zeta + K\zeta = \text{const}. \tag{7–35}$$

We see that the configuration of the free surface for geometrically similar vessels depends on only one parameter $B=\varrho g l/\alpha$, which is called the Bond number. Low Bond numbers correspond to low-gravity conditions; large Bond numbers correspond to the case when the gravitational forces perceptibly exceed the surface tension forces. Cases when the Bond numbers are very small or very large are the simplest.

[7-2.2] Weightless Fluid

Since in this case $B=0$, Eq. (7–35) yields

$$K\zeta = \text{const}.$$

Thus the free surface of the weightless fluid is a surface of constant curvature; *i.e.*, it is a part of a sphere. Its radius and the position of the center of the sphere are determined from the isoperimetric condition and from the requirement that the contact angle be equal to some specified value. The same conditions define the line of intersection of the surfaces Σ and S^0.

Let, for example, the region occupied by the fluid have an axis of symmetry; then the curve Γ (the line of intersection of surfaces S^0 and Σ) is a plane curve (Fig. 31). We can specify the equation of the surface Σ in the form $r=F(h)$, where r is the distance of the point from the axis of symmetry and h is the height, measured from the bottom of the vessel. Since the center of the

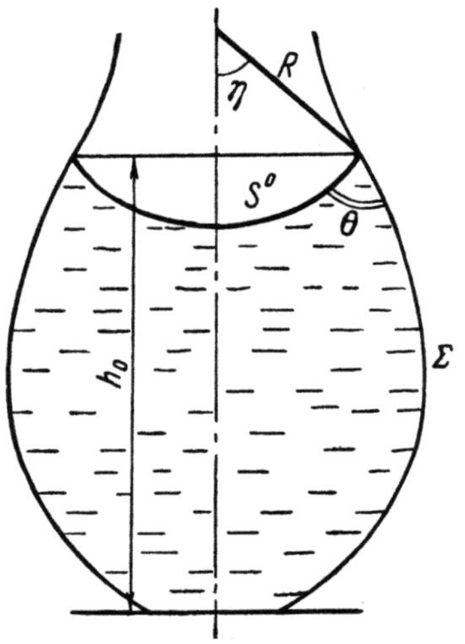

Fig. 31.

spherical surface S^0 is on the axis of symmetry, the problem reduces to finding two parameters: the radius of the sphere R and the height h_0 (see Fig. 31). To define them we have two equations:
the isoperimetric condition:

$$\pi \int_0^{h_0} F^2(h)\, dh - \tfrac{1}{3}\pi R^3 (1 - \cos\alpha)^2 \times (2 + \cos\alpha) = \tau ; \qquad (7\text{–}36)$$

and the condition of the "contact angle":

$$F'(h_0) = \cot(\eta + \theta) .$$

The first term in (7–36) is the volume of the body of revolution bounded by the surface Σ and the plane $h=h_0$. The second term is the volume of a spherical segment with a subtended angle of 2η and with a radius R. In the case under study the problem reduces to solving two transcedental equations, since the angle η is uniquely determined by the magnitude of h_0 and R:

$$\eta = \arcsin \frac{F(h_0)}{R} .$$

If the surface Σ is a sphere with radius R^* (Fig. 32), the problem is simplified greatly:

$$\tfrac{1}{3}\pi R^{*3}\,[4-(1-\cos\beta)^2(2+\cos\beta)] - \tfrac{1}{3}\pi R^3(1-\cos\eta)(2+\cos\eta) = \tau\,.$$

Here the angle β is defined uniquely by the angle α and the radius R

$$R^*\sin\beta = R\sin\eta\,,$$

the angle η, in turn, is related to the contact angle

$$\theta+\eta+\beta \equiv \theta+\eta+\arcsin\frac{R}{R^*}\sin\eta = \pi\,.$$

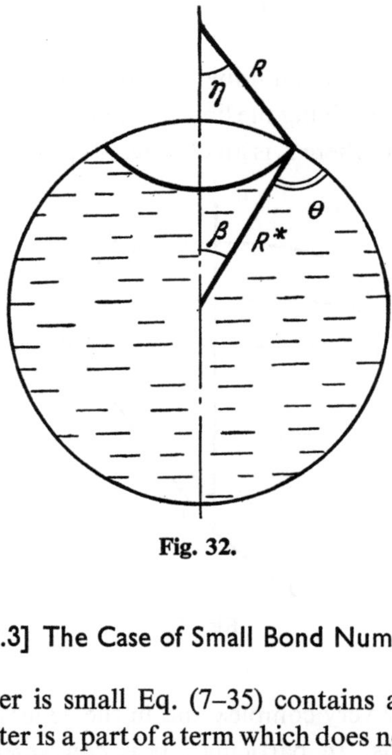

Fig. 32.

[7-2.3] The Case of Small Bond Numbers

If the Bond number is small Eq. (7–35) contains a small parameter; in addition, this parameter is a part of a term which does not contain derivatives. Consequently, the Poincaré method can be used for constructing Eq. (7–32).

Thus we shall assume that B in Eq. (7–35) is a first-order infinitesimal. We set

$$\zeta = \zeta_0 + fB\,,$$

where f is of the order of unity. The operator K is an analytical operator. We denote its linear part by L. Then

$$K\zeta = K\zeta_0 + BLf + O(B^2)\,.$$

We shall regard the function ζ_0 as the specified solution of the equation $K\zeta_0 = \text{const}$. On this basis we rewrite Eq. (7–35) in the form

$$Lf = \varphi\,, \tag{7–37}$$

where

$$\varphi = -\zeta_0 + O(B)\,.$$

We denote the line of intersection of the surfaces Σ and $z = \zeta_0$ by Γ_0. Then, repeating the deliberations that brought us to Eq. (7–26), we find the condition that the function f should satisfy on Γ_0:

$$\frac{\partial f}{\partial n_\Gamma} \sin\gamma + f\,(K_S \cos\gamma - K_\Sigma) = 0\,. \tag{7–38}$$

Thus in the case of small Bond numbers the determination of the equilibrium configuration reduces to solving the linear boundary-value problem described by Eqs. (7–37) – (7–38), where L is an elliptical-type operator.

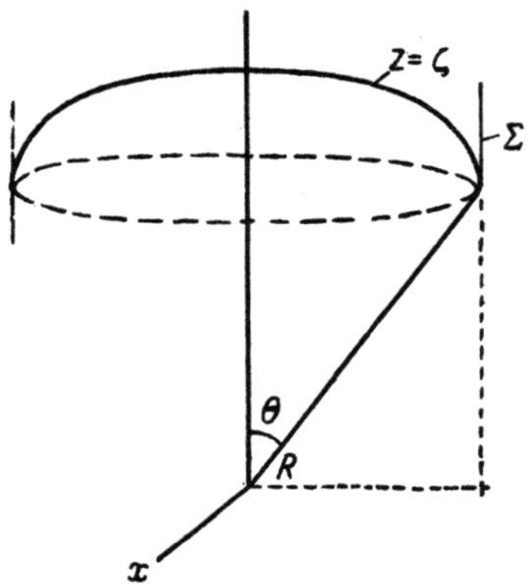

Fig. 33.

This problem is still very complex and in the general case must be solved by numerical methods. In particular, it is convenient to use variational methods for this purpose. According to the general theory, the problem (7–37) – (7–38) reduces to the variational problem for the functional

$$(Lf\,, f) + (f\,, \varphi)\,.$$

As we shall see in the following section, the operator L is a Legendre operator, and consequently the coordinate functions of the Ritz method can be constructed by resorting to spherical functions.

[7-2.4] Problems in Spherical Coordinates

We shall now consider in more detail the case when the line Γ_0 is a circle of radius R^*. This case corresponds, for example, to axially symmetrical vessels when the axis is identical with the direction of the gravity force. In this case it is convenient to use spherical coordinates. We place the origin at the center of the surface ζ_0, which is a part of a spherical surface of radius R (Fig. 33):

$$\left.\begin{aligned} x &= r \sin\vartheta \cos\psi\,, \\ y &= r \sin\vartheta \sin\psi\,, \\ z &= r \cos\vartheta\,. \end{aligned}\right\} \tag{7–39}$$

We shall describe the surface S^0 by the expression

$$r = \varphi(\vartheta, \psi) = R + Bf(\vartheta, \psi)\,. \tag{7–40}$$

Here, without any loss of generality, we can assume that $R=1$ (*i.e.*, we can use the radius as a characteristic scale). Using Eq. (7–40), we can represent the expression for $K\zeta$ in the form

$$K\zeta = 2 + BLf + O(B^2)\,, \tag{7–41}$$

where

$$Lf = -\frac{1}{\sin^2\vartheta}\frac{\partial^2 f}{\partial\psi^2} + \frac{1}{\sin\vartheta}\frac{\partial}{\partial\vartheta}\left(\sin\vartheta\,\frac{\partial f}{\partial\vartheta}\right) + 2f\,.$$

Equation (7–35) then takes on the form

$$B(1 + Bf)\cos\vartheta \mp 2 \mp BLf = \text{const}\,. \tag{7–42}$$

Since the above equation should be satisfied when $B=0$, the constant is equal to two. Assuming that the function Bf is of the order of the Bond number, and dropping second-order infinitesimals, we finally obtain

$$Lf = \pm\cos\vartheta\,. \tag{7–43}$$

The sign is selected on the basis of the sign of the Bond number, *i.e.*, by the position of the fluid above or below the free surface.

The form of the right-hand side of Eq. (7–43) shows that we should seek the solution in the form

$$f = f(\vartheta)\,.$$

Then we can rewrite (7–43) as

$$Lf \equiv \frac{1}{\sin\vartheta}\frac{d}{d\vartheta}\left(\sin\vartheta\,\frac{df}{d\vartheta}\right) + 2f = \pm\cos\vartheta\,. \tag{7–44}$$

The general solution of the above equation can be obtained in quadratures.

By setting $\cos\vartheta = \xi$ we can reduce this equation to the form

$$Lf \equiv \frac{d}{d\xi}\left[(1-\xi^2)\frac{df}{d\xi}\right] + 2f = \pm\,\xi\,. \tag{7–45}$$

The equation $Lf=0$ is Legendre's equation. Its general solution has the form

$$f = C_1 P_1 + C_2 Q_1\,,$$

where P_1 and Q_1 are Legendre functions of the first and second kind:

$$P_1 = \xi\,, \quad Q_1 = \xi \ln\frac{1+\xi}{1-\xi} - 2\,.$$

Using the method of variation of constants, the general solution of Eq. (7–45) can be represented as

$$f = D_1\xi + D_2\left(\xi\ln\frac{1+\xi}{1-\xi} - 2\right) \pm$$
$$\pm\frac{1}{2}\int_0^{\xi}\left[\left(\xi\ln\frac{1+\xi}{1-\xi} - 2\right)\xi_1 - \left(\xi_1\ln\frac{1+\xi_1}{1-\xi_1} - 2\right)\xi\right](1-\xi_1^2)\,d\xi_1\,. \tag{7–46}$$

The constants D_1 and D_2 in the above expression are determined from the boundary conditions, which, as is easy to check, have in the given case the form

$$\begin{aligned} \frac{df}{d\xi} + \omega f &= 0 \quad \text{when} \quad \xi = \xi_0\,, \\ \frac{df}{d\xi} - \omega f &= 0 \quad \text{when} \quad \xi = -\,\xi_0\,. \end{aligned} \tag{7–47}$$

Here $\omega = \cos\theta - c$. The dimensionless parameter c is equal to the ratio of the radius of the surface S^0 to the radius of curvature of the surface Σ, at the point of contact.

[7-2.5] The Case of Large Bond Numbers

We now pass on to the case when the surface tension forces at the point of contact are small in comparison with the gravitational attraction, *i.e.*, when the dimensions of the vessel are sufficiently large, or when the gravitational attraction is strong. Denoting $B = 1/\varepsilon^2$, we rewrite Eq. (7–35) as

$$\zeta = \varepsilon^2 K\zeta + h\,, \tag{7–48}$$

where h is the isoperimetric constant.

The above equation contains a small parameter in the term containing higher-order derivatives. It therefore should be expected that the solution of the boundary-value problem for Eq. (7–48) for small ε will have the property of a boundary-layer type function; *i.e.*, the function $\zeta(x, y)$ will differ from a constant only in some narrow region adjoining the contour. These considerations are fully substantiated experimentally (and by everyday observation). If the dimensions of the vessel are appreciable, the free surface of the fluid is practically horizontal.

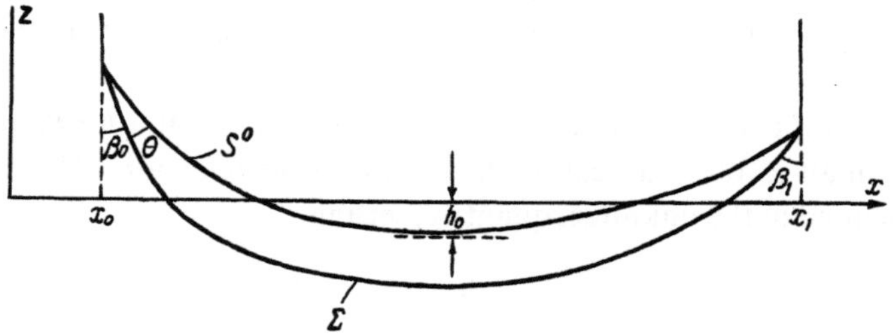

Fig. 34.

We first consider the two-dimensional case. Then the problem reduces to determining the function $\zeta(x)$ that satisfies the following equation and boundary conditions (Fig. 34):

$$\zeta = \pm \frac{\varepsilon^2 \zeta''}{(1 + \zeta'^2)^{3/2}} + h, \tag{7–49}$$

$$\begin{aligned} \zeta' &= -\cot(\beta_0 + \theta) \quad \text{when} \quad x = x_0, \\ \zeta' &= \cot(\beta_1 + \theta) \quad \text{when} \quad x = x_1. \end{aligned} \tag{7–50}$$

Here ζ is the elevation of the free surface above the level of the fluid when $\varepsilon = 0$; β_0 and β_1 are the angles of inclination of the contour Σ to the vertical at the points of contact x_0 and x_1.

The plus sign in Eq. (7–49) is used in the case when the fluid is beneath the free surface (as in Fig. 34), and the minus sign when the fluid is above it.

The boundary-value problem described by Eqs. (7–49) – (7–50) has the peculiar feature that the points of contact are not previously specified. In addition, the constant h is unknown. To determine them, we have the three conditions

$$\begin{aligned} \zeta(x_0) &= F_0(x_0), \\ \zeta(x_1) &= F_1(x_1), \\ V &= 0, \end{aligned}$$

where V is the volume bounded by the surface $z = \zeta(x)$, the walls of the vessel, and the straight line $z=0$; $z=F_0(x)$ and $z=F_1(x)$ are the equations of the walls of the vessel, and the last condition is the isoperimetric condition.

Remark

The methods of boundary-layer theory are based on a change of variables that makes it possible to determine second-degree terms (higher-order terms with respect to ε) and to replace the starting equation with a simpler one, which is called degenerate (or the equation of the boundary layer). It is easy to check that Eq. (7–49) cannot be simplified as $\varepsilon \to 0$, since it becomes identical with the degenerate equation; *i.e.*, it is itself the equation of the boundary layer.

When $\varepsilon=0$ the solution of (7–49) has the form $\zeta=0$ and $h=0$. Thus h is small for small ε. We shall call it the *displacement thickness.* This constant can be included in the unknown quantity, setting

$$\zeta = \eta + h\,, \tag{7–51}$$

since $Kh=0$. After this replacement, Eq. (7–49) has the form

$$\eta = \pm \frac{\varepsilon^2 \eta''}{(1+\eta'^2)^{3/2}}\,. \tag{7–52}$$

The main assertion of boundary-layer theory (which is not proved here) consists in the following. The solution of the boundary-value problem (7–49) can be represented

$$\zeta = h + \eta_1 + \eta_2 + \text{higher-order terms} \tag{7–53}$$

where η_1 and η_2 are some solutions of Eq. (7–52) satisfying the conditions

$$\begin{aligned} &\eta_1 \to 0, \quad \eta_1' \to 0 \quad \text{as} \quad x \to +\infty\,, \\ &\eta_2 \to 0, \quad \eta_2' \to 0 \quad \text{as} \quad x \to -\infty\,. \end{aligned} \tag{7–54}$$

This assertion has a simple physical meaning: for small ε the behavior of the function in the vicinity of the point x_0 does not affect the behavior of the function when $x=x_1$, and conversely.

Equation (7–52) has the first integral

$$\frac{\eta^2}{2} = \mp \frac{\varepsilon^2}{(1+\eta'^2)^{1/2}} + C_1\,. \tag{7–55}$$

We now construct the function η_1, which should satisfy the first of conditions (7–54). Since far from the boundary the surface is always situated beneath the free surface, we should use Eq. (7–55) with the upper sign, and thus $C_1=\varepsilon^2$. We note that Eq. (7–55) can be used to calculate the value of $\eta=\eta^*$ for which (7–55) changes sign. It is obvious that this point corresponds to the tangent $\eta' = \infty$, and we find that $|\eta^*| = \sqrt{2}\varepsilon$.

We now solve (7–55) for η':

$$\eta' = \pm \frac{\eta\sqrt{4\varepsilon^2 - \eta^2}}{\eta^2 - 2\varepsilon^2}. \tag{7–56}$$

whence it follows immediately that $|\eta| \leqslant 2\varepsilon$.

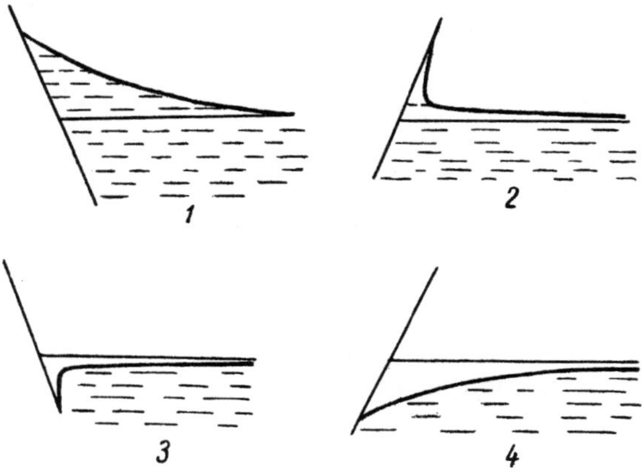

Fig. 35.

We now consider any of the four possible cases that may arise (Fig. 35). We see that the signs of η and η' are opposite when $|\eta| < \eta^*$ and are identical when $|\eta| > \eta^*$. On this basis we can conclude that in each of these cases the plus sign should be used in Eq. (7–56).

Equation (7–56) is easy to integrate:

$$x = C - \sqrt{4\varepsilon^2 - \eta^2} - \frac{\varepsilon}{2}\ln\frac{1 - \sqrt{1 - (\eta/2\varepsilon)^2}}{1 + \sqrt{1 - (\eta/2\varepsilon)^2}}.$$

To determine the constant of integration it is necessary, first, to use the conditions at the wall (7–50) and the circumstance that the number x_0 satisfies the boundary equation. Thus, the complete solution of the problem, including evaluation of the thickness of the displacement layer, also requires the solution of a system of transcedental equations.

Although the solution of Eq. (7–52) is written in quadratures, its analysis in each individual case is quite laborious. Hence, foregoing consideration of the general case, we shall consider the elementary case characterized by angles $\beta_0 + \theta$ and $\beta_1 + \theta$ that are close to $\pi/2$. Now it is possible to disregard η'^2 in (7–49), and we will have

$$\eta_1 = C_1 e^{-x/\varepsilon}, \quad \eta_2 = C_2 e^{x/\varepsilon}.$$

In this case, all of the calculations can easily be completed.

We now consider, as an example, the case shown in Fig. 36. The constants C_1 and C_2 are found from the condition

$$(\eta_1')_{x=x_0} = m_1\,, \quad (\eta_2')_{x=x_1} = m_2\,.$$

The numbers m_1 and m_2 are easy to find when the contact angle and the angles of inclination of the vessel walls are specified:

$$\eta_1 = -\,\varepsilon m_1 e^{\frac{x_0-x}{\varepsilon}}\,, \quad \eta_2 = \varepsilon m_2 e^{\frac{x-x_1}{\varepsilon}}\,. \tag{7–57}$$

Fig. 36.

The isoperimetric condition can be written

$$\int\limits_{x_0}^{x_1} (\eta_1 + \eta_2 + h)\, dx + I_0 + I_1 = 0\,,$$

where I_0 and I_1 are the areas of the triangles ABx_0 and CDx_1, respectively. Setting $\int_{x_0}^{x_1} \eta_1 dx = \int_{x_0}^{\infty} \eta_1 dx$, $\int_{x_0}^{x_1} \eta_2 dx = \int_{-\infty}^{x_1} \eta_2 dx$, we obtain

$$\varepsilon^2 \{m_2 - m_1 + \tfrac{1}{2}(m_1^2 \cos\beta_0 + m_2^2 \cos\beta_1)\} = -\,h(x_1 - x_0)\,.$$

Since $x - x_0 = l + O(\varepsilon)$, where l is the width of the vessel, we finally obtain the equation for the displacement thickness:

$$h = -\frac{\varepsilon^2}{l}\left\{m_2 - m_1 + \frac{m_1^2\cos\beta_0 + m_2^2\cos\beta_1}{2}\right\}. \tag{7–58}$$

Let us consider another similar problem. We assume that the walls are vertical and that the contact angle is the same and is close to $\pi/2$ at both walls (see Figs. 37 a and b). Then Eq. (7–52) contains still another unknown parameter, the value of the derivative at the points of contact between the free surface and the walls. If this parameter is small, the configuration of the free surface can be represented as a power series in this parameter. To construct the solution of Eq. (7–52) in this form, we replace this equation by the expression

$$\eta = \frac{\varepsilon^2 \eta''}{(1 + \mu\eta'^2)^{3/2}}\,,$$

and we shall seek its solution in the form

$$\eta = \eta^0 + \mu\eta^{(1)} + \cdots .$$

Upon constructing the functions $\eta^{(0)}, {}^{(}\eta^{)1}$,..., we find the solution by setting $\eta = 1$ in it. The equations that satisfy the functions $\eta^{(i)}$ have the form

$$\begin{aligned} &\varepsilon^2 \eta^{(0)''} = \eta^{(0)}, \\ &\varepsilon^2 \eta^{(1)''} = \eta^{(1)} + \tfrac{3}{2}\eta^{(0)}\eta^{(1)'2}, \\ &\cdots\cdots\cdots\cdots \end{aligned}$$

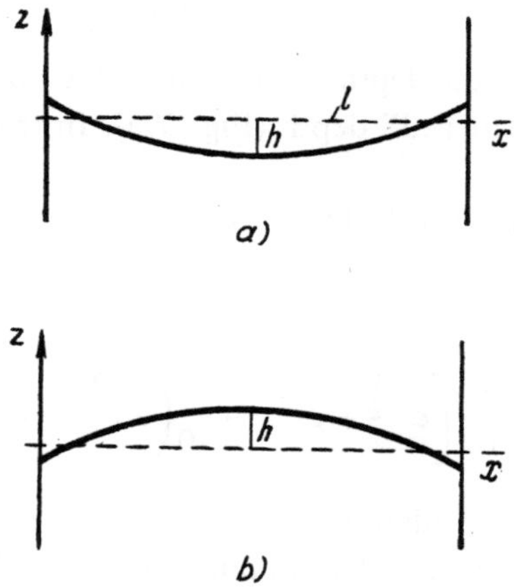

Fig. 37.

The function $\eta^{(0)}$ should satisfy the conditions

$$\eta^{(0)'}(0) = \beta, \quad \eta^{(0)}(\infty) = 0.$$

Hence

$$\eta^{(0)} = -\,\varepsilon\beta e^{-x/\varepsilon}.$$

We subject the function $\eta^{(1)}$ to the condition

$$\eta^{(1)'}(0) = 0, \quad \eta^{(1)}(\infty) = 0.$$

After obvious computations we obtain

$$\eta^{(1)} = \frac{3\beta^3\varepsilon}{16}\left(e^{-x/\varepsilon} - \frac{3}{16}e^{-3x/\varepsilon}\right)$$

etc.

If two terms are retained, the approximate solution has the from

$$\eta = -\,\varepsilon\beta e^{-x/\varepsilon} + \frac{3\beta^3\varepsilon}{16}\left(e^{-x/\varepsilon} - \frac{3}{16}e^{-3x/\varepsilon}\right).$$

Since we are considering a problem in which the plane $x = l/2$ is a plane of symmetry, the displacement thickness will be defined as

$$-h = \frac{2}{l}\int_0^{l/2} \eta(x)\,dx\,.$$

Setting approximately $e^{-l/\varepsilon}=0$, we finally have

$$h = \varepsilon^2\beta\left(1 - \frac{\beta^2}{8}\right).$$

If $\beta<0$, then $h<0$; *i.e.*, the depth of the fluid in the middle of the vessel will be less than its "undisturbed" depth (Fig. 37a). In the opposite case, $h>0$ (Fig. 37b).

The general expression for the free surface in this case will be

$$\zeta = \varepsilon^2\beta\left(1 - \frac{\beta^2}{8}\right) - \varepsilon\beta \times \times \left\{e^{-x/\varepsilon} + e^{\frac{x-l}{\varepsilon}} + \frac{3\beta^2}{16}\left[e^{-\frac{x}{\varepsilon}} + e^{\frac{x-l}{\varepsilon}} - \frac{3}{10}\left(e^{-\frac{3x}{\varepsilon}} + e^{\frac{3(x-l)}{\varepsilon}}\right)\right]\right\}. \tag{7–58$'$}$$

We now pass on to consideration of the three-dimensional case. For large Bond numbers the problem of determining the free surface also reduces to a boundary-layer problem (7–49).

In order to prove this, we introduce an orthogonal coordinate grid ξ, η on the surface S so that the boundary of S, *i.e.*, the line Γ, is the line $\xi=0$. We consider some point $P\in\Gamma$. The peculiar feature of the boundary layer is that the curvature of the line $\xi=0$ at point P should be appreciably less than the curvature of the line $\eta=$const, which also passes through this point. Its curvature should be the greatest from among other normal sections. Thus the line $\beta=$const will be one of the principal cross sections. Then the curve $\xi=0$ will be the second normal section in this point. We write the operator K in this coordinate system

$$K\zeta = \frac{\zeta_{\xi\xi}}{(1-\zeta_\xi^2)^{1/2}} + \frac{\zeta_{\eta\eta}}{(1-\zeta_\eta^2)^{1/2}}.$$

If the characteristic scale of the variation in ζ in the direction η is ε, while in the direction ξ it is of the order of 1, then performing the standard transformations of boundary-layer theory, we arrive at

$$K\zeta = \frac{\zeta_{\xi\xi}}{(1-\zeta_\xi^2)^{1/2}} + O(\varepsilon^2).$$

Changing to our previous variables, we again obtain Eq. (7–49).

Remark

The theory of the boundary layer in the problems of the theory of equilibrium configurations, which is developed in this section, is not entirely rigorous. Moreover, all of the deliberations concerning the boundary layer become invalid when the curvature of the surface Σ is large. Substantiation of the use of the boundary-layer theory encounters the same difficulties encountered by the boundary-layer theory when applied to a viscous fluid. The source of these difficulties is the substantial nonlinearity of the problem.

[7-3] THE THEORY OF SMALL VIBRATIONS

[7-3.1] Solvability of Linear Problems

We now assume that the problem of determining the equilibrium configuration of the free surface, *i.e.*, of the surface S^0, has been solved. Then we can state the problem of small vibrations of the fluid subjected to gravity and surface tension forces. It was shown in Sec. 1 of this chapter that this problem reduces to determining a function $\tilde{\varphi}(x, y, z, t)$ that is harmonic in region τ (see Fig. 38) and a function $\tilde{\psi}(x, y, z, t)$ on the basis of the conditions

$$\text{on}\quad S^0:\quad \frac{\partial\tilde{\varphi}}{\partial t} + \kappa\tilde{\psi} + \frac{\alpha}{\varrho} L\tilde{\psi} = 0, \tag{7–59}$$

$$\text{on}\quad S^0:\quad \frac{\partial\tilde{\psi}}{\partial t} + \frac{\partial\tilde{\varphi}}{\partial n} = 0, \tag{7–60}$$

$$\text{on}\quad \Sigma:\quad \frac{\partial\tilde{\varphi}}{\partial n} = 0, \tag{7–61}$$

$$\text{on}\quad \Gamma:\quad \frac{\partial\tilde{\psi}}{\partial n_\Gamma} + \omega\tilde{\psi} = 0. \tag{7–62}$$

The problem under study can be reduced to a single operator equation. For this it is again necessary to use the Neumann operator H. The expression

$$\varphi(P) = Hf(Q), \quad P \in \tau, \quad Q \in S^0, \quad \int_{S^0} f\, dS = 0$$

will denote that the function $f(Q)$ is matched with the function φ (harmonic in τ), the normal derivative of which is zero on Σ, and f on S. The operator H, as was shown in Chapter 6, is positive, fully continuous and self-adjoint.

From (7–60) we have

$$\tilde{\varphi} = -H\frac{\partial\tilde{\psi}}{\partial t}.$$

Consequently, we can eliminate $\tilde{\varphi}$ from (7–59)

$$M\tilde{\psi} \equiv -H\frac{\partial^2\tilde{\psi}}{\partial t^2} + \kappa\tilde{\psi} - \frac{\alpha}{\varrho}L\tilde{\psi} = 0. \qquad (7\text{–}62')$$

The operator M is defined on a class of functions satisfying condition (7–62). The expression $M\tilde{\psi}=0$ is some integro-differential equation.

The basic problem stated in the theory of small vibrations is that of finding the principal modes of vibrations and the frequencies of these vibrations. We set $\tilde{\psi}=\psi\cos\mu t$; then

$$\mu^2 H\psi = A\psi, \qquad (7\text{–}63)$$

where $A=\alpha/\varrho L-\kappa I$, and I is a unit operator.

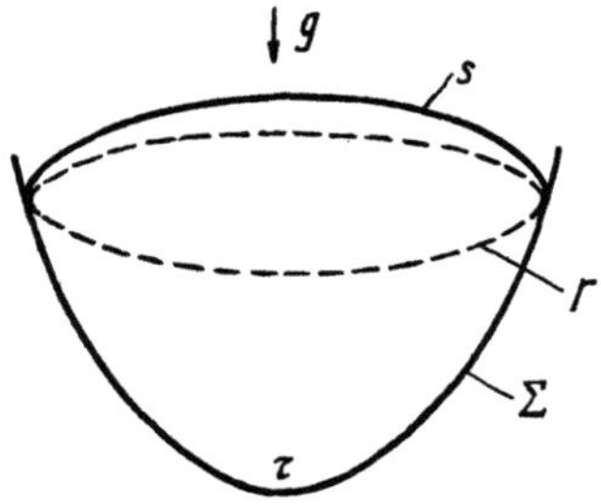

Fig. 38.

The operator L is self-adjoint. This fact is almost obvious, since the operator L describes a conservative system. Hence we shall restrict ourselves to clarifying this fact, using the particular case when the operator L is defined on a set of functions, which depend only on the angle ϑ and $\omega\equiv 0$. In this case

$$L\psi = -\frac{1}{R^2}\left[\frac{1}{\sin\vartheta}\frac{d}{d\vartheta}\left(\sin\vartheta\frac{d\psi}{d\vartheta}\right) + 2\psi\right],$$

$$\psi'(\vartheta_0) = \psi'(\vartheta_1) = 0,$$

where ϑ_0, ϑ_1 are the values of the angle ϑ corresponding to the walls of the vessel. Upon defining the scalar product

$$(f, g) = \int_{\vartheta_0}^{\vartheta_1} \sin\vartheta f(\vartheta)\, g(\vartheta)\, d\vartheta,$$

we shall prove the self-adjointness of L by the following chain of expansions:

$$(Lf, g) = -\frac{1}{R^2}\int_{\vartheta_0}^{\vartheta_1} \sin\vartheta \left[\frac{1}{\sin\vartheta}\frac{d}{d\vartheta}\left(\sin\vartheta\frac{df}{d\vartheta}\right) + 2f\right] g\,d\vartheta =$$

$$= \frac{1}{R^2}\int_{\vartheta_0}^{\vartheta_1} \left(\frac{df}{d\vartheta}\frac{dg}{d\vartheta}\sin\vartheta - 2fg\sin\vartheta\right) d\vartheta =$$

$$= -\frac{1}{R^2}\int_{\vartheta_0}^{\vartheta_1} \left(f\frac{d}{d\vartheta}\left(\sin\vartheta\frac{dg}{d\vartheta}\right) + 2fg\sin\vartheta\right) d\vartheta =$$

$$= -\frac{1}{R^2}\int_{\vartheta_0}^{\vartheta_1} \sin\vartheta \left[\frac{1}{\sin\vartheta}\frac{d}{d\vartheta}\left(\sin\vartheta\frac{dg}{d\vartheta}\right) + 2g\right] f\,d\vartheta = (f, Lg).$$

The proof of the self-adjointness of L in the general case is just as elementary. To realize this, we must replace integration by parts by application of Green's formula. The self-adjointness of the operator A follows immediately from the self-adjointness of L.

We shall consider only the case when A is a positive definite operator. Since the expression $(A\psi, \psi)$ is equal to twice the potential energy, our assumptions will mean that in the equilibrium position the potential energy of the system has an isolated minimum.

On the basis of these assumptions, we arrive at the standard eigenvalue problem (7–63), where H and A are self-adjoint operators; H, in addition, is fully continuous and positive, and A is positive definite. On the basis of known theorems, we can formulate the following corollaries.

1. The spectrum of problem (7.63) is discrete, of finite multiplicity, the only limiting point of which is $\mu = \infty$. This means that, under the conditions being studied, there is an infinite set of natural frequencies μ_n such that $\mu_n \to \infty$ when $n \to \infty$, and each natural frequency has corresponding to it a finite number of possible modes of principal vibrations.
2. The spectrum is situated in its entirety on the positive part of the real axis; *i.e.*, each principal vibration is stable.
3. The system of eigenfunctions is complete in the Freidrichs norm.
4. All of the principal vibrations and natural frequencies can be obtained by the Ritz method, which is convergent.

This last circumstance opens up certain prospects for constructing an effective numerical method. The Ritz method can be successfully used in all cases for which we can insert the volume τ into some region for which a solution is known. For this enveloping region it is sufficient to consider the problem in which $\alpha=0$, since this auxiliary problem should only make it possible to define a complete system of coordinate functions. In many cases it is possible to use the method of perturbation theory developed by Petrov[69].

The theorem formulated above contains directions for the method of studying the equilibrium configuration. This problem reduces to the study of positive definiteness of the operator A. If the mass force vector is directed as shown in Fig. 38, the position of the free surface is stable; *i.e.*, all of the principal vibrations are periodic motions of the fluid. If the direction of the vector $\boldsymbol{g}$ is reversed, the free surface of the fluid (which will be flat) will not be stable in the absence of surface tension forces. However, when certain relationships exist between the intensity of the mass forces and the surface tension coefficient, this position may be stable. The equilibrium configurations of a fluid subjected to surface tension forces and to gravity forces were studied by Myshkis and his students[80].

[7-3.2] The Boundary Condition in the Linear Problem

In the above we have utilized condition (7–62). This is inconvenient in performing specific calculations, since the direction $\boldsymbol{n}_\Gamma$ is the direction of the normal to the contour Γ, which lies in a plane tangent to the undisturbed surface ζ_0. We shall write this boundary condition in terms of derivatives of the function ζ_0. For this we shall repeat the deliberations presented in Sec. 1 of this chapter in the particular case when the free surface can be specified by a single-valued function of the Cartesian coordinates.

Let $z=\zeta_0(x, y)$ be the equation of the undisturbed surface S^0, and let $z=\zeta(x, y)+f(x, y)$ be the equation of the disturbed surface S. The quantity f is considered a first-order infinitesimal. Further, $\Phi_0\equiv z-\zeta(x, y)$;

$$\Phi \equiv z - \zeta(x, y) - f(x, y).$$

Condition (7–62) can be written as

$$(\boldsymbol{n}_S \cdot \boldsymbol{n}_\Sigma)_B = \cos\theta. \tag{7–64}$$

Here n_Σ is the normal to the surface Σ, and n_S is the normal to the surface S at the points of contact

$$\boldsymbol{n}_S = \frac{\nabla\Phi}{|\nabla\Phi|} = \frac{\boldsymbol{z}^0 - \zeta_x\boldsymbol{x}^0 - \zeta_y\boldsymbol{y}^0 - f_x\boldsymbol{x}^0 - f_y\boldsymbol{y}^0}{\sqrt{(\boldsymbol{z}^0 - \boldsymbol{x}^0(\zeta_x + f_x) - \boldsymbol{y}^0(\zeta_y + f_y))^2}}.$$

We draw a vertical plane; this may be done arbitrarily. We then draw planes parallel to the zOy plane. We match point B of contour Γ (see Fig. 39) with point A of contour Γ_0, which lies in a single vertical plane.

We express the values of the scalar product (7–64) by the quantity $(\boldsymbol{n}_{s^0} \cdot \boldsymbol{n}_\Sigma)_A$ and derivatives of the function f, where $\boldsymbol{n}_{S^0}$ is the unit vector of the normal to the surface S^0. We note that, unlike the construction made in Sec. 1 of this chapter, the plane of the drawing in Fig. 39 does not coincide with the plane drawn through the vectors $\boldsymbol{n}_{S^0}$ and $\boldsymbol{n}_\Sigma$ at point A.

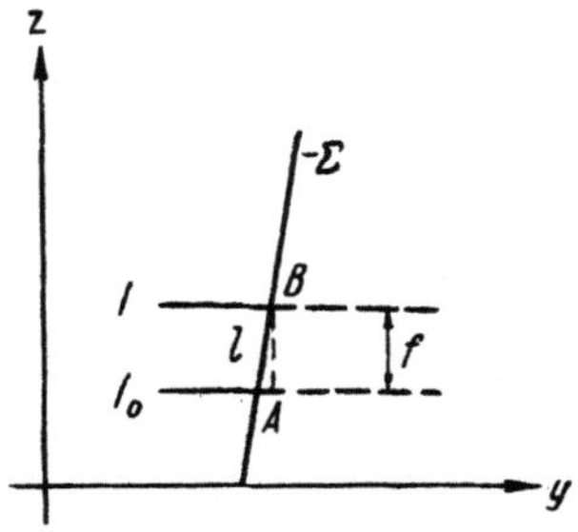

Fig. 39.

We now transform the expression for $\boldsymbol{n}_S$, retaining in it only first-order infinitesimals. Since

$$\begin{aligned}|\nabla\Phi| &= \sqrt{1 + (\zeta_x + f_x)^2 + (\zeta_y + f_y)^2} = \\ &= \sqrt{1 + \zeta_x^2 + \zeta_y^2 + 2\zeta_x f_x + 2\zeta_y f_y + \cdots} = \\ &= \sqrt{|\nabla\Phi_0|^2 + 2\nabla\zeta\nabla f + \cdots},\end{aligned}$$

the expression for $\boldsymbol{n}_S$ can be reduced to the form

$$\boldsymbol{n}_S = \frac{1}{|\nabla\Phi_0|}\left\{\nabla\Phi_0 - \nabla f - \frac{\nabla\Phi_0\,(\nabla\zeta\cdot\nabla f)}{|\nabla\Phi_0|\,|\nabla\Phi_0|}\right\} + O(f^2).$$

But

$$\frac{\nabla\Phi_0}{|\nabla\Phi_0|} = \boldsymbol{n}_{S^0}.$$

Thus

$$\boldsymbol{n}_S = \boldsymbol{n}_{S^0} - \frac{1}{|\nabla\Phi_0|}\left\{\nabla f + \frac{\boldsymbol{n}_{S^0}(\nabla\zeta\cdot\nabla f)}{|\nabla\Phi_0|}\right\} + O(f^2).$$

We now calculate the scalar product $(\boldsymbol{n}_S \cdot \boldsymbol{n}_\Sigma)_A$. Here the surface Σ will be specified by the expression

$$\Phi^* \equiv z - F(x, y) = 0.$$

In addition, we make use of the fact that $(\boldsymbol{n}_{S^0} \cdot \boldsymbol{n}_\Sigma) = \cos\theta$. Then

$$(\boldsymbol{n}_S \cdot \boldsymbol{n}_\Sigma)_A = \cos\theta + \frac{\nabla f}{|\nabla\Phi_0|}\left\{\frac{\nabla F}{|\nabla\Phi^*|} + \frac{\nabla\zeta}{|\nabla\Phi_0|}\cos\theta\right\}.$$

The above expression can also be rewritten as

$$(\boldsymbol{n}_S \cdot \boldsymbol{n}_\Sigma)_A = \cos\theta + a\,\frac{\partial f}{\partial x} + b\,\frac{\partial f}{\partial y}, \tag{7–65}$$

where

$$a = \frac{1}{\sqrt{1+\zeta_x^2+\zeta_y^2}}\left\{\frac{\dfrac{\partial F}{\partial x}}{\sqrt{1+F_x^2+F_y^2}} + \frac{\dfrac{\partial \zeta}{\partial x}\cos\theta}{\sqrt{1+\zeta_x^2+\zeta_y^2}}\right\},$$

$$b = \frac{1}{\sqrt{1+\zeta_x^2+\zeta_y^2}}\left\{\frac{\dfrac{\partial F}{\partial y}}{\sqrt{1+F_x^2+F_y^2}} + \frac{\dfrac{\partial \zeta}{\partial y}\cos\theta}{\sqrt{1+\zeta_x^2+\zeta_y^2}}\right\}.$$

Let us now return to expression (7–64), which we rewrite in the form

$$(\boldsymbol{n}_S \cdot \boldsymbol{n}_\Sigma)_B = (\boldsymbol{n}_S \cdot \boldsymbol{n}_\Sigma)_A + l\left\{\left(\frac{\partial \boldsymbol{n}_S}{\partial l}\cdot \boldsymbol{n}_\Sigma\right) + \left(\frac{\partial \boldsymbol{n}_\Sigma}{\partial l}\cdot \boldsymbol{n}_S\right)\right\} + \cdots = \cos\theta\,.$$

The first term on the left-hand side was just calculated. Let us now determine $\partial \boldsymbol{n}_\Sigma/\partial l$. We consider the plane curve AB, which lies in a vertical section, perpendicular to the Ox axis. The equation of the surface Σ was specified in the form

$$z = F(x, y)\,.$$

Consequently the equation for AB is

$$z = F(x_A, y)\,,$$

where x_A is the value of the abscissa of point A, through which the section was drawn. Then

$$\begin{gathered}\frac{\partial}{\partial l} = \frac{\partial}{\partial y}\cdot\frac{1}{\sqrt{1+F_y^2}} + \frac{\partial}{\partial z}\,\frac{F_y}{\sqrt{1+F_y^2}}\,,\\ l = f\,\frac{F_y}{\sqrt{1+F_y^2}}\,.\end{gathered} \tag{7–66}$$

Disposing of the expressions for $\boldsymbol{n}_S$ and $\boldsymbol{n}_\Sigma$ and of Eqs. (7–66), we can calculate the scalar product contained in the braces. They will contain second derivatives of the functions F and ζ.

Collecting all of the results, substituting them into (7–64), and restricting ourselves to terms linear in f, we obtain the condition at point A in the form

$$a\,\frac{\partial f}{\partial x} + b\,\frac{\partial f}{\partial y} + \nu f = 0\,, \tag{7–67}$$

where

$$\nu = \frac{F_y}{\sqrt{1 + F_y^2}} \left\{ \left(\frac{\partial \boldsymbol{n}_{S0}}{\partial l} \cdot \boldsymbol{n}_\Sigma \right) + \left(\frac{\partial \boldsymbol{n}_\Sigma}{\partial l} \cdot \boldsymbol{n}_{S0} \right) \right\}.$$

In the two-dimensional case these expressions simplify somewhat:

$$a = 0,$$

$$b = \frac{1}{\sqrt{1 + \zeta'^2}} \left[\frac{F'}{\sqrt{1 + F'^2}} + \frac{\zeta' \cos\theta}{\sqrt{1 + \zeta'^2}} \right],$$

$$\nu = \frac{F'}{\sqrt{1 + F'^2}} \left\{ \frac{K_S F'}{\sqrt{1 + F'^2}} + \frac{K_\Sigma \zeta'}{\sqrt{1 + \zeta'^2}} - \right. \tag{7–68}$$

$$\left. - \frac{\zeta'' \zeta'}{\sqrt{(1 + \zeta'^2)^3} \cdot \sqrt{1 + F'^2}} - \frac{F'' F'}{\sqrt{(1 + F'^2)^3} \cdot \sqrt{1 + \zeta'^2}} \right\}.$$

If the walls are vertical, $dl = dz$, and $\partial \boldsymbol{n}_\Sigma / \partial l = \partial \boldsymbol{n}_S / \partial l = 0$, since $\boldsymbol{n}_\Sigma$ and $\boldsymbol{n}_S$ depend only on x and y. In this case condition (7–64) becomes

$$a \frac{\partial f}{\partial x} + b \frac{\partial f}{\partial y} = 0. \tag{7–69}$$

If the vessel is axially symmetrical, the curve Γ_0 will be plane and the normals $\boldsymbol{n}_S$ and $\boldsymbol{n}_\Sigma$ will lie in the plane of the vertical section containing the axis of symmetry. This is precisely the plane which we shall use to establish a one-to-one correspondence between points of the contours Γ and Γ_0. The normal to Γ_0, lying in the horizontal plane, is denoted by $\boldsymbol{n}$. All three of the vectors $\boldsymbol{n}_S$, $\boldsymbol{n}$, and $\boldsymbol{n}_\Sigma$ lie in the same vertical plane; the latter two are identical:

$$\boldsymbol{n}_\Sigma = \boldsymbol{n} = \boldsymbol{y}^0 .$$

In addition, the vector $\nabla \zeta$ also lies on this vertical plane,

$$\nabla \zeta = c\boldsymbol{n} ,$$

Thus Eq. (7–64) can be rewritten in the form

$$\boldsymbol{n}_S = \boldsymbol{n}_{S0} - \frac{1}{|\nabla \Phi_0|} \left\{ \nabla f + \frac{c\boldsymbol{n} \dfrac{\partial f}{\partial n}}{|\nabla \Phi_0|} \right\}$$

and, consequently,

$$(\boldsymbol{n}_S \cdot \boldsymbol{n}_\Sigma)_A = \cos\theta + b \frac{\partial f}{\partial n},$$

where

$$b = \frac{1}{|\nabla \Phi_0|} \left\{ 1 + \frac{c}{|\nabla \Phi_0|} \right\}.$$

Thus, for axially symmetrical vessels with vertical walls, condition (7–69) takes the form

$$\frac{\partial f}{\partial n} = 0 . \tag{7–70}$$

Remark

It was of importance in deriving condition (7–70) that the walls of the vessel be vertical only in the vicinity of the line of contact Γ_0. In addition, the curve Γ_0 should be a plane curve. The assumption of axial symmetry which we have made is sufficient for this purpose, but it is not necessary.

[7-3.3] Vibration of a Fluid Between Two Vertical Walls

Let us consider the two-dimensional problem of the vibration of a fluid between two vertical walls. In this case the surface is a single-valued function

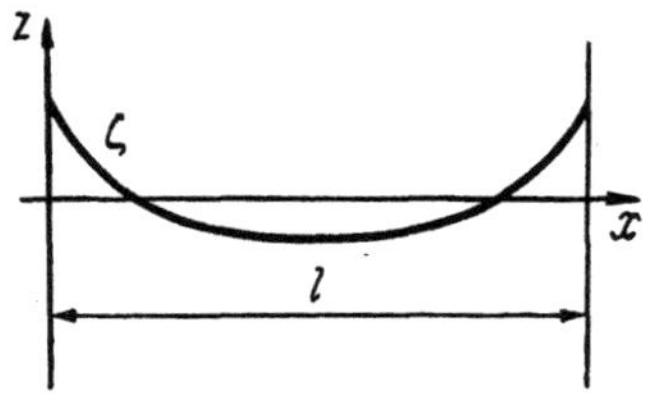

Fig. 40.

of the abscissa x (Fig. 40). Since the problem is two-dimensional, according to (7–34), the expression for $K\tilde{\zeta}$ has the form

$$K\tilde{\zeta} = + \frac{\tilde{\zeta}_{\tilde{x}\tilde{x}}}{(1 + \tilde{\zeta}_{\tilde{x}}^2)^{3/2}} \cdot$$

Here $\tilde{z} = \tilde{\zeta}(\tilde{x}, \tilde{t})$ is the equation of the free surface. We set $\tilde{\zeta} = \tilde{\zeta}^0(\tilde{x}) + \tilde{f}(\tilde{x}, \tilde{t})$ and we separate the linear part of the operator K:

$$L\tilde{f} = \tilde{f}_{\tilde{x}\tilde{x}}(1 + K_1(\tilde{\zeta}_0)) + \tilde{\zeta}_{\tilde{x}} K_2(\tilde{\zeta}_0),$$

$$K_1(\tilde{\zeta}^0) = \left(\frac{1}{(1 + \tilde{\zeta}_x^{0 2})^{3/2}} - 1\right),$$

$$K_2(\tilde{\zeta}^0) = - \frac{3\tilde{\zeta}^0_{\tilde{x}\tilde{x}} \tilde{\zeta}_x}{(1 + \tilde{\zeta}_x^{0 2})^{5/2}} \cdot$$

Here the sign should always be taken as positive, since the fluid is beneath the free surface.

Further, the potential of the gravity forces on the free surface has the form of $g\tilde{\zeta}$. Hence the linearized problem given by Eqs. (7–59)–(7–62) can be formulated as

$$\begin{aligned} &\text{on } S^0: \quad \frac{\partial \tilde{\varphi}}{\partial t} + g\tilde{f} - \frac{\alpha}{\varrho} L\tilde{f} = 0, \\ &\text{on } S^0: \quad \frac{\partial \tilde{f}}{\partial t} = \frac{\partial \tilde{\varphi}}{\partial \tilde{z}} - \tilde{\zeta}^0_{\tilde{x}} \tilde{\varphi}_{\tilde{x}}, \\ &\text{on } \Sigma: \quad \frac{\partial \tilde{\varphi}}{\partial n} = 0, \\ &\left.\begin{aligned} x &= 0 \\ x &= l \end{aligned}\right\} \frac{\partial \tilde{f}}{\partial x} = 0. \end{aligned} \tag{7–71}$$

We introduce the dimensionless coordinates

$$\begin{aligned} &\tilde{x} = xl, \quad \tilde{f} = \tilde{f}^* l, \quad \tilde{\zeta}^0 = \zeta^0 l, \\ &\tilde{t} = tT, \quad \tilde{\varphi} = \varphi^* \frac{l^2}{T}. \end{aligned}$$

As the characteristic linear dimension we take the channel width l, and T is some characteristic time. Further, we set

$$f^* = \sin \omega t \cdot f, \quad \varphi^* = \cos \omega t \cdot \varphi.$$

Problem (7–71) now takes on the form

$$\begin{aligned} &\text{on } S^0: \quad -\mathrm{S}\omega\varphi + f - \varepsilon^2 L f = 0, \\ &\text{on } S^0: \quad \omega f = \frac{\partial \varphi}{\partial z} - \zeta^0_x \frac{\partial \varphi}{\partial x}, \\ &\text{on } \Sigma: \quad \frac{\partial \varphi}{\partial n} = 0, \\ &\left.\begin{aligned} x &= 0 \\ x &= 1 \end{aligned}\right\} f' = 0. \end{aligned} \tag{7–72}$$

Here $\varepsilon^2 = \alpha/\varrho g l^2$ is the reciprocal of the Bond number and $\mathrm{S} = l/T^2 g$ is the Strouhal number. These conditions must be supplemented by the requirement that the system must be at rest as $z \to -\infty$. If we use the Fourier method, the problem (7–72) can be reduced to an infinite system of algebraic equations.

We seek the function f in the form

$$f = \sum_{n=1}^{\infty} q_n \cos n\pi x, \tag{7–73}$$

thus automatically satisfying the last of conditions (7–72).

The function $\varphi(x, z)$ is sought in the form

$$\varphi = \sum_{n=1}^{\infty} p_n \cos n\pi x \, e^{n\pi z}. \tag{7–73'}$$

This function satisfies the condition of rest as $z \to -\infty$ and the condition of smoothness of the surface $\sum$, and is harmonic in the region $0<x<1$.

We introduce the expansions

$$\cos n\pi x e^{n\pi\zeta^0} + \zeta_x^0 \sin n\pi x e^{n\pi\zeta^0} = \sum_{m=1}^{\infty} \alpha_{nm} \cos m\pi x,$$

$$\cos n\pi x e^{n\pi\zeta^0} = \sum_{m=1}^{\infty} \beta_{nm} \cos m\pi x,$$

$$L \cos n\pi x = \sum_{m=1}^{\infty} \gamma_{nm} \cos m\pi x, \quad n = 1, 2, \ldots.$$

Substituting these expressions into the first two of Eqs. (7–72) and equating the coefficients of the cosines of angle with the same multiplicity, we obtain the infinite system of algebraic equations

$$\left. \begin{aligned} -\mathrm{S}\omega \sum_{m=1}^{\infty} p_m \beta_{mn} + q_n &= \varepsilon^2 \sum_{m=1}^{\infty} q_m \gamma_{mn}, \\ \omega q_n = n\pi \sum_{m=1}^{\infty} p_m \alpha_{mn}, &\quad n = 1, 2, \ldots \end{aligned} \right\} \tag{7–74}$$

The above system is a system of homogeneous equations for q_n and p_n; by equating the determinant of this system to zero we obtain the frequency equation.

If the series (7–73) and (7–74) are replaced by sums with a finite number of terms, we will obtain a finite system that can studied by known methods.

If the surface ζ^0 is close to the plane $z=0$, the mechanics of perturbation theory can be put to effective use. The quantity ζ^0 is small when the Bond number is large, *i.e.*, when ε is small. In addition, this condition is also satisfied when the contact angle is close to $\pi/2$, *i.e.*, when $\zeta^{0\prime}$ is small; the Bond number in this case can be arbitrary. To construct the perturbation theory for large Bond numbers, we replace the first two conditions of system (7–72) by

$$\left. \begin{aligned} \text{on} \quad S^0: &\quad -\mathrm{S}\omega(\varphi)_0 + f = \mu\Phi, \\ \text{on} \quad S^0: &\quad \omega f = \left(\frac{\partial\varphi}{\partial z}\right)_0 + \mu\Psi, \end{aligned} \right\} \tag{7–75}$$

where

$$\Phi = \varepsilon^2 L f + \mathrm{S}\omega\eta, \quad \eta = (\varphi)_\zeta - (\varphi)_0,$$

$$\Psi = -\zeta_x^0 \left(\frac{\partial\varphi}{\partial x}\right)_\zeta + \psi, \quad \psi = \left(\frac{\partial\varphi}{\partial z}\right)_\zeta - \left(\frac{\partial\varphi}{\partial z}\right)_0.$$

If the contact angle is close to $\pi/2$, we can rewrite the first of conditions (7–75) as

$$-\,S\omega(\varphi)_0 + f - \varepsilon^2 f_{xx} = \mu\Phi^*\,, \qquad (7\text{–}75')$$

where

$$\Phi^* = \varepsilon^2\left(K_1(x)\,f_{xx} + K_2(x)\,f_x\right).$$

The second of conditions (7–75) is then left unchanged. The subscript $(\;)_0$ means that the value of the function is taken at points of the straight line $z=0$, and the subscript ζ means that the value of the function was calculated on the free surface in the state of equilibrium [*i.e.*, when $z=\zeta^0(x)$]. When $\mu=1$, conditions (7–75) and (7–75′) become conditions (7–72).

We shall seek the solutions of the above problem in the form of the series

$$\begin{aligned} \varphi &= \varphi_0 + \mu\varphi_1 + \cdots, \\ f &= f_0 + \mu f_1 + \cdots, \\ \omega &= \omega_0 + \mu\omega_1 + \cdots. \end{aligned} \qquad (7\text{–}76)$$

Upon finding the unknowns f_i, φ_i, and ω_i and substituting them into the above series, we set $\mu=1$.

Substituting series (7–76) into (7–75) and equating the coefficients of the same powers of μ, we obtain the following conditions that should be satisfied by $\varphi_0, \varphi_1, f_0, f_1, \omega_0, \omega_1$, *etc.*:

$$\begin{aligned} -\,S\omega_0\varphi_0 + f_0 &= 0\,, \\ \omega_0\varphi_0 &= \frac{\partial\varphi_0}{\partial z}\,, \end{aligned} \qquad (7\text{–}77)$$

$$\begin{aligned} -\,S\omega_0\varphi_1 + f_1 &= \Phi_0 + s\omega_1\varphi_0\,, \\ \omega_0 f_1 &= \frac{\partial\varphi_1}{\partial z} + \Psi_0 - \omega_1 f_0 \end{aligned} \qquad (7\text{–}78)$$

etc.

The functions φ_0, f_0, φ_1, and f_1 are calculated in these expressions at points of the straight line $z=0$. Furthermore, Φ_0 and Ψ_0 are the values of the functions Φ and Ψ calculated under the condition that $\varphi=\varphi_0, f=f_0$ and $z=0$.

For contact angles close to $\pi/2$, the first equations of the systems (7–77) and (7–78) are replaced by

$$-\,S\omega_0\varphi_0 + f_0 - \varepsilon^2 f_{0xx} = 0\,, \qquad (7\text{–}77')$$

$$-\,S\omega_0\varphi_1 + f_1 - \varepsilon^2 f_{1xx} = \Phi_0^* + S\omega_1\varphi_0\,. \qquad (7\text{–}78')$$

The functions f_0 and φ_0 will be sought in the form of (7–73) and (7–73′):

$$f_0 = \sum_{n=1}^{\infty} a_n \cos n\pi x\,,$$

$$\varphi_0 = \sum_{n=1}^{\infty} A_n \cos n\pi x e^{n\pi z}\,.$$

Substituting these expressions into condition (7–77), we obtain the following system of homogeneous equations for a_n and A_n:

$$- S\omega_0 A_n + a_n = 0 ,$$
$$\omega_0 a_n = n\pi A_n , \quad n = 1, 2, \ldots .$$

Eliminating A_n we obtain

$$a_n\left(- \frac{S\omega_0^2}{n\pi} + 1 \right) = 0 . \tag{7–79}$$

The above system can have nontrivial solutions if and only if ω_0 is one of the natural frequencies

$$\omega_0^2 = \omega_m^2 = \frac{m\pi}{S} . \tag{7–80}$$

In this case all of the

$$a_n = 0 , \quad \text{if} \quad n \neq m .$$

The quantity a_m is arbitrary. This reflects the fact that the amplitude of the waves in the linear problem is arbitrary. Without loss of generality, we can assume $a_m = 1$. Thus,

$$f_0 = \cos m\pi x , \quad \varphi_0 = \frac{\omega_0}{m\pi} \cos m\pi x e^{m\pi z} . \tag{7–81}$$

The above equation and Eq. (7–80) show that in the zeroth approximation the vibration under study is identical with the principal vibration of a fluid not subjected to surface tension forces.

We obtain similar results when the contact angle is close to $\pi/2$:

$$\begin{aligned} f_0 &= \cos m\pi x , \\ \varphi_0 &= \frac{\omega_0}{m\pi} \cos m\pi x e^{m\pi z} , \\ \omega_0^2 &= \frac{m\pi}{s} (1 + \varepsilon^2 m^2 \pi^2) . \end{aligned} \tag{7–82}$$

Thus, in the zeroth approximation, the mode of vibration of any number will be the same for a fluid that is not subjected to surface tension forces, but the frequency of the principal vibrations will be higher. The difference in magnitude of the frequencies will increase with an increase in the number of the principal vibration. In the present case the square of the natural frequency will increase as the cube of the number of the principal vibration, while for a fluid not subjected to surface tension forces it increases in proportion to the number of the principal vibration.

We now move on to finding the following approximation. We set

$$f_1 = \Sigma b_n \cos n\pi x\,,$$
$$\varphi_1 = \Sigma B_n \cos n\pi x e^{n\pi z}\,.$$

Substituting these series into the system (7–77), equating the coefficients of cosines of the same multiplicity, and using (7–80), we obtain

$$-\, S\omega_0 B_n + b_n = \Phi_n\,, \quad n \neq m\,, \tag{7–83a}$$

$$-\, S\omega_0 B_m + b_m = \Phi_m + \omega_1 \frac{S\omega_0}{m\pi}\,, \tag{7–83b}$$

$$\omega_0 b_n = B_n n\pi + \Psi_n\,, \quad n \neq m\,, \tag{7–83c}$$

$$\omega_0 b_m = B_m m\pi + \Psi_m - \omega_1\,, \tag{7–83d}$$

where Φ_n and Ψ_n are the Fourier coefficients of the functions Φ_0 and Ψ_0,

$$\Phi_n = \tfrac{1}{2}\int_0^1 \Phi_0 \cos n\pi x dx\,, \quad \Psi_n = \tfrac{1}{2}\int_0^1 \Psi_0 \cos n\pi x dx\,.$$

Determining B_m from (7–83d) and substituting into (7–83b), we obtain

$$b_m\left(-\frac{S\omega_0^2}{m\pi} + 1\right) + \frac{S\omega_0}{m\pi}\Psi_m - \frac{S\omega_0}{m\pi}\omega_1 = \Phi_m + \frac{S\omega_0\omega_1}{m\pi}\,. \tag{7–84}$$

The multiplier of b_m is zero; hence, in order for Eq. (7–84) to be meaningful, ω_1 should be such that the sum of all of the remaining terms of (7–84) is zero.

We now write the expressions for Φ_m and Ψ_m:

$$\Phi_m = -\, \varepsilon^2 m^2 \pi^2 (1 + c_m) - \varepsilon^2 m\pi g_m + \frac{S\omega_0^2 d_m}{m\pi}\,,$$
$$\Psi_m = \omega_0 e_m + \omega_0 d_m\,,$$

where c_m, d_m, e_m, and g_m are the coefficients of the expansions

$$k_1(\zeta^0) \cos m\pi x = \sum c_n \cos n\pi x\,,$$
$$(e^{m\pi\zeta^0} - 1) \cos m\pi x = \sum d_n \cos n\pi x\,,$$
$$\zeta_x^0 \sin m\pi x e^{m\pi\zeta^0} = \sum e_n \cos n\pi x\,,$$
$$k_2(\zeta^0) \sin m\pi x = \sum g_n \cos n\pi x\,.$$

We can now rewrite Eq. (7–84) in the form

$$\frac{2S\omega_0}{m\pi}\omega_1 = \varepsilon^2 m^2 \pi^2 (1 + c_m) + \varepsilon^2 m\pi g_m + e_m$$

or

$$\omega_1 = \varepsilon^2 \frac{m^3\pi^3(1+c_m) + m^2\pi^2 g_m}{2S\omega_0} + \frac{\omega_0}{2} e_m. \qquad (7\text{–}84')$$

In the case when ε is arbitrary, but the slope of the arbitrary free surface at the walls is small (the contact angle is close to $\pi/2$), we have a formula for the correction similar to (7–84′):

$$\frac{2S\omega_0}{m\pi}\omega_1 = \varepsilon^2 m^2\pi^2 c_m + \varepsilon^2 m\pi g_m + \frac{S\omega_0^2}{m\pi} e_m. \qquad (7\text{–}84'')$$

We recall that, in contrast to (7–84′), ω_0 is given here by the expression $\omega_0^2 = m\pi/S(1+\varepsilon^2 m^2\pi^2)$.

We have thus calculated how much the natural frequency has changed as a result of the action of surface tension forces. The term $(\varepsilon^2 m^3\pi^3/2S)(1+c_m)$ is always positive. It characterizes the change in the "rigidity" of the free surface as a consequence of surface tension. However, it would be wrong to conclude that the surface tension forces increase the natural frequencies, since the quantities g_m and e_m can be either positive or negative and depend on the magnitude of the contant angle. Thus the action of surface tension forces may cause either increases or decreases in the natural frequencies.

To determine the mode of the principal vibrations, we must solve the systems (7–83a) and (7–83c). These are systems of nonhomogeneous algebraic equations. As a result we establish that the free surface has the form

$$\zeta_0 = a_m \cos m\pi x + \sum_{n \neq m} b_n \cos n\pi x;$$

i.e., on the principal wave, which gives the mode of vibration of a fluid not subjected to surface tension forces, are superimposed waves of small amplitude b_n, uniquely determined by the amplitude of the principal vibration.

The following approximations are calculated according to the same procedure. Equation (7–84) shows that, in order to determine the correction to the natural frequency, it is sufficient to evaluate the functions e_m, c_m, and g_m. If we are speaking of calculating the first natural frequencies, these functionals can be calculated approximately.

We first consider the case when the contact angle is close to $\pi/2$. The free surface ζ^0 is defined by Eq. (7–58′), where β is a small parameter. Then $\zeta_{0x} = O(\beta)$, and consequently $c_m = O(\beta^2)$, $g =_u O(\beta^2)$. Then Eq. (7–84′) yields

$$\omega_1 = \tfrac{1}{2}\omega_0 e_m + O(\beta^2). \qquad (7\text{–}84''')$$

We now evaluate e_m using the representation of the free surface in the form of (7–58′). First we note that

$$\zeta_{0x} \sin m\pi x e^{m\pi\zeta_0} = \zeta_{0x} \sin m\pi x + O(\beta^2).$$

Hence

$$e_m = 2\int_0^1 \zeta_{0x} \sin m\pi x \cos m\pi x dx = \int_0^1 \zeta_{0x} \sin 2m\pi x dx =$$

$$= -2m\pi \int_0^1 \zeta_0 \cos 2m\pi x dx$$

or, substituting (7–58′) and calculating with the same accuracy, we obtain

$$e_m = \varepsilon\beta \int_0^1 (e^{-\frac{x}{\varepsilon}} + e^{\frac{x-1}{\varepsilon}})\, dx = 2\varepsilon^2\beta[1 - e^{-\frac{1}{\varepsilon}}] \approx 2\varepsilon^2\beta.$$

We thus see that the sign of the correction to the frequency depends on the sign of β. If the slope of the tangent is negative, *i.e.*, if the contact angle is less than $\pi/2$, the rigidity is reduced—the natural frequencies become smaller. If the contact angle is larger than $\pi/2$, the rigidity always increases.

We now move on to analyzing cases of small ε. We first consider the coefficient c_m. We note that, since the function ζ_{0x} remains finite over the intervals $(0, \varepsilon)$ and $(1-\varepsilon, 1)$, while outside these intervals it is close to zero, the function

$$K_1 = \frac{1}{(1+\zeta_{0x}^2)^{3/2}} - 1$$

also has these properties. Consequently, for any bounded function $F(x)$ the estimate

$$\int_0^1 F(x)\, K_1(x)\, dx = O(\varepsilon)$$

is valid. Hence, for a given n, $c_n = O(\varepsilon)$. On this basis we can drop the term $\varepsilon^2 m^3\pi^3 c_m/2S\omega_0$ in Eq. (7–84″), since it will be a third-order infinitesimal. The identical procedure can be applied to $\varepsilon^2 g_m$.

We now pass on to calculating e_m. We note immediately that for a given m, $e^{m\pi\zeta^0} = 1 + O(\varepsilon)$, since $\zeta^0 = O(\varepsilon)$. On this basis

$$e_m = 2\int_0^1 \zeta_x^0 \sin m\pi x \cos m\pi x dx = \int_0^1 \zeta_{0x} \sin 2m\pi x dx =$$

$$= -2m\pi \int_0^1 \zeta^0 \cos 2m\pi x dx.$$

Further, according to (7–54) we obtain

$$e_m = -2m\pi \int_0^1 (\eta_1 + \eta_2) \sin 2m\pi x dx .$$

The function η_1 decreases rapidly and is nonzero only inside the segment $(0, \varepsilon)$; hence

$$\int_0^1 \eta_1 \cos 2m\pi x dx \sim \int_0^\infty \eta_1 (x) dx .$$

Similarly

$$\int_0^1 \eta_2 \cos 2m\pi x dx \sim \int_{-\infty}^0 \eta_2 (x) dx \cdot (-1)^{2m} .$$

These integrals are equal to each other by their sign and their absolute magnitude. On the basis of this

$$e_m = -4m\pi \int_0^\infty \eta_1 (x) dx .$$

If the contact angle is smaller than $\pi/2$, then $\eta_1 > 0$ and $e_m < 0$. In the opposite case, $e_m < 0$. To evaluate this integral it is convenient to change from integration with respect to x to integration with respect to η_1 (Fig. 41):

$$e_m = -4m\pi \int_0^{\eta^*} x(\eta) d\eta$$

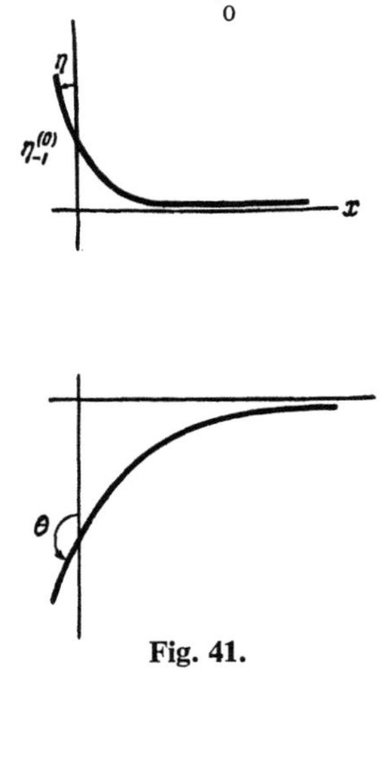

Fig. 41.

or, integrating by parts,

$$e_m = -4m\pi x(\eta)\,\eta|_0^{\eta^*} + 4m\pi \int_0^{\eta^*} \frac{\eta d\eta}{\eta_x}.$$

But the first term is zero. Using Eq. (7–57) we have

$$e_m = 4m\pi \int_0^{\eta^*} \frac{\eta^2 - 2\varepsilon^2}{\sqrt{4\varepsilon^2 - \eta^2}}\, d\eta = -2m\pi\eta^* \sqrt{4\varepsilon^2 - \eta^{*2}}.$$

On the basis of the above, the arithmetic value of the root is always used here.

We recall further [see (7–50)] that at the point $x=0$

$$\eta' = -\cot\theta,$$

where θ is the contact angle. We thus immediately have from (7–56) that

$$\eta^{*2} = 2\varepsilon(1 - \sin\theta).$$

Further, $\eta^* > 0$ if $\theta \in (0, \pi/2)$, and $\eta^* < 0$ if $\theta \in (\pi/2, \pi)$; hence

$$\eta^* = \sqrt{2}\,\varepsilon\sqrt{1 - \sin\theta} \quad \text{if} \quad \theta \in \left(0, \frac{\pi}{2}\right),$$

$$\eta^* = -\sqrt{2}\,\varepsilon\sqrt{1 - \sin\theta} \quad \text{if} \quad \theta \in \left(\frac{\pi}{2}, \pi\right).$$

Therefore, finally, for all of the $\theta \in (0, \pi)$

$$e_m = -4m\pi\varepsilon^2 \cos\theta.$$

We now rewrite Eq. (7–84) as

$$\omega_1 = \frac{\omega_0}{2}\, m\pi\varepsilon^2 (m\pi - 4\cos\theta) + O(\varepsilon^3). \qquad (7\text{–}85)$$

The above expression shows that for all of the frequencies, starting with the second, the correction for the frequency is positive; here this correction increases rapidly with an increase in the natural frequency number. As to the first natural frequency, it may be either positive or negative, depending on the cosine of the contact angle. The critical value of the angle θ^* is defined as

$$\theta^* = \arccos\frac{\pi}{4} \sim 38°18'.$$

If $0 > \theta^*$, then the first frequency also increases; *i.e.*, the rigidity of the system as a whole increases.

[7-3.4] Arbitrary Cylinders

We now consider a more general case—the vibrations of a fluid in a vessel with a vertical generatrix—assuming as before that the line of contact Γ between the free surface and the walls of the cylinder is a plane curve.

Selecting, in one manner or another, the characteristic scales, we arrive at a problem similar to that defined by Eq. (7–72)

$$\begin{aligned} &\text{on} \quad S^0: \quad -S\omega\varphi + f - \varepsilon^2 Lf = 0, \\ &\text{on} \quad S^0: \quad \omega f = \frac{\partial\varphi}{\partial z} - (\nabla\zeta^0\nabla\varphi), \\ &\text{on} \quad \Sigma: \quad \frac{\partial\varphi}{\partial n} = 0, \\ &\text{on} \quad \Gamma: \quad \frac{\partial f}{\partial n_\Gamma} = 0. \end{aligned} \tag{7–86}$$

The gradient in this system of conditions is calculated on the basis of the variables x and y. The third of conditions (7–86) is satisfied not only at the side walls, but also at the bottom where $z = -h$. We call h the depth of the fluid. It is measured from the plane $z=0$, which is identical with the free surface if $\varepsilon=0$. The normal $\boldsymbol{n}_\Gamma$ to the curve Γ lies in a horizontal plane.

We denote the eigenfunctions of the problem by $\chi_n(x, y)$:

$$\Delta\chi_n + \lambda^2\chi_n = 0, \quad P \in S_0, \tag{7–87}$$

$$\frac{\partial\chi_n}{\partial n_\Gamma} = 0, \quad P \in \Gamma, \tag{7–88}$$

where S_0 is the cylinder cross section produced by the plane $z=0$. The Laplacian in (7–87) is calculated on the basis of the variables x and y.

Since the function f identically satisfies the condition (7–88), to it is natural seek the configuration of the free surface in the form

$$f_1 = \Sigma q_n\chi_n(x, y). \tag{7–89}$$

On the other hand, a function that is harmonic in τ and that satisfies the condition $\partial\varphi/\partial n = 0$ on Σ can be represented in the form

$$\varphi = \sum p_n\chi_n(x, y)\cosh\lambda_n(z + h). \tag{7–90}$$

The expansions (7–89) and (7–90) are meaningful, since the system of functions $\{\chi_n\}$ is complete and orthogonal on S_0, *i.e.*, on a plane figure bounded by curve Γ. The functions χ_n will be regarded as normalized by the conditions

$$\int_{S_0} \chi_n^2 \, dS = 1.$$

In a manner similar to that used in the preceding subsection, we can reduce problem (7–86) to the infinite system of algebraic equations

$$\begin{aligned} \mathrm{S}\omega \sum_{m=1}^{\infty} p_m\beta_{mn} + q_n &= \varepsilon^2 \sum_{m=1}^{\infty} q_m\gamma_{mn}, \\ \omega q_n &= \sum_{m=1}^{\infty} p_m\alpha_{mn}, \quad n, m = 1, 2, \dots, \end{aligned} \tag{7–74}$$

where the coefficients α_{mn}, β_{mn}, and γ_{mn} are given by

$$\begin{aligned} \alpha_{mn} &= \int_{S_0} \{\lambda_m\chi_m(x, y)\sinh\lambda_m(\zeta^0 + h) - \\ &\quad - (\nabla\zeta^0\nabla\chi_m)\cosh(\zeta^0 + h)\}\,\chi_n(x, y)\,dydx, \\ \beta_{mn} &= \int_{S_0} \chi_m(x, y)\cosh\lambda_m(\zeta^0 + h)\,\chi_n(x, y)\,dxdy, \\ \gamma_{mn} &= \int_{S_0} L\chi_m(x, y)\,\chi_n(x, y)\,dxdy. \end{aligned} \tag{7–91}$$

Thus the problem of vibrations of a fluid in a cylindrical vessel for an arbitrary Bond numbers can be reduced to an infinite system of algebraic equations. The use of the Galerkin method makes it possible to reduce the problem to the study of a finite number of equations such as (7–74).

This result is easily extended to vessels of arbitrary shape, if the walls are shaped so that the line of contact is a plane curve and the condition $\partial f/\partial n = 0$ is satisfied along it.

Let us also consider the case of large Bond numbers. Since we intend to construct only the first two terms of the series of pertubation theory, we immediately transform the coefficients (7–91). Here the numbers m and n will be regarded (as in the preceding section) as fixed and smaller than some number $N \ll 1/\varepsilon^2$. We transform the coefficients α_{mn}, β_{mn}, and γ_{mn}, retaining numbers of the order of $O(\varepsilon^2)$ in them:

$$\begin{aligned} \alpha_{mn} &= \int_{S_0} \{\lambda_m\chi_m\sinh\lambda_m h + \lambda_m^2\chi_m\zeta^0\cosh\lambda_m h - \\ &\quad - (\nabla\zeta^0\nabla\chi_m)\cosh\lambda_m h\}\,\chi_n(x, y)\,dxdy, \\ \beta_{mn} &= \int_{S_0} \chi_m(x, y)\,[\cosh\lambda_m h + \lambda_m\zeta^0\sinh\lambda_m h]\,\chi_n(x, y)\,dxdy, \\ \gamma_{mn} &= \int_{S_0} \Delta\chi_m\chi_n\,dxdy. \end{aligned}$$

We can thus write

$$\begin{aligned} \alpha_{mn} &= \lambda_m \sinh \lambda_m h \delta_n^m + \alpha_{mn}^*, \\ \beta_{mn} &= \cosh \lambda_m h \delta_n^m + \beta_{mn}^*, \\ \gamma_{mn} &= -\lambda_m^2 \delta_n^m . \end{aligned} \tag{7–92}$$

Here δ_n^m is the Kronecker delta,

$$\alpha_{mn}^* = \int\limits_{S_0} \cosh \lambda_m h \left[\zeta^0 \lambda^2 \chi_m - (\nabla \zeta^0 \nabla \chi_m)\right] \chi_n(x, y)\, dx dy ,$$

$$\beta_{mn}^* = \int\limits_{S_0} \lambda_m \sinh \lambda_m h \cdot \zeta^0 \chi_m(x, y)\, \chi_n(x, y)\, dx dy .$$

In calculating γ_{mn}, use was made of the fact that the function χ_m satisfies the equation $\Delta\chi_m - \lambda^2\chi_m = 0$.

In these equations we have only retained the first terms of expansions in terms of ε; *i.e.*, the functions that enter the integrands are first-order infinitesimals, but the integrals proper are second-order infinitesimals, since the function ζ^0 is nonzero only in a region with width ε.

We now replace the system (7–74) by the system

$$\begin{aligned} -\mathrm{S}\omega p_n \cosh \lambda_n h + q_n &= \mu \{-\lambda_n^2 \varepsilon^2 + \mathrm{S}\omega \sum_m p_m \beta_{mn}^*\}, \\ \omega q_n &= \lambda_n \sinh \lambda_n h \cdot p_n + \mu \sum_m p_m \alpha_{mn}^*, \quad n, m = 1, 2, \ldots, N, \end{aligned} \tag{7–93}$$

which reverts to (7–74) when $\mu = 1$. We shall seek the solution of (7–93) in the form

$$\begin{aligned} p_n &= p_{n0} + \mu p_{n1}, \\ q_n &= q_{n0} + \mu q_{n1}, \\ \omega_n &= \omega_0 + \mu \omega_1 . \end{aligned}$$

Substituting these expressions into (7–93), we find the systems of equations

$$\begin{aligned} -\mathrm{S}\omega_0 p_{n0} \cosh \lambda_n h + q_{n0} &= 0, \\ \omega_0 q_{n0} &= \lambda_n \sinh \lambda_n h p_{n0}, \end{aligned} \tag{7–94}$$

$$\begin{aligned} -\mathrm{S}\omega_0 p_{n1} \cosh \lambda_n h + q_{n1} &= -\lambda_n^2 \varepsilon^2 q_{n0} + \\ &\quad + \mathrm{S}\omega_0 \sum_m p_{m0} \beta_{mn}^* - \mathrm{S}\omega_1 \cosh \lambda_n h \cdot p_{n0}, \\ \omega_0 q_{n1} &= \lambda_n \sinh \lambda_n h p_{n1} + \sum_m p_{m0} \alpha_{mn}^* - \omega_1 q_{n0} . \end{aligned} \tag{7–95}$$

Eliminating p_{n_0} from (7–94), we obtain

$$\left(\frac{-\mathrm{S}\omega_0^2}{\lambda_n \tanh \lambda_n h} + 1\right) q_{n0} = 0,$$

whence it follows immediately that the system (7–94) has nontrivial solutions of the form

$$q_{n0} = 0 \quad \text{if} \quad n \neq m, \quad \omega_0^2 = \frac{\lambda_n \tanh \lambda_n h}{S}. \tag{7–96}$$

The quantity q_m is arbitrary and we assume that it is equal to unity. Thus

$$p_m = \frac{\omega_0}{\lambda_m \sinh \lambda_m h}, \qquad p_n = 0 \quad \text{if} \quad n \neq m. \tag{7–96′}$$

Using the above expressions and Eqs. (7–96), we rewrite the system (7–95) as

$$- S\omega_0 p_{n1} \cosh \lambda_n h + q_{n1} = \left[- \lambda_n^2 \varepsilon^2 - \frac{S\omega_1 \cosh \lambda_n h \cdot \omega_0}{\lambda_m \sinh \lambda_m h} \right] \delta_n^m + \frac{S\omega_0^2}{\lambda_m \sinh \lambda_m h} \beta_{mn}^*,$$

$$\omega_0 q_{n1} = \lambda_n \sinh \lambda_n h \cdot p_{n1} + \frac{\omega_0}{\lambda_m \sinh \lambda_m h} \alpha_{mn}^* - \omega_1 \delta_n^m,$$

$$n = 1, 2, \ldots, N.$$

We now perform calculations for $n = m$. We use the second equation to find p_m, substitute the value found into the first equation, and make use of the expression for ω_0^2:

$$\frac{S\omega_0^2 \cdot \alpha_{mm}^*}{\lambda_m^2 \tanh \lambda_m h \cdot \sinh \lambda_m h} + \lambda_m^2 \varepsilon^2 - \frac{S\omega_0^2 \cdot \beta_{mm}^*}{\cosh \lambda_m h \cdot \tanh \lambda_m h} = \frac{2S\omega_1 \omega_0}{\lambda_m \tanh \lambda_m h}.$$

This expression can be regarded as the equation for determining the "frequency correction," *i.e.*, ω_1. It can also be rewritten in the form

$$\omega_1 = \frac{\omega_0}{2} \left\{ \varepsilon^2 \lambda_m^2 + \frac{\alpha_{mm}^*}{\lambda_m \sinh \lambda_m h} - \frac{\beta_{mm}^*}{\cosh \lambda_m h} \right\}. \tag{7–97}$$

Setting $n \neq m$, we obtain by the same procedure a nonhomogeneous equation for q_n. Upon determining these quantities, we shall determine the Fourier coefficients of the expansion for the free surface. Together with Eq. (7–97), these results complete the problem.

The quantities α_{mm}^* and β_{mm}^* can be expressed in terms of the contact angle. We first perform one auxiliary transformation:

$$\begin{aligned} \zeta^0 \lambda_m^2 \chi_m \chi_n - (\nabla \zeta^0 \nabla \chi_m)\, \chi_m &= - \{\zeta^0 \chi_n \nabla \chi_m + (\nabla \zeta^0 \nabla \chi_m)\, \chi_n\} = \\ &= - \{\operatorname{div}(\zeta^0 \chi_n \cdot \nabla \chi_m) - \zeta^0 (\nabla \chi_n \nabla \chi_m)\}. \end{aligned} \tag{7–98}$$

Its validity is checked by direct computation. On the basis of the above equation, the expression for α^*_{mm} can be rewritten as

$$\alpha^*_{mn} = -\cosh\lambda_m h \{\int\limits_{S_0} \operatorname{div}(\zeta^0 \chi_n \chi_m)\, dS - \int\limits_{S_0} \zeta^0 (\nabla\chi_m \nabla\chi_n)\, dS\}.$$

But the first of these integrals is zero. In order to satisfy ourselves that this is so, we should apply Green's formula to it and use the condition that $\partial\chi_n/\partial n_\Gamma = 0$.

Further, ζ^0 can be represented in the form of the sum

$$\zeta^0 = \eta + \eta_0,$$

where η_0 is the displacement thickness, and η is a boundary-layer-type function. We calculate

$$\int\limits_{S_0} \eta_0 (\nabla\chi_m \nabla\chi_n)\, dS = \eta_0 \int\limits_{\Gamma} \chi_m \frac{\partial\chi_n}{\partial n_\Gamma}\, dl = 0.$$

Thus

$$\alpha^*_{mn} = \cosh\lambda_m h \int\limits_{S_0} \eta (\nabla\chi_m \nabla\chi_n)\, dS.$$

Since the function η is nonzero only in a narrow region of width ε adjoining contour Γ, we can approximately evaluate this integral as

$$\alpha^*_{mn} \sim \cosh\lambda_m h \int\limits_{\Gamma} \nabla\chi_m \nabla\chi_n dl \cdot \int\limits_0^\infty \eta\, dn_\Gamma,$$

where n is the internal normal to the contour Γ. Since $\partial\chi_n/\partial n_\Gamma = 0$,

$$\alpha^*_{mn} = \cosh\lambda_m h \int\limits_{\Gamma} \frac{\partial\chi_m}{\partial l} \cdot \frac{\partial\chi_n}{\partial l}\, dl \cdot \int\limits_0^\infty \eta\, dn_\Gamma. \tag{7–99}$$

The inner integral was evaluated in the preceding section [see derivation of Eq. (7–85)]:

$$\int\limits_0^\infty \eta\, dn_\Gamma = \varepsilon^2 \cos\theta.$$

Thus, finally,

$$\alpha^*_{mn} = \varepsilon^2 \cosh\lambda_m h \int\limits_{\Gamma} \frac{\partial\chi_n}{\partial l} \frac{\partial\chi_m}{\partial l}\, dl \cdot \cos\theta.$$

Similarly,

$$\beta^*_{mn} = \varepsilon^2 \lambda_m \sinh \lambda_m h \int_\Gamma \chi_n \chi_m dl\,.$$

Substituting into (7–97), we find the following equation, which expresses the dependence of the frequency on the magnitude of the surface tension and of the contact angle (7–99)

$$\omega_1 = \varepsilon^2 \frac{\omega_0}{2} \left\{ \lambda_m^2 + \left[\frac{1}{\lambda_m \tanh \lambda_m h} \int_\Gamma \left(\frac{\partial \chi_m}{\partial l} \right)^2 dl - \lambda_m \tanh \lambda_m h \int_\Gamma \chi_m^2 dl \right] \cos \theta \right\}.$$

The above procedure can be used for arbitrary regions, provided the walls are vertical and the line of contact is a plane curve[94]. For inclined walls the phenomenon becomes appreciably more complicated. In the case described above the frequency correction is of the order of ε^2. For arbitrary walls this correction contains a term of the order of ε, multiplied by the cosine of the angle of inclination. In our case this correction was zero, since the walls were assumed to be vertical.

[7-4] PROBLEMS OF DYNAMICS OF A BUBBLE

[7-4.1] Some Features of the Problem

Up to now we have considered only the case when the free surface S intersects the vessel surface Σ. However, it may turn out that these surfaces do not intersect, *i.e.*, a bubble is located inside the fluid. The problems of the dynamics of a bubble and of a body containing in a cavity a fluid with a bubble have a large number of anomalies, some of which are considered here (see [61]).

We assume that mass forces are not acting and that the fluid is at rest; then the bubble is at rest and is spherical since it is bounded by a surface with a constant average curvature and not containing infinitely removed points.

We fix to the center of the bubble the spherical coordinate system

$$x = r \sin \vartheta \cos \psi\,,$$
$$y = r \sin \vartheta \sin \psi\,,$$
$$z = r \cos \vartheta\,.$$

We now consider some surface S_ζ, given by the expression

$$\zeta \equiv \zeta(\vartheta, \psi) = r(\vartheta, \psi) - R\,. \qquad (7\text{–}100)$$

We shall assume that the surface S_ζ is close to a sphere of radius R. We calculate its average curvature $K(\zeta)$, restricting ourselves to first-order infinitesimals with respect to ζ:

$$K(\zeta) = \frac{2}{R} + L\zeta .$$

The operator L is defined as

$$L = -\frac{1}{R^2}\left[\frac{1}{\sin^2 \vartheta}\frac{\partial^2}{\partial \psi^2} + \frac{1}{\sin \vartheta}\frac{\partial}{\partial \vartheta}\left(\sin \vartheta \frac{\partial}{\partial \vartheta}\right) + 2\right]. \qquad (7\text{–}101)$$

We now clarify some properties of this operator. We consider the equation

$$L\zeta = 0 . \qquad (7\text{–}102)$$

It can be shown by a simple check that Eq. (7–101) has the nontrivial solutions

$$\begin{aligned} \zeta_1 &= CP_1(\cos\vartheta) = c\cos\vartheta , \\ \zeta_2 &= CP_1^1(\cos\vartheta)\cos\psi = c\sin\vartheta\cos\psi , \\ \zeta_3 &= CP_1^1(\cos\vartheta)\sin\psi = c\sin\vartheta\sin\psi . \end{aligned} \qquad (7\text{–}103)$$

Here $P_i(x)$ denote Legendre polynomials, the P_i^1 denote the associated Legendre functions of the first kind and C is an arbitrary constant.

It follows from the theory of Legendre polynomials that Eq. (7–102) has no single-valued and bounded solutions except for (7–103). This can be easily proven directly on the basis of the properties of the associated Legendre functions. In fact, any solution of Eq. (7–102) that is periodic with respect to ψ can be represented in the form of a linear combination of functions

$$\Phi_1^n(\vartheta)\,{\sin \atop \cos}\, n\psi , \quad n = 1, 2, \dots ,$$

where the functions $\Phi(n/1)$ satisfy the expression

$$\frac{1}{\sin\vartheta}\frac{d}{d\vartheta}\left(\sin\vartheta \frac{d}{d\vartheta}\Phi_1^n(\vartheta)\right) + \left(2 - \frac{n^2}{\sin^2\vartheta}\right)\Phi_1^n(\vartheta) = 0 .$$

When $n=0$, this equation is a particular case of the Legendre equation. Its linearly independent solutions will be P_1 and Q_1, where P_1 is a Legendre polynomial, and Q_1 is a Legendre function of the first kind (the function Q_1 is not bounded). When $n\neq 0$ this equation defines the associated Legendre functions $P_1^{(n)}$ and $Q_1^{(n)}$. It is known from the theory of these functions (see [81]), that $P_m^{(n)}$ and $Q_m^{(n)}$ possess certain features: the function $Q_m^{(n)}$ for any relationship between n and m, and the function $P_m^{(n)}$ under the condition that $n>m$. This inequality is valid precisely in the case under study.

Thus the functions (7–103) are the only bounded solutions of Eq. (7–102). From this it follows that the equation

$$L\zeta = f \tag{7–104}$$

has bounded solutions only in the space of functions that are orthogonal to the functions ζ_1, ζ_2, and ζ_3, since Fredholm's alternative applies to the operator L; *i.e.*, either $Lf=0$ and Eq. (7–104) has no bounded solutions, or $Lf \neq 0$ and Eq. (7–104) has bounded solutions.

The solutions (7–103) have a simple physical meaning. They describe infinitesimal displacements of the sphere as a whole along Ox, the Oy, and Oz axes. Let us consider, for example, the function ζ_1 that describes an infinitesimal displacement along the Oz axis. In fact, if we move the sphere along the Oz axis as a single entity by an amount c, its equation will become

$$x^2 + y^2 + (z - c)^2 = R^2$$

or

$$r^2 \sin^2 \vartheta + (r \cos \vartheta - c)^2 = R^2 .$$

Dropping second-order infinitesimals, we obtain

$$r^2 = R^2 + 2cR \cos \vartheta$$

and, finally,

$$r = R + c \cos \vartheta .$$

Comparing this to (7–100), we find that $\zeta = \zeta_1 = c \cos \vartheta$, *etc.*

Let us now assume that the fluid is subjected to a homogeneous field

$$U = az ,$$

where a is the strength of the field, which is regarded as low (of the order of ζ). The linearized dynamic condition will be

$$\varepsilon \cos \vartheta - L\zeta = 0 . \tag{7–105}$$

The above equation is of the type of Eqs. (7–103), where $f = \zeta_1$. Thus, no matter how low the strength a of the mass forces, Eq. (7–105) has no bounded solutions. This means that the bubble cannot be in equilibrium inside the fluid if the intensity of the mass force field is nonzero. In this case the bubble, in addition to vibrational motion, will inevitably start to perform translational motion.

[7-4.2] Elementary Problem of Vibrations of a Bubble

We now consider an elementary case of the dynamic problem—the natural vibrations of a spherical bubble situated inside a spherical shell

(Fig. 42). We shall assume that the bubble (surface S) and the shell Σ are concentric. Thus, reference will be had here to vibrations of a spherical layer of an ideal fluid subjected solely to surface tension forces (we have just established that purely vibrational motions of the bubble are impossible if the fluid is subjected to mass forces).

We shall consider the problem in dimensionless variables, taking R_1, the radius of the inside sphere, as the characteristic linear scale. We denote $q=R_2/R_1>1$.

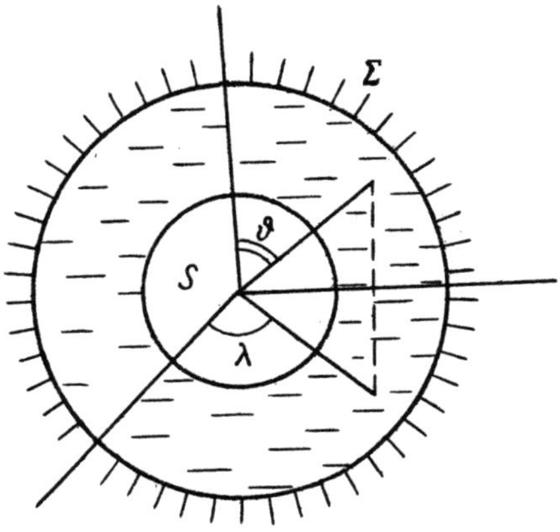

Fig. 42.

We have decided to consider free vibrations, *i.e.*, to seek solutions of the form $\varphi=e^{\sigma t}\Phi$. This means that all of the functions that we shall consider from now on are automatically assumed not to be time-dependent.

The motion is, as always, regarded as potential and the velocity potential Φ inside the spherical layer τ satisfies Laplace's equation, which we now write in spherical coordinates as

$$\frac{\partial}{\partial r}\left(r^2\frac{\partial\Phi}{\partial r}\right)+\frac{1}{\sin\vartheta}\frac{\partial}{\partial\vartheta}\left(\sin\vartheta\frac{\partial\Phi}{\partial\vartheta}\right)+\frac{1}{\sin^2\vartheta}\frac{\partial^2\Phi}{\partial\psi^2}=0. \qquad (7\text{–}106)$$

Since the outside shell is impermeable, we should have when $r=q$

$$\left(\frac{\partial\Phi}{\partial r}\right)_{r=q}=0. \qquad (7\text{–}107)$$

The particular solutions of Eq. (7–106) that are bounded and single-valued functions of the variables r, ϑ, and ψ, satisfying condition (7–107), have the form

$$\Phi^{(1),(2)}_{n,m}=c\left[\frac{n+1}{n}q^{-(2n+1)}r^{n}+r^{-(n+1)}\right]P_n^m(\cos\vartheta)\,{\sin\atop\cos}\,m\psi, \qquad (7\text{–}108)$$

where c is an arbitrary constant, n is an arbitrary positive integer, $m=0, 1, 2, \ldots, n$, and $P_m^{(n)}(x)$ are spherical functions.

We now introduce the expression $\zeta=f(\vartheta, \psi)$, which is the equation of the free surface. It should satisfy the kinematic condition

$$\left(\frac{\partial \Phi}{\partial r}\right)_{r=1} = f. \tag{7–109}$$

The linearized dynamic condition is

$$(\sigma^2 \Phi = Lf)_{r=1}, \tag{7–110}$$

where the expression for Lf was set up in the preceding part of this section. Using one of the particular solutions of (7–108) as Φ, we substitute it into (7–109). From this condition we determine f and substitute the expression thus obtained into (7–110). As a result, dividing through by c, we can obtain an expression for σ^2, the frequency of the natural vibrations:

$$\sigma_n^2 = (n-1)(n+2) \frac{1 - q^{-(2n+1)}}{1 + \dfrac{n+1}{n} q^{-(2n+1)}}. \tag{7–111}$$

This expression shows that the frequency of vibrations is independent of m, whence it follows that all of the principal vibrations, the modes of which are described by the functions $\Phi_{n,m}^{(1),(2)}$ $(m=1, 2, \ldots, n)$, have the same frequency. Thus the problem under study has a multiple spectrum, and the eigenfunction σ_n is of multiplicity $2n+1$.

We note that $\sigma_0=\sigma_1=0$. We see on the basis of the above that the natural frequencies correspond to nontrivial solutions of the equation $L\zeta=0$; *i.e.*, they correspond to secular rather than vibrational motion of the bubble.

The lowest natural frequency σ_2 corresponds to $n=2$.

We calculate the first natural frequencies

$$\sigma_2^2 = 12 \frac{1 - q^{-5}}{1 + \frac{3}{2} q^{-5}}, \quad \sigma_3^2 = 40 \frac{1 - q^{-7}}{1 + \frac{4}{3} q^{-7}},$$

$$\sigma_4^2 = 90 \frac{1 - q^{-9}}{1 + \frac{5}{4} q^{-9}}, \quad \textit{etc}.$$

Thus the natural frequencies depend to a large extent on the depth only in the case of low depths, and this only for natural frequencies of the first numbers. As early as $q=2$ (when the thickness of the layer is equal to the radius of the bubble) we have

$$\frac{\sigma_2^2}{\sigma_{2\infty}^2} = 0.961, \quad \frac{\sigma_3^2}{\sigma_{3\infty}^2} = 0.9909, \quad \textit{etc}.$$

Here $\sigma_{i\infty}$ denotes the natural frequency of the layer on the assumption that $q=\infty$.

A well-known problem is that of capillary waves on the surface of a spherical volume of fluid. In this problem the natural frequencies are given by[16] (Landau + Lifshitz)

$$\sigma_n^{*2} = n(n-1)(n+2).$$

Equating this expression to (7–110), we obtain

$$\sigma_n^2 = \sigma_n^{*2}\,\frac{n+1}{n}\,\frac{1-q^{-(2n+1)}}{1+\dfrac{n+1}{n}\,q^{-(2n+1)}}.$$

We finally write the expression for the natural frequency in the dimensional form

$$\omega_n^2 = \frac{\alpha}{\varrho R^3}\,\sigma_n^2,$$

where σ_n^2 is a number given by Eq. (7–111).

Problems of the dynamics of a bubble are of great interest, even in the simplified form considered here. It should be remembered that conditions of complete weightlessness are the usual ones during a space flight.

We have considered the most elementary problem when the bubble and the shell are concentric, and we have obtained a result in closed form. It should be noted that this solution is the only known exact solution. There are a number of approaches for approximately solving this problem that, in one way or another, use the smallness of the bubble's dimensions and the fact that it is appreciably removed from the wall of the cavity.

[7-5] ASYMPTOTIC BEHAVIOR FOR "SHALLOW WATER"

The difficulites encountered in solving the problems of dynamics of a body containing a fluid add interest to those instances when the problem can be simplified in some manner. One such possibility arises if the depth of the fluid is small (Fig. 43) in comparison with the linear dimensions of its free surface. Then the problem can be substantially simplified by using asymptotic methods of narrow strips[82].

To reduce computations and to demonstrate more graphically the procedure in applying the asymptotic approximation to the problem, we shall consider the elementary case of a weightless fluid inside of an axially symmetrical vessel. The free surface in the equilibrium position is a sphere. The equation of the surface of the fluid in the equilibrium position and the equations of the walls of the vessel will be written in the form

$$\begin{aligned} r &= R, \\ r &= R + z(\vartheta) = R + \varepsilon f. \end{aligned} \tag{7–112}$$

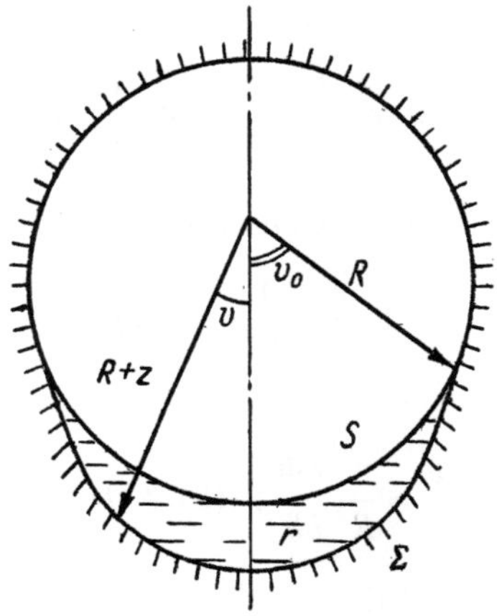

Fig. 43.

The problem consists in finding a function $\Phi(t, r, \vartheta)$ satisfying the Laplace equation in τ:

$$\frac{\partial}{\partial r}\left(r^2 \frac{\partial \Phi}{\partial r}\right) + \frac{1}{\sin\vartheta}\frac{\partial}{\partial \vartheta}\left(\sin\vartheta \frac{\partial \Phi}{\partial \vartheta}\right) = 0, \tag{7–113}$$

and a function $\psi(t, \vartheta)$ based on the conditions

$$\frac{\partial \Phi}{\partial n} = 0 \quad \text{when} \quad r = R + z, \tag{7–114}$$

$$\frac{\partial \Phi}{\partial t} + \alpha L\psi = 0 \quad \text{when} \quad r = R, \tag{7–115}$$

$$\frac{\partial \psi}{\partial t} = -\frac{\partial \Phi}{\partial r} \quad \text{when} \quad r = R, \tag{7–116}$$

$$\frac{\partial \psi}{\partial r} = 0 \quad \text{when} \quad r = R,\ \vartheta = \vartheta_0. \tag{7–117}$$

Here ϑ_0 is the value of the angle ϑ at the point of contact between the undisturbed free surface and the walls.

We shall first consider the auxiliary problem of determining the function $\Phi(t, r, \vartheta)$ satisfying Eq. (7–113), condition (7–114), and the condition

$$\Phi(t, R, \vartheta) = a(t, \vartheta), \tag{7–118}$$

where a is some specified function of its variables. We construct an asymptotic solution of this problem as $\varepsilon \to 0$. We set $r = R + \varepsilon\xi$ in (7–113):

$$\frac{\partial}{\partial \xi}\left[(R+\varepsilon\xi)^2 \frac{\partial \Phi}{\partial \xi}\right] = -\varepsilon^2 A\Phi. \tag{7–119}$$

The letter A is here used to denote an operator:

$$A\Phi = \frac{1}{\sin\vartheta}\frac{\partial}{\partial\vartheta}\left(\sin\vartheta \frac{\partial \Phi}{\partial \vartheta}\right). \tag{7–119'}$$

Using the above notation, we write an expression for L [cf. (7–101)]:

$$L = -\frac{1}{R^2}(A+2).$$

We can write (7–114) in the form

$$\frac{\partial \Phi}{\partial \xi} = \frac{\varepsilon^2}{R^2}\frac{\partial \Phi}{\partial \vartheta} f'(\vartheta) + O(\varepsilon^3).$$

We construct the solution of the auxiliary problem in the form of the series

$$\Phi = \Phi_0 + \varepsilon\Phi_1 + \varepsilon^2\Phi_2 + \cdots. \tag{7–120}$$

For Φ_0 we obtain

$$\frac{\partial^2 \Phi_0}{\partial \xi^2} = 0, \quad P \in \tau,$$

$$\frac{\partial \Phi_0}{\partial \xi} = 0 \quad \text{when} \quad \xi = f, \quad \Phi_0 = a \quad \text{when} \quad \xi = 0.$$

Consequently, $\Phi_0 = a(t, \vartheta)$.

Performing identical procedures, we find $\Phi_1 = 0$. The function Φ_2 satisfies the conditions

$$\begin{gathered}\frac{\partial^2 \Phi_2}{\partial \xi^2} = -\frac{1}{R^2} A\Phi_0 = -\frac{1}{R^2} Aa(t,\vartheta), \quad P \in \tau, \\ \Phi_2 = 0 \quad \text{when} \quad \xi = 0, \\ \frac{\partial \Phi_2}{\partial \xi} = \frac{1}{R^2} f'(\vartheta)\frac{\partial a}{\partial \vartheta} \quad \text{when} \quad \xi = f.\end{gathered} \tag{7–121}$$

The solution of the above problem is given by the formula

$$\Phi_2 = \frac{\xi}{R^2}\left[\left(f - \frac{\xi}{2}\right)Aa + f'(\vartheta)\frac{\partial a}{\partial \vartheta}\right].$$

Calculation of the remaining terms of the expansion (7–120) is just as simple. The series (7–120) is divergent in the general case. Reference [82] gives *a* *priori* conditions* that must be met for the series to be asymptotic. We shall regard them as satisfied and shall restrict ourselves to calculating Φ_0 and Φ_2. Returning to the previous variables, we have

$$\Phi = a(t, \vartheta) + \frac{r-R}{R^2} \times \left[\left(z(\vartheta) - \frac{r-R}{2}\right) Aa + z'(\vartheta)\frac{\partial a}{\partial \vartheta}\right]. \tag{7–122}$$

Using the above expression, we rewrite (7–115) and (7–116)

$$\begin{aligned} \frac{\partial a}{\partial t} &= \frac{\alpha}{R^2}(A\psi + 2\psi), \\ \frac{\partial \psi}{\partial t} &= -\frac{1}{R^2}\left(zAa + z'\frac{\partial a}{\partial \vartheta}\right). \end{aligned} \tag{7–123}$$

Thus the problem has been reduced to finding two functions that do not depend on the radius vector.

To find the natural frequencies, we set

$$a = b(\vartheta)\, e^{\omega t}, \quad \psi = \beta(\vartheta)\, e^{\omega t},$$

and then it follows from (7–123) that

$$\omega b = \frac{\alpha}{R^2}(A+2)\beta, \quad \omega\beta = -\frac{1}{R^2}(zAb + z'b').$$

This problem is equivalent to the eigenvalue problem

$$\alpha R^{-4}(A+2)(zA\beta + z'\beta') = -\omega^2\beta.$$

The problem obtained as a result of the asymptotic treatment is simpler than the original problem, if only because the unknown function depends only on the variable ϑ.

Appendix

Formulating the Problem of the Motion of a Fluid Subjected to Gravity Forces and Surface Tension Forces

Chapter 1 contained a derivation of the equations of motion of a rigid body containing a fluid. As a particular case, we obtained the formulation

* The meaning of these conditions is that the derivatives of all of the functions with respect to ϑ are of higher order, the higher the order of the derivative

$$\frac{\partial^k \varphi}{\partial \vartheta^k} = o\left(\frac{\partial^{k-1}\varphi}{\partial \vartheta^{k-1}}\right).$$

of the boundary-value problem (7–1)–(7–8) that describes the motion of the fluid in a motionless vessel. We now present a more detailed derivation of these conditions on the basis of Hamilton's principle, which clarifies some details of the deliberations presented in Chapter 1.

According to Hamilton's principle, for virtual displacements

$$\delta L = 0, \tag{7–124}$$

where δL is the variation of the action integral

$$L = \int_0^{t_1} (T - \Pi)\, d\tau dt, \tag{7–125}$$

defined for a set of motions that transfer the system from one fixed state to another during time t_1. The motion should be constrained by the incompressibility condition. Introducing the Lagrange multiplier p, we consider, instead of Eq. (7–124), the equation

$$\int_0^{t_1} (\delta T - \delta\Pi)\, dtL + \int_0^{t}\int_\tau p \operatorname{div} \delta\boldsymbol{r} d\tau dt = 0. \tag{7–126}$$

We establish the validity of two auxiliary relationships:

$$\delta \int_\tau \Phi\varrho d\tau = \int_\tau \delta\Phi\varrho d\tau, \tag{7–127}$$

$$\int_0^t \Big(\int_\tau \Phi\varrho d\tau\Big)\, dt = \int_\tau \Big(\int_0^t \Phi dt\Big)\, \varrho d\tau. \tag{7–128}$$

The first follows directly from the continuity equation

$$\frac{d}{dt}(\varrho d\tau) = 0$$

and is valid for incompressible as well as for compressible fluids. In precisely the same manner

$$\int_0^t \Big(\int_\tau \Phi\varrho d\tau\Big)\, dt = \lim_{n\to\infty} \int_0^t \sum_{i=1}^{n} \Phi_i \varrho_i \Delta\tau_i dt.$$

But $\varrho_i \Delta\tau_i = \text{const}$; hence

$$\int_0^t \Big(\int_\tau \Phi\varrho d\tau\Big)\, dt = \lim_{n\to\infty} \sum_{i=1}^{n} \varrho_i \Delta\tau_i \int_0^t \Phi_i dt = \int_\tau \Big(\int_0^t \Phi dt\Big)\, \varrho d\tau.$$

We now transform (7–126); since

$$T = \frac{1}{2}\int_\tau v^2 d\tau,$$

according to (7–127)

$$\delta T = \varrho \int_\tau (\boldsymbol{v}, \delta \boldsymbol{v})\, d\tau = \varrho \int_\tau \left(\boldsymbol{v}, \frac{d\delta \boldsymbol{r}}{dt}\right) d\tau.$$

Further, $\Pi = \Pi_1 + \Pi_2$, whereupon

$$\delta \Pi_1 = - \int_\tau (\boldsymbol{F}\delta \boldsymbol{r})\, d\tau; \tag{7–129}$$

i.e., the variation of the potential energy of the mass forces is equal in magnitude but is of opposite sign to the work performed by the mass forces $\boldsymbol{F} = -g\boldsymbol{z}^0$ over the possible displacement $\delta \boldsymbol{r}$. Further,

$$\delta \Pi_2 = \alpha_1 \delta \Sigma_1 + \alpha_2 \delta \Sigma_2 + \alpha \delta S. \tag{7–130}$$

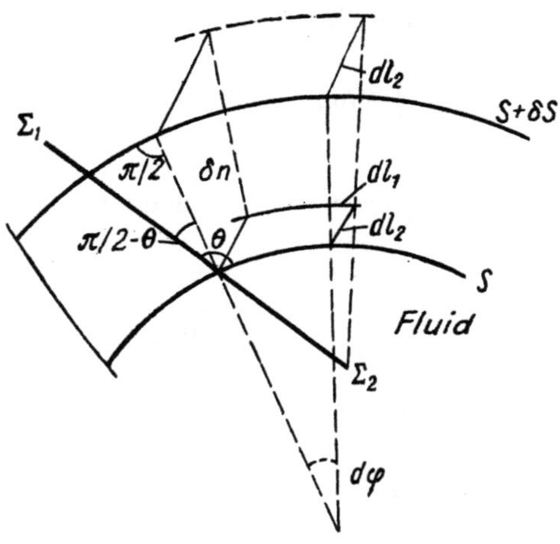

Fig. 44.

We examine Fig. 44, where δn denotes the segment of the normal contained between the surface S and the variational surface $S' = S + \delta S$. The quantity δn is thought of as positive if δn is a segment of the external normal. We evaluate δS:

$$\delta S = \int_S (dS' - dS) + \int_\Gamma \delta n \cot \theta dl.$$

Here Γ denotes the curve along which the surfaces Σ and S intersect. To within higher-order infinitesimals,

$$dS = dl_1 \cdot dl_2,$$

where dl_1 and dl_2 are elements of the arcs of circles of the principal normal sections of the surface S. In precisely the same manner,

$$dS' = dl'_1 \cdot dl'_2 .$$

But, as is easy to see,

$$dl'_i = dl_i\left(1 + \frac{\delta n}{R_i}\right), \quad i = 1, 2,$$

where R_i are the principal radii of curvature. In fact, to within first-order infinitesimals, $dl_i = R_i d\varphi$ and $dl'_i = (R_i + \delta n)\, d\varphi$. From this, eliminating $d\varphi$, we obtain the above. This expression makes it possible to obtain immediately the rule of signs derived in Chapter 1. If $\delta n > 0$ and the center of curvature lies on the same side of surface S as the fluid, then $dl'_i > dl_i$, and consequently $R_i > 0$. In the opposite case the radius of curvature should be taken with the minus sign. Thus, to within first-order infinitesimals,

$$\delta dS = dS' - dS = K dS \delta n ,$$

and we have the following expression for the variation:

$$\delta S = \int_S K \delta n dS + \int_\tau \cot \theta \delta n dl . \tag{7–131}$$

The quantities $\delta \Sigma_1$ and $\delta \Sigma_2$ are of the same magnitude but of opposite sign. As follows from Fig. 44,

$$\delta \Sigma_1 = - \delta \Sigma_2 = \int_\Gamma \frac{\delta n}{\sin \theta} dl . \tag{7–132}$$

It only remains to transform the last term in (7–126). First we note that

$$\int_\tau p \operatorname{div} \delta \boldsymbol{r} d\tau = \int_\tau \operatorname{div}(p \delta \boldsymbol{r})\, d\tau - \int_\tau (\nabla p \delta \boldsymbol{r})\, d\tau .$$

Making use of the fact that $\delta n = 0$ on Σ, we finally obtain

$$\int p \operatorname{div} \delta \boldsymbol{r} d\tau = \int_S p \delta n dS - \int_\tau (\nabla p \delta \boldsymbol{r})\, d\tau .$$

Collecting all the results thus obtained, and using the fact that the variations δr, δn, and δl are arbitrary, we obtain (7–3) – (7–8).

8

Vibrations of a Viscous Fluid and of a Body Containing a Viscous Fluid

In this chapter we make the following step on the path of complicating our models: we shall consider problems in which the fluid is regarded as viscous.

Dropping the assumption that the fluid is ideal appreciably complicates the problem. For a long time this theory provided qualitative results only. It is only quite recently that it has been possible to develop an apparatus for effectively constructing solutions. However, this involves constructing asymptotic solutions and can be used for considering only extreme values of Reynolds numbers. For large Reynolds numbers, the method of asymptotic integration of linearized, nonsteady-state Navier-Stokes equations developed in Refs. [83–85] gave the first solutions of problems of the vibrations of a fluid in vessels.

In Refs. [86, 87], the asymptotic formalism was subjected to further refinement, and these works present examples of asymptotic solutions in the dynamics of a rigid body containing a viscous fluid. In particular, they considered the motion of an artificial earth satellite along an elliptical orbit, on the assumption that inside of it is situated a spherical cavity containing a viscous fluid. The dissipation of energy results in a gradual rounding out of the orbit. The asymptotic method makes it possible to determine the radius of the limiting circular orbit.

The second extreme case, that of low Reynolds numbers, was considered in Refs. [88, 89]. These authors succeeded in examining in detail the transition processes related to the loss of stability of a gyroscope with an ellipsoidal cavity completely filled with a viscous fluid.

The limitation of the theory describing the motion of a body containing a viscous fluid, *i.e.*, that it is only possible to construct asymptotic solutions, reflects the general situation existing in the hydrodynamics of viscous fluids.

Despite all this, there is a wide class of problems of practical interest that can be studied by the above methods. This justifies the theory we are about to develop.

In this chapter we shall first consider several problems of vibrations of a viscous fluid, and then we shall give an example of solving a problem of the dynamics of a body containing a viscous fluid. We do not consider general problems because they are cumbersome, and we restrict ourselves to a detailed presentation of illustrative examples that demonstrate the methods of constructing solutions and the basic properties of the motions.

We shall consider only one problem: the plane vibrations of a rectangular vessel under the action of the restoring force of an elastic spring. For the case of an ideal fluid this problem was considered as early as 1951[93,94]. These were apparently the first publications on the dynamics of a fluid-containing body that assumed that the fluid has weight and a free surface. Despite the simplicity of this problem, its solution has made it possible to clarify a large number of characteristics of the motions under study, in particular, to clarify the structure of the spectrum. Thus it is natural to consider this problem first.

The theory of asymptotic solutions still contains many problems that require additional study. First of all, any results that would give a rigorous substantiation of the theory are lacking. There are also a number of difficulties in the very statement of the problem (conditions under which the problem may be linearized, the behavior of the solution at the point of contact between the free surface and the vessel shell). Finally, in the actual procedure of calculating the elements of the fluid's motion, it has not as yet been possible to attain the degree of standardization that is needed for mass calculations and that is developed in the theory of vibrations of an ideal fluid.

[8-1] ELEMENTARY PROBLEMS OF THE VIBRATIONS OF A VISCOUS FLUID WITH A LARGE REYNOLDS NUMBER

[8-1.1] Statement of the Problem

Let us consider the problem of two-dimensional vibrations of a viscous fluid contained in a vessel. The wave amplitude is assumed to be small, making it possible to linearize the problem. The coordinate system and the notation are shown in Fig. 45.

We describe the motion of the fluid by the linearized Navier-Stokes equations

$$\begin{aligned} \frac{\partial v_1}{\partial t} &= -\frac{1}{\varrho}\frac{\partial p}{\partial x} + \nu \Delta v_1 , \\ \frac{\partial v_2}{\partial t} &= -\frac{1}{\varrho}\frac{\partial p}{\partial z} + \nu \Delta v_2 - g \end{aligned} \tag{8–1}$$

and by the continuity equation

$$\frac{\partial v_1}{\partial x} + \frac{\partial v_2}{\partial z} = 0 . \tag{8–2}$$

Here ν is the kinematic viscosity coefficient, and Δ is the Laplacian operator

$$\Delta = \frac{\partial^2}{\partial x^2} + \frac{\partial^2}{\partial z^2} .$$

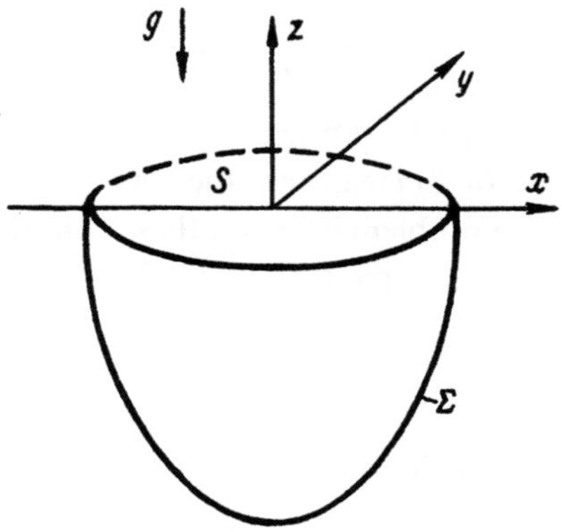

Fig. 45.

We introduce dimensionless variables, using the width l of the free surface of the vessel as the characteristic linear scale, and $1/\omega_0$ as the characteristic time; here ω_0 is the vibrational frequency of the first tone of free vibrations of an ideal heavy fluid

$$v = \tilde{v} l \omega_0 , \quad x = \tilde{x} l , \quad z = \tilde{z} l , \quad t = \frac{t}{\omega_0} , \quad p = \tilde{p} \varrho l^2 \omega_0^2 . \tag{8–3}$$

Here the sign ~ denotes that the variables are dimensionless. Performing the substitution (8–3) and dropping the above sign, we reduce the system (8.1) to the form

$$\begin{aligned} \frac{\partial v_1}{\partial t} &= -\frac{\partial p}{\partial x} + \frac{1}{\mathsf{R}} \Delta v_1 , \\ \frac{\partial v_2}{\partial t} &= -\frac{\partial p}{\partial z} + \frac{1}{\mathsf{R}} \Delta v_2 - \frac{1}{\mathsf{F}} , \end{aligned} \tag{8–4}$$

where R and F are the Reynolds and Froude numbers, respectively:

$$\mathsf{R} = \frac{\omega_0 l^2}{\nu} , \quad \mathsf{F} = \frac{\omega_0^2 l}{g} .$$

The form of Eq. (8–2) is retained.

The solution of Eqs. (8–4) must satisfy the following boundary conditions:
a) condition of "adhesion" on the surface Σ

$$v_1 = v_2 = 0; \tag{8–5}$$

b) condition of equality to zero of the tangential stresses

$$v_{1z} + v_{2x} = 0; \tag{8–6}$$

c) condition of equality to zero of the normal stresses

$$p = \frac{2}{\mathsf{R}} v_{2z} . \tag{8–7}$$

The last two conditions must be satisfied on the free surface.

Conditions (8–6) and (8–7) mean that the wave surface $z = f(x, t)$ must be free of stresses. Since the problem is linear, these conditions must be satisfied on the straight line $z = 0$. The free surface $z = f(x, t)$ can be determined from the kinematic condition

$$\frac{\partial f}{\partial t} = v_2 . \tag{8–8}$$

We shall seek the solution in the form

$$\boldsymbol{v} = \nabla\varphi + \nabla^*\psi , \tag{8–9}$$

where the operator ∇^* is defined as

$$\nabla^* = \frac{\partial}{\partial z} \boldsymbol{x}^0 - \frac{\partial}{\partial x} \boldsymbol{z}^0 .$$

We shall require that the function φ be harmonic. Satisfaction of this requirement means that the field $\boldsymbol{v}$ will be sinusoidal, and consequently Eq. (8–2) is automatically satisfied.

Using representation (8–9), we rewrite Eq. (8–4) in the form

$$\frac{\partial}{\partial x}\left(\frac{\partial\varphi}{\partial t} + p + \frac{1}{\mathsf{F}} z\right) = -\frac{\partial}{\partial z}\left(\frac{\partial\psi}{\partial t} - \frac{1}{\mathsf{R}} \Delta\psi\right),$$

$$\frac{\partial}{\partial z}\left(\frac{\partial\varphi}{\partial t} + p + \frac{1}{\mathsf{F}} z\right) = \frac{\partial}{\partial x}\left(\frac{\partial\psi}{\partial t} - \frac{1}{\mathsf{R}} \Delta\psi\right).$$

We define the function p such that

$$\frac{\partial\varphi}{\partial t} + p + \frac{1}{\mathsf{F}} z = 0 . \tag{8–10}$$

Then the function ψ will have to satisfy the thermal conductivity equation

$$\frac{\partial\psi}{\partial t} = \frac{1}{\mathsf{R}} \Delta\psi . \tag{8–11}$$

If we perform the replacement (8–9) in the boundary conditions (8–5)–(8–8) and use Eq. (8–10), the problem will reduce to determining the harmonic function φ, the function ψ satisfying the thermal conductivity equation, and the function $f(x, t)$, together with the boundary conditions:

a) condition of "adhesion" at surface Σ

$$\begin{aligned} \varphi_x + \psi_z &= 0, \\ \varphi_z - \psi_x &= 0; \end{aligned} \tag{8–12}$$

b) condition of equality to zero of the tangential stresses on the free surface

$$2\varphi_{zx} + \psi_{zz} - \psi_{xx} = 0; \tag{8–13}$$

c) condition of equality to zero of the normal stresses on the free surface

$$\mathsf{F}\frac{\partial \varphi}{\partial t} + f = \frac{2\mathsf{F}}{\mathsf{R}}\frac{\partial}{\partial z}(\psi_x - \varphi_z); \tag{8–14}$$

d) kinematic condition on the free surface

$$\frac{\partial f}{\partial t} = \varphi_z - \psi_x. \tag{8–15}$$

We transform Eq. (8.13), eliminating $\psi_{zz} = \mathsf{R}\psi_t - \psi_{xx}$:

$$\psi_t = \frac{2}{\mathsf{R}}\frac{\partial}{\partial x}(\psi_x - \varphi_z). \tag{8–16}$$

Now we eliminate the function f from (8–14). For this we differentiate (8–14) with respect to t and use (8–15). In addition, we make use of the fact that $\varphi_{zz} = -\varphi_{xx}$. Then

$$\mathsf{F}\varphi_{tt} + \varphi_z = \frac{\partial}{\partial x}\left[\psi + 2\frac{\mathsf{F}}{\mathsf{R}}\frac{\partial}{\partial t}(\varphi_x + \psi_z)\right]. \tag{8–17}$$

Thus the problem reduces to determining the two functions φ and ψ, one of which is harmonic while the other satisfies the thermal conductivity equation, and boundary conditions (8–12), (8–16), and (8–17).

The representation (8–19) was apparently first derived by Lamb. Boundary conditions in the form (8–16) and (8–17) were also used by Sretenskiy[90] who also showed that the new problem thus obtained is equivalent to the original problem.

We shall study free vibrations, using this term to denote solutions of the problem stated in the form

$$\varphi = e^{\sigma t}\Phi(x, z), \quad \psi = e^{\sigma t}\Psi(x, z). \tag{8–18}$$

Here $\sigma = \alpha + i\Omega$ is a complex number. We shall call $|\alpha|$ the damping factor and β the instantaneous frequency of the vibrational process described by this solution.

By virtue of its linearity, the problem under study can have solutions of the form of (8–18); here both the real and imaginary parts of (8–18) satisfy the boundary-value problem. The main problem consists in finding the dependence of α and β on the viscosity and on the Froude number for a vessel of the specified shape.

The functions Φ and Ψ satisfy the expressions

$$\Delta\Phi = 0, \tag{8–19}$$

$$\sigma\Psi = \frac{1}{\mathsf{R}}\,\Delta\Psi. \tag{8–20}$$

The boundary conditions on Σ are again (8–12). Conditions (8–16) and (8–17), after substitution (8–18), take on the form

$$\sigma\Psi = \frac{2}{\mathsf{R}}\frac{\partial}{\partial x}(\Psi_x - \Phi_z), \tag{8–21}$$

$$\sigma^2\mathsf{F}\Phi + \Phi_z = \frac{\partial}{\partial x}\left[\Psi + 2\,\frac{\mathsf{F}\sigma}{\mathsf{R}}(\Phi_x + \Psi_z)\right]. \tag{8–22}$$

Thus the functions Φ and Ψ satisfying Eqs. (8–19) and (8–20) should be determined together with boundary conditions (8–12), (8–21), and (8–22).

For high Reynolds numbers the function Ψ, characterizing the vortex component of the velocity field, will be a boundary-layer type function, since Eq. (8–20) contains a small parameter with a higher-order derivative, and degeneration of the equation as $1/\mathsf{R}\to 0$ is regular (for more details see Ref. [91]). This function will compensate for the mismatch between boundary conditions that arises when the problem is solved within the framework of an ideal fluid, and it will drop exponentially along the normal to the boundary of the region. Physically it is also obvious that the vortex component of the velocity field differs appreciably from zero only in the vicinity of the boundaries where, as a result of the boundary conditions, appreciable velocity gradients arise.

[8-1.2] Lamb's Problem

Lamb was apparently the first to state the problem of damping of standing gravity waves on the surface of an unbounded fluid under the action of viscous forces; he gave a solution for the general case of an arbitrary fluid.*

* These results were published in the first edition (1897) of his *Mathematical Theory of the Motion of Fluids.*

Here we consider this problem for the case of low viscosity. Simultaneously, we also study the structure of the boundary layer near the free surface of the fluid.

The fluid is assumed to be unbounded; hence we replace condition (8–12) by the requirement that the solution must be bounded as $z \to -\infty$. In this case it is natural to take the function Φ in the form

$$\Phi = A \cos \omega x e^{\omega z}, \tag{8–19'}$$

where ω is some real number.

In this problem the boundary layer arises around the free surface since in the general case it is impossible to select a harmonic function that will simultaneously satisfy conditions (8–21) and (8–22). Following the standard methods of boundary-layer theory, we replace $\eta = z\sqrt{\mathsf{R}}$, in (8–20), after which it takes the form

$$\sigma \Psi = \Psi_{\eta\eta} + \frac{1}{\mathsf{R}} \Psi_{xx}.$$

Dropping the term containing $1/\mathsf{R}$,* we obtain

$$\Psi_{\eta\eta} - \sigma \Psi = 0. \tag{8–23}$$

The above equation is an ordinary, second-order differential equation with constant coefficients. Its linearly independent solutions are

$$\Psi = Ce^{\pm\sqrt{\sigma}\eta}.$$

We choose the root with positive real part. The constant of integration will be some (as yet unknown) function of x.

Returning to the old variables, we write the general solution of Eq. (8–20) as

$$\Psi = C(x)\, e^{\sqrt{\sigma \mathsf{R}} z} + C_1(x)\, e^{-\sqrt{\sigma \mathsf{R}} z}. \tag{8–24}$$

By assumption, the solution should be bounded as $z \to -\infty$. Consequently, $C = 0$.

To determine the function $C(x)$ we resort to condition (8–21). Using Eqs. (8–19') and (8–24), we find for the function $C(x)$ the ordinary differential equation

$$C'' - \tfrac{1}{2}\sigma \mathsf{R} C = -\,\omega^2 A \sin \omega x. \tag{8–25}$$

The above expression contains a large parameter. Consequently, its general solution can be represented in the form

$$C = \frac{2\omega^2 A}{\sigma \mathsf{R}} \sin \omega x + D_1 e^{\sqrt{\frac{1}{2}\sigma \mathsf{R}} x} + D_2 e^{-\sqrt{\frac{1}{2}\sigma \mathsf{R}} x} + O\left(\frac{1}{\sqrt{\mathsf{R}}}\right).$$

* That is, we seek the solution in the form of the series $\Psi = \Psi_0 + \frac{1}{\mathsf{R}} \Psi + \cdots$ and retain only the first term.

Since $C(x)$ should be bounded as $x \to +\infty$, the constants of integration D_1 and D_2 should be assumed to be zero.

Thus we have the following asymptotic representation for the function Ψ:

$$\Psi(x, z) = \frac{2\omega^2 A}{\sigma \mathsf{R}} \sin \omega x \exp\{\sqrt{\sigma \mathsf{R} z}\}. \tag{8–26}$$

Upon constructing the function Ψ in the above form we shall satisfy condition (8–21), requiring the absence of tangential stresses, to within a quantity of the order of $1/\mathsf{R}$. It now remains only to satisfy the requirement that the normal stresses must be absent. This can be achieved by proper selection of σ.

We substitute for Ψ its expression (8–26) in (8–22) and for Φ its expression (8–19′). Retaining quantities not higher than the first order relative to $1/\mathsf{R}$, we arrive at the following expression for σ:

$$\mathsf{F}\sigma^2 + \omega = \frac{2\omega^2}{\mathsf{R}}\left[\frac{\omega}{\sigma} - \mathsf{F}\sigma\right]. \tag{8–27}$$

We now examine the asymptotic behavior of the roots of the above equation. Two of its roots will be close to the frequencies of free vibrations of an ideal fluid. These roots are represented in the form

$$\sigma = \pm i\sqrt{\frac{\omega}{\mathsf{F}}} + \frac{\sigma_1}{\mathsf{R}} + O\left(\frac{1}{\mathsf{R}^2}\right).$$

We immediately calculate that

$$\sigma_1 = -2\omega^2. \tag{8–28}$$

The third root of the equation will be represented in the form

$$\sigma = \frac{2\omega^2}{\mathsf{R}} + O\left(\frac{1}{\mathsf{R}^2}\right).^* \tag{8–29}$$

We are considering problems of the motion of a low-viscosity fluid. Hence it is natural to seek solutions describing vibrations that are close to those of an ideal fluid. Solutions such as (8–29) have no physical meaning within the framework of this theory.

Equation (8–28) yields an expression for the damping factor

$$|\alpha| = \frac{2\omega^2}{\mathsf{R}}. \tag{8–29′}$$

Thus the damping factor, to within $O(1/\mathsf{R})$, is inversely proportional to the Reynolds number. The vibrational frequency Im $\sigma = \Omega$, within the frame-

* To find this expression it is sufficient to set $\sigma = \lambda/R$ in Eq. (8–27).

work of this approximation, does not depend on the Reynolds number (*i.e.*, on the viscosity) and is equal to the frequency of the corresponding vibrations of an ideal fluid.

We replace ω by the dimensionless wavelength $\lambda=2\pi/\omega$; then for the instantaneous frequency of vibrations Ω and for the damping factor $|\alpha|$ we will have

$$\Omega = \sqrt{\frac{2\pi}{\lambda \mathsf{F}}}, \quad |\alpha| = \frac{8\pi^2}{\lambda^2 \mathsf{R}}. \tag{8–30}$$

For examining this result it is convenient to again return to dimensional variables

$$t^* = \frac{t}{\omega_0}, \quad \lambda^* = \lambda l, \quad \Omega^* = \Omega\frac{t}{t^*}, \quad |\alpha^*| = |\alpha|\frac{t}{t^*}.$$

Here the superscript* denotes dimensional variables, and l and $1/\omega_0$ are the characteristic length and time scales. After this substitution, Eq. (8–30) takes the form

$$\Omega^* = \sqrt{\frac{2\pi g}{\lambda^*}}, \quad |\alpha^*| = \nu\frac{8\pi^2}{\lambda^{*2}}. \tag{8–31}$$

It is interesting that the above expression is identical with Lamb's known formula, which was obtained without any assumptions about the magnitude of the Reynolds number. In this case the asymptotic process of solution yields, at the first stage, exact values of the damping factor and of the frequency.

For illustration we present the values of the damping factor as a function of the wavelength for water ($\nu=0.018$) and kerosene ($\nu=0.029$)

λ^*, cm	$\lvert\alpha^*\rvert$, $\sec^{-1}$	
	water	kerosene
0.1	141.978	196.40
0.2	35.495	49.10
1.0	1.420	1.96
2.0	0.355	0.49
10.0	0.014	0.019
50.0	0.0006	0.0008

We note that the expansion of the function Ψ starts with terms of the order of $O(1/\mathsf{R})$. Consequently, to satisfy the requirement of absence of tangential stresses, we must take into account a term of the order of $O(1/\mathsf{R})$. In other words, if this condition is not satisfied we are permitting an error of order of $O(1/\mathsf{R})$. This remark will subsequently be utilized.

[8-1.3] Standing Waves Between Two Vertical Walls

We now consider transverse vibrations of a fluid of infinite depth in a channel with vertical walls. Taking the channel width l as the characteristic dimension, we shall regard it as being equal to unity in dimensionless variables. (Fig. 46).

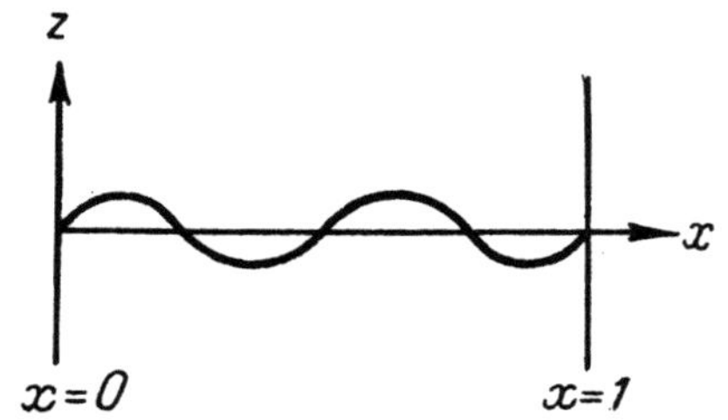

Fig. 46.

In contrast to the problem considered above, we are here faced with the existence of three boundary layers: near the wall $x=0$, near the wall $x=1$, and near the free surface. Accordingly we shall represent the function Ψ in the form of the three components $\Psi=\Psi_1+\Psi_2+\Psi_3$. The function Ψ_3, which describes the boundary layer near the free surface, was constructed in the preceding subsection. We now construct the function Ψ_1, which compensates for the mismatch in the boundary conditions at the line $x=0$. According to the general procedure, we set $\xi=x\sqrt{\mathsf{R}}$, so that the function Ψ_1 will satisfy the equation

$$\sigma\Psi_1=\Psi_{1\xi\xi}+\frac{1}{\mathsf{R}}\Psi_{1zz}.$$

Dropping the second term in this equation, we find its particular solutions. Returning to the variable x, we can write these solutions in the form

$$\Psi_1=C(z)\,e^{\pm\sqrt{\sigma\mathsf{R}}x}.$$

As before, we shall use the root for which $\mathrm{Re}\sqrt{\sigma}>0$. The function Ψ_1 should satisfy the boundedness condition as $x\to+\infty$; hence we have the following representation for it:

$$\Psi_1\sim C(z)\,e^{-\sqrt{\sigma\mathsf{R}}x}. \qquad (8\text{–}32)$$

To determine the function $C(z)$, we use the second of boundary conditions (8–12):

$$C=-\frac{1}{\sqrt{\sigma\mathsf{R}}}\Phi_z.$$

Thus, finally,

$$\Psi_1 \sim -\frac{1}{\sqrt{\sigma \mathrm{R}}}(\Phi_z)_{x=0}\exp\{-\sqrt{\sigma \mathrm{R}x}\}. \tag{8–33}$$

The function Ψ_2, which compensates for the mismatch in the boundary conditions on the line $x=1$, can be constructed similarly.

In selecting particular solutions we must take into account the fact that this function must be bounded as $x\to-\infty$. Repeating our deliberations, we obtain

$$\Psi_2 \sim \frac{1}{\sqrt{\sigma \mathrm{R}}}(\Phi_z)_{x=1}\exp\{\sqrt{\sigma \mathrm{R}}(x-1)\}. \tag{8–34}$$

As is shown by the above expression and by Eq. (8–33), the structure of the boundary layer at the solid walls differs appreciably from the structure of the boundary layer at the wave surface. The expansion of function Ψ_3 starts from terms of the order of $O(1/\mathrm{R})$, while the expansions of functions Ψ_1 and Ψ_2 start with terms of the order of $O(1/\sqrt{\mathrm{R}})$. Thus, if we wish to restrict ourselves in our calculations to terms of the order of $O(1/\sqrt{\mathrm{R}})$, we should assume that the function Ψ_3 is zero. Then the requirement that no tangential stresses must be present will not be met, but the error will be of the same order as the error which is produced if we restrict ourselves in the representations of functions Ψ_1 and Ψ_2 to the first terms of the asymptotic expansions.

Here we shall restrict ourselves everywhere to terms of the order of $O(1/\sqrt{\mathrm{R}})$. On this basis, the asymptotic representation of the function Ψ is taken in the form

$$\Psi = -\frac{1}{\sqrt{\sigma \mathrm{R}}}\{(\Phi_z)_{x=0}\,e^{-\sqrt{\sigma \mathrm{R}x}} + (\Phi_z)_{x=1}\,e^{\sqrt{\sigma \mathrm{R}}(x-1)}\}. \tag{8–35}$$

We now transform the first of conditions (8–12). For this we evaluate

$$(\Psi_{1z})_{x=0} = -\frac{1}{\sqrt{\sigma \mathrm{R}}}\{\Phi_{zz}\}_{x=0} = \frac{1}{\sqrt{\sigma \mathrm{R}}}(\Phi_{xx})_{x=0},$$

$$(\Psi_{2z})_{x=1} = \frac{1}{\sqrt{\sigma \mathrm{R}}}\{\Phi_{zz}\}_{x=1} = -\frac{1}{\sqrt{\sigma \mathrm{R}}}\{\Phi_{xx}\}_{x=1}.$$

These conditions will thus take on the form

$$\begin{aligned} &\Phi_x + \frac{1}{\sqrt{\sigma \mathrm{R}}}\Phi_{xx} = 0 \quad \text{when} \quad x = 0, \\ &\Phi_x - \frac{1}{\sqrt{\sigma \mathrm{R}}}\Phi_{xx} = 0 \quad \text{when} \quad x = 1. \end{aligned} \tag{8–36}$$

We now transform condition (8–22), replacing the function Ψ in it by its expression given by (8–35). Retaining in it terms of the order of $O(1/\sqrt{\mathrm{R}})$, we have

$$\sigma^2 \mathrm{F}\Phi + \Phi_z = (\Phi_z)_{x=0}\, e^{-\sqrt{\sigma \mathrm{R}}x} + (\Phi_z)_{x=1}\, e^{\sqrt{\sigma \mathrm{R}}(x-1)}. \tag{8–37}$$

The problem has reduced to determining the function Φ, which is harmonic in τ and of number σ, using conditions (8–36) and (8–37).

Thus the asymptotic method developed here again reduces the problem (as in the case of an ideal fluid) to finding a single harmonic function. However, as we shall presently see, the boundary conditions are such that the problem ceases to be self-adjoint, *i.e.*, its eigenvalues turn out to be complex.

Remark

We have considered in detail the procedure for reducing the problem to a harmonic problem only for the case of vibrations of a fluid inside a channel with vertical walls. However, with insignificant complication of the calculations, we shall arrive at the same results in the arbitrary case of a two-dimensional problem. It is not too difficult to show that the problem of free two-dimensional vibrations of a low-viscosity fluid in a channel of arbitrary cross section reduces [within the framework of the asymptotic theory only terms $O(1/\sqrt{\mathrm{R}})$ are considered] to finding a function Φ, which is harmonic in region τ, according to the conditions

$$\Phi_n + \frac{1}{\sqrt{\sigma \mathrm{R}}}\Phi_{nn} = 0 \quad \text{on} \quad \Sigma, \tag{8–38}$$

$$\sigma^2 \mathrm{F}\Phi + \Phi_z = (\Phi_z)_{x=a}\, e^{-\sqrt{\sigma \mathrm{R}}(x-a)} \cos\alpha_a + (\Phi_z)_{x=b}\, e^{\sqrt{\sigma \mathrm{R}}(x-b)} \cos\alpha_b .$$

Here $x=a$ and $x=b$ are the points of intersection of the contour of the channel with the Ox axis, α_a and α_b are angles made by the contour of the channel with the axis Ox, and $\boldsymbol{n}$ is the unit vector of the internal normal to Σ. We shall see below what are the complications introduced by the fact that the problem is three-dimensional.

Let us now return to the determination of the harmonic function Φ, using conditions (8–36) and (8–37). We denote the harmonic functions $\varphi_n = \psi_n(x)\, e^{\omega_n z}$ by $\varphi_n(x, z)$, where the function $\psi_n(x)$ satisfies the equation

$$\psi_n'' = -\,\omega_n^2 \psi_n$$

and conditions (8–36). This system is complete and orthogonal. We normalize it by the condition $(\psi_n, \psi_n) = \int_0^1 \psi_n^2 dx = 1$.

The functions ψ_n have the form

$$\psi_n = C_n(\cos \omega_n x + B_n \sin \omega_n y),$$

where $B_n = \omega_n/\sqrt{\sigma \mathsf{R}}$, C_n is a norming multiplier, and ω_n is a root of the transcendental equation

$$\sin \omega_n + \frac{2\omega_n}{\sqrt{\sigma \mathsf{R}}} \cos \omega_n + \frac{\omega_n}{\sigma \mathsf{R}} \sin \omega_n = 0 .$$

Retaining terms up to $O(1/\sqrt{\mathsf{R}})$ inclusive, we obtain

$$\omega_n = n\pi + \frac{2n\pi}{\sqrt{\sigma \mathsf{R}}} .$$

We designate further

$$\frac{1}{\sqrt{\sigma \mathsf{R}}} \alpha_n = (e^{-\sqrt{\sigma \mathsf{R}} \cdot x}, \psi_n), \quad \frac{1}{\sqrt{\sigma \mathsf{R}}} \beta_n = (e^{+\sqrt{\sigma \mathsf{R}}(x-1)}, \psi_n) .$$

If we use Galerkin's method and seek the function φ in the form $\Phi = \sum_{n=1}^{N} a_n \varphi_n$, we will obtain the following equations for the coefficients a_n:

$$\sigma^2 \mathsf{F} a_n + \omega_n a_n = \sum_{m=1}^{N} a_m \omega_m \psi_m(0) \frac{\alpha_n}{\sqrt{\sigma \mathsf{R}}} + \sum_{m=1}^{N} a_m \omega_m \psi_m)1) \frac{\beta_n}{\sqrt{\sigma \mathsf{R}}}, \qquad (8\text{–}38')$$
$$n = 1, 2, \ldots, N .$$

Our problem has thus reduced to solving the above system of equations, which in the general case is a system of interrelated equations but contains small parameters.

To obtain an approximate solution we can apply the method of perturbation theory:

$$a_n = a_{n0} + \varepsilon a_{n1} ,$$
$$\sigma = \sigma_0 + \varepsilon \sigma_1 ,$$

where $\varepsilon = 1/\sqrt{\mathsf{R}}$. It is easy to see that

$$\sigma_0 = \sqrt{\frac{m\pi}{\mathsf{F}}} ;$$

i.e., in the zero approximation we obtain the frequency of the mth principal vibration of an ideal fluid.

Using the estimate for ω_n obtained above, and evaluating the Fourier coefficients α_n and β_n, we can set up an equation for a_{n1}. Dropping intermediate operations (for more details see Ref. [84]), we shall present only the final result. The frequency of free vibrations that is close to the frequency of the mth principal vibration is given by

$$\sigma = -\sqrt[4]{\frac{m\pi}{4\mathsf{R}^2\mathsf{F}}} + i \left\{ \sqrt{\frac{m\pi}{\mathsf{F}}} - \sqrt[4]{\frac{m\pi}{4\mathsf{R}^2\mathsf{F}}} \right\} + O\left(\frac{1}{\mathsf{R}}\right). \qquad (8\text{–}39)$$

This expression shows that the friction at the walls not only damps the wave but also changes the instantaneous frequency.

We now calculate the damping factor $|\mathrm{Re}\,\sigma|$ in dimensional variables:

$$|\mathrm{Re}\,\sigma| = \sqrt[4]{\frac{g\nu^2 m\pi}{4l^5}}.$$

In the problem of standing waves, the wave length λ is related to the distance between the walls by the expression

$$\lambda = \frac{2l}{m}.$$

Hence the damping decrement will be related to the wave length by the expression

$$|\mathrm{Re}\,\sigma| = \frac{1}{l}\sqrt[4]{\nu^2\frac{\pi g}{2\lambda}}. \tag{8–40}$$

It is interesting to compare calculations performed using the above equation with the results of Lamb's formula. This makes it possible to evaluate the contribution made by the boundary layers at the free surface and at the walls.

Since, for the boundary layer adjoining the free surface, the damping factor* is inversely proportional to the square of the wavelength, while for the boundary layer at the wall it is proportional only to the fourth root of the wavelength, then for a fixed viscosity the damping produced by the first boundary layer can sometimes be greater than that produced by the second boundary layer. We present some results for the first tone of the principal vibrations. If the width of the vessel is 25 cm (wave length $\lambda = 50$ cm), the damping factor for standing waves in water resulting from friction at the walls will be greater by a factor of 500 than the damping factor resulting from the boundary layer at the free surface. If the width of the vessel is of the order of 2.5 cm, the damping factor resulting from the walls will now be only 50 times greater than the damping factor due to the surface. In a vessel approximately 0.7 cm wide the contribution made by both boundary layers becomes of the same order.

The contribution of the boundary layer at the free surface becomes increasingly greater with an increase in the number of the tone of the principal vibration. We denote the ratio between the two damping factors by γ. Then, for the vessel 25 cm wide $\gamma = 500$ for the first tone, $\gamma = 50$ for the second tone, and $\gamma = 5$ for the tenth. These numbers can serve as a basis for estimating the applicability of the above scheme of computations in specific individual cases.

* This is the real part of the complex quantity σ.

[8-1.4] Forced Vibrations of a Viscous Fluid

We now consider the problem of forced vibrations of a viscous fluid induced by a force $\boldsymbol{Q}$ of specified frequency σ. We shall, for simplicity, regard this force as potential, *i.e.*,

$$\boldsymbol{Q}(t, x, z) = e^{\sigma t}\nabla U(x, z), \tag{8–41}$$

where σ is a purely imaginary quantity.

In the beginning of this section we examined the statement of the problem of vibration of a viscous fluid subjected to the force of gravity. Now we consider the more general case when the fluid is subjected to both gravity forces and conservative forces of the form of (8–41). Obviously all the deliberations presented above also remain valid for the present case. In particular, the analog of the Cauchy-Lagrange equation, which gives us the pressure distribution, has the form

$$\frac{\partial\varphi}{\partial t} + p + \frac{1}{\mathsf{F}} z - e^{\sigma t} U = 0. \tag{8–42}$$

The requirement that normal stresses must be absent on the free surface [condition (8–17)] is now written in the form

$$\mathsf{F}\varphi_{tt} + \varphi_z = \frac{\partial}{\partial x}\left[\psi + \frac{2\mathsf{F}}{\mathsf{R}}\frac{\partial}{\partial t}(\varphi_x + \varphi_z)\right] + \sigma \mathsf{F} e^{\sigma t} U. \tag{8–43}$$

The requirement that tangential stresses must be absent is not written out, since we conduct all discussion to a precision within $O(1/\sqrt{\mathsf{R}})$. The error produced by neglecting this condition is of the order of $O(1/\mathsf{R})$, as was established above.

We shall seek the vibrations of the fluid that take place at the frequency of the disturbing force. We shall seek the solution in the form

$$\varphi = e^{\sigma t}\Phi(x, z), \quad \psi = e^{\sigma t}\Psi(x, z). \tag{8–44}$$

Substituting the above expression into (8–43), we obtain

$$\sigma^2 \mathsf{F}\Phi + \Phi_z = \frac{\partial}{\partial x}\left[\Psi + \frac{2\mathsf{F}\sigma}{\mathsf{R}}(\Phi_x - \Psi_z)\right] + \sigma \mathsf{F} U. \tag{8–45}$$

Restricting ourselves to quantities of the order of $1/\sqrt{\mathsf{R}}$, we shall represent the stream function in the form

$$\Psi = -\frac{1}{\sqrt{\sigma\mathsf{R}}}(\Phi_z)_{x=0}\, e^{-\sqrt{\sigma\mathsf{R}}x} + \frac{1}{\sqrt{\sigma\mathsf{R}}}(\Phi_z)_{x=1}\, e^{\sqrt{\sigma\mathsf{R}}(x-1)}.$$

Dropping numbers of order higher than $\mathsf{R}^{-1/2}$ in Eq. (8–45), we will have

$$\sigma^2 \mathsf{F}\Phi + \Phi_z = (\Phi_z)_{x=0}\, e^{-\sqrt{\sigma\mathsf{R}}x} + (\Phi_z)_{x=1}\, e^{\sqrt{\sigma\mathsf{R}}(x-1)} + \sigma^2 \mathsf{F} U. \tag{8–46}$$

The function Φ on the walls of the channel should satisfy conditions (8–36). With this as a basis, we shall seek it in the form of the series

$$\Phi = \sum_{n=1}^{N} a_n \varphi_n ,$$

where

$$\varphi_n = \psi_n(x)\, e^{\omega_n z} .$$

The functions ψ_n and the number ω_n were derived in the preceding subsection; they satisfy the equation

$$\psi_n'' + \omega_n^2 \psi_n = 0 ,$$

where ω_n is the root of the transcendental equation

$$\sin \omega_n + \frac{2\omega_n}{\sqrt{\sigma \mathrm{R}}} \cos \omega_n + \frac{\omega_n}{\sigma \mathrm{R}} \sin \omega_n = 0 . \tag{8–47}$$

Designating $u_n = (U, \psi_n)$, and using the standard procedure, we arrive at the following system of algebraic equations:

$$\sigma^2 \mathrm{F} a_n + \omega_n a_n = \sum_{m=1}^{N} a_m \omega_m \psi_m(0) \frac{\alpha_n}{\sqrt{\sigma \mathrm{R}}} + \sum_{m=1}^{N} a_m \omega_m \psi_m(1) \frac{\beta_n}{\sqrt{\sigma \mathrm{R}}} + \sigma \mathrm{F} u_n . \tag{8–48}$$

The above system is nonhomogeneous. The number σ is specified. Since σ is a purely imaginary quantity, it cannot be a root of the characteristic determinant, since we established above that its roots always have a positive real part. Consequently, Eq. (8–48) has a solution and this solution is unique. This assertion is valid no matter what the number σ.

If the fluid had been ideal, Eq. (8–48) would have had a solution for all values of σ, with the exception of resonance values. We use this term to denote the values of σ that are identical to the natural frequencies of vibration of the fluid. This is the basic difference in the character of forced vibrations of ideal and viscous fluids.

[8-2] VIBRATION OF AN OPEN VESSEL CONTAINING A VISCOUS FLUID

[8-2.1] Statement of the Problem

Apparently the most elementary problem of vibration of an open vessel containing a viscous fluid is that of a rectangular vessel (Fig. 47). We shall assume that the vessel is capable of performing vibrations along the Ox axis under the action of springs. We shall assume that the problem is two-dimensional, although for a viscous fluid this assumption is much less well substantiated than in the case of an ideal fluid, for which this problem is solved in Refs. [93, 94], since here we disregard the energy dissipation brought about by friction at the ends. We do this only to reduce the calculations, since we are primarily interested in making clear the method of study itself.

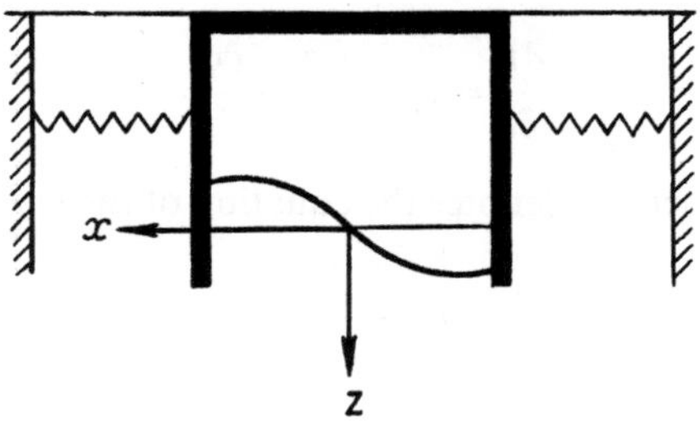

Fig. 47.

We fix the moving coordinate system Oxz to the vessel, as is shown in Fig. 47. This coordinate system will perform only translational motion. We write the momentum vector

$$\boldsymbol{K} = M\boldsymbol{v}_0 + \varrho \int_\tau \boldsymbol{v} d\tau . \tag{8–49}$$

Here M is the mass of the vessel and the fluid, $\boldsymbol{v}_0$ is the velocity with which the coordinate origin O moves, and $\boldsymbol{v}$ is the velocity vector of the fluid particles in the moving coordinate system. Integration in Eq. (8–49) will be extended over to the entire volume occupied by the fluid. However, by virtue of the linearity of the problem, we can assume that integration is performed only over the volume occupied by the fluid in the equilibrium position.

Following the methods presented in the preceding section, we set

$$\boldsymbol{v} = \boldsymbol{V} + \boldsymbol{u} ,$$

where $\boldsymbol{V}$ is the potential vector $\boldsymbol{V} = \nabla\varphi$, and $\boldsymbol{u}$ is the solenoidal vector $\boldsymbol{u} = \nabla^*\psi$. The function φ is harmonic in τ, and the function ψ satisfies the thermal conductivity equation

$$\frac{\partial \psi}{\partial t} = \nu \Delta \psi .$$

The adhesion condition $\boldsymbol{V} + \boldsymbol{u} = 0$ is satisfied on the wetted surface Σ. Projecting the vector in the left-hand side of this condition on the normal and on the tangent to the contour in sequence, we have

$$v_n \equiv \frac{\partial \varphi}{\partial n} + \frac{\partial \psi}{\partial l} = 0\,,$$
$$v_l \equiv \frac{\partial \varphi}{\partial l} - \frac{\partial \psi}{\partial n} = 0\,. \tag{8–50}$$

On the free surface ($z = 0$) we should have the kinematic relationship

$$\frac{\partial f}{\partial t} = v_z = \frac{\partial \varphi}{\partial z} - \frac{\partial \psi}{\partial x}, \tag{8–51}$$

where $z = f(t, x)$, as before, denotes the equation of the free boundary.

We evaluate

$$\int_\tau v_x d\tau = \int_\tau \boldsymbol{v} \nabla x d\tau\,.$$

Since $\boldsymbol{v}$ is a solenoidal vector, we can utilize Green's formula:

$$\int_\tau v_x d\tau = \int_{\Sigma + S} v_n x d\tau\,.$$

Making use of conditions (8–50) and (8–51), we obtain

$$\int_\tau v_x d\tau = \int_S \frac{\partial f}{\partial t} x d\tau\,.$$

Thus the projection of the momentum vector on the Ox axis has the form

$$K_x = M v_{0x} + \varrho \int_S \frac{\partial f}{\partial t} x dS\,. \tag{8–52}$$

The above expression is identical to the analogous expression for an ideal fluid.

We shall describe the motion of the fluid by the momentum equation. Since the vessel performs translational motion parallel to the Ox axis, it suffices to write only one scalar equation for the projection of the momentum vector on the Ox axis. We shall assume that the only external force acting on the vessel is the restoring force of the spring Mk^2x, where x is the deviation of the vessel from the equilibrium position. On these assumptions, the momentum equation can be written as

$$Mx'' + \varrho \int_S \frac{\partial^2 f}{\partial t^2} x dS + Mk^2 x = 0\,. \tag{8–53}$$

Using dimensionless variables, let us rewrite this equation in the form

$$mx'' + \int_S \frac{\partial^2 f}{\partial t^2} x dS + k^2 x = 0. \tag{8–53'}$$

The meaning of the designations is obvious. For example, m is the ratio of the mass of the vessel to the mass of one fluid, k is the frequency of a spring plus mass system, with a mass equal to the mass of the fluid and vibrating under the action of a spring of the given characteristic, *i.e.*, Mk^2.

The function f is determined by solving the hydrodynamic problem of vibrations of a fluid in an immobile vessel under the action of gravity and inertia forces. The potential of the gravity force is z/F, while the potential of the inertia force is $x''x$. Repeating the deliberations of Sec. 1, we set up an expression analogous to (8–10):

$$\varphi_t + p + \frac{1}{\mathsf{F}} f - x''x = \frac{2}{\mathsf{R}} (\psi_{xz} - \varphi_{zz}). \tag{8–54}$$

The condition of absence of normal stresses (8–7) will now have the form

$$\varphi_t + \frac{1}{\mathsf{F}} f - x''x = \frac{2}{\mathsf{R}} (\psi_{xz} - \varphi_{zz}). \tag{8–55}$$

The condition of absence of tangential stresses is not written out since the error thus produced is of the order of $O(1/\mathsf{R})$. With this as a basis we drop terms of the order of $O(1/\mathsf{R})$ in (8–55):

$$\varphi_t + \frac{1}{\mathsf{F}} f - x''x = 0. \tag{8–56}$$

Restricting ourselves to consideration only of free vibrations, we set

$$\begin{aligned} x &= e^{\sigma t} A, \quad \varphi = e^{\sigma t} \Phi, \\ f &= e^{\sigma t} q, \quad \psi = e^{\sigma t} \Psi. \end{aligned} \tag{8–57}$$

The momentum equation, the kinematic condition, and the requirement that normal stresses must be absent take the following form, after substitution (8–57):

$$\sigma^2 (mA + \int_S qx dS) + k^2 A = 0, \tag{8–58}$$

$$\sigma q = \Phi_z - \Psi_x, \tag{8–59}$$

$$\sigma\Phi + \frac{1}{\mathsf{F}} q = \sigma^2 A x. \tag{8–60}$$

The function ψ is a boundary-layer type function. It describes the boundary layer on the side walls ($x=0$ and $x=1$) and on the bottom ($z=-h$). If the depth of the fluid is appreciable, the contribution made by this boundary layer is not great, and we shall disregard it in order to simplify our calculations. Then we can write the flow function in the form

$$\Psi = -\frac{1}{\sqrt{\sigma \mathrm{R}}}\{(\Phi_z)_{x=0}\, e^{-\sqrt{\sigma \mathrm{R}}x} + (\Phi_z)_{x=1}\, e^{\sqrt{\sigma \mathrm{R}}(x-1)}\}. \tag{8–35}$$

Using the above expression, we can eliminate the function Ψ from Eq. (8–59):

$$\sigma q = \Phi_z - \{(\Phi_z)_{x=0}\, e^{-\sqrt{\sigma \mathrm{R}}x} + (\Phi_z)_{x=1}\, e^{\sqrt{\sigma \mathrm{R}}(x-1)}\}. \tag{8–61}$$

Thus the study of free vibrations of a vessel containing a viscous fluid is reduced to determining the function $q(x)$, the harmonic function $\Phi(x, y)$, and the number σ (the number A, as we shall presently see, can be arbitrary). In addition, Φ should satisfy conditions (8–36) on the side walls. Since we have decided not to take into account the boundary layer at the bottom, we require when $z=-h$ that the function $\Phi_z=0$ should satisfy the impermeability condition

$$\Phi_z = 0. \tag{8–62}$$

[8-2.2] The Galerkin Method

We shall seek the solution of the above problem by the Galerkin method. For this we write

$$\begin{aligned} \Phi &= \sum_{n=1}^{N} a_n \psi_n(x) \cosh \omega_n (z+h), \\ q &= \sum_{n=1}^{N} b_n \psi_n(x), \end{aligned} \tag{8–63}$$

The functions ψ_n were derived in the preceding section. We can always regard them as normalized. The function Φ represented in the form (8–63) is harmonic and satisfies the boundary conditions on the bottom and at the walls.

We denote by α_n the Fourier coefficients of the series

$$x = \sum \alpha_n \psi_n,$$

i.e.,

$$\alpha_n = (x, \psi_n).$$

We note further that if $V(x)$ is a function that is bounded together with its derivative,

$$\int_0^1 e^{-\sqrt{\sigma \mathrm{R}}x}\, V(x)\, dx = \frac{V(0)}{\sqrt{\sigma \mathrm{R}}} + O\left(\frac{1}{\mathrm{R}}\right) \tag{8–64}$$

and, similarly,

$$\int_0^1 e^{\sqrt{\sigma \mathrm{R}}(x-1)} V(x)\, dx \approx \frac{V(1)}{\sqrt{\sigma \mathrm{R}}}.$$

To use Galerkin's method, we substitute (8–63) into Eqs. (8–60) and (8–61). Multiplying these equations in sequence by $\psi_1, \psi_2, \ldots, \psi_N$, and making use in the process of Eq. (8–64), we obtain

$$\sigma b_n = a_n \omega_n \sinh \omega_n h - \frac{1}{\sqrt{\sigma \mathrm{R}}} \sum_{k=1}^{N} a_k (\psi_k(0) + \psi_k(1)), \qquad (8\text{–}65)$$

$$\sigma a_n \cosh \omega_n h + \frac{1}{\mathrm{F}} b_n = \sigma^2 A \alpha_n .$$

which should be supplemented by Eq. (8–58), which is now written as

$$\sigma^2 (mA + \sum_{n=1}^{N} b_n \alpha_n) + k^2 A = 0 . \qquad (8\text{–}66)$$

Thus the problem has been reduced to solving some homogeneous system of algebraic equations. Having determined the roots $\sigma_1, \sigma_2, \ldots, \sigma_N$ of the characteristic equation, we can determine the constants a_n, b_n, and A. One of these numbers (for example, A) can be chosen arbitrarily. We shall assume that A, which is called the amplitude of the vibrations, is our arbitrary quantity.

The system of equations (8–65) contains a small parameter, *i.e.*, the quantity $1/\sqrt{\mathrm{R}}$. This makes it possible to construct an approximate solution of the problem. Since the system (8–65) – (8–66) proper is approximate, and we have dropped from it all terms of an order higher than $O(1/\sqrt{\mathrm{R}})$, all of the calculations should be performed to within that quantity.

If, in the first of Eqs. (8–65) we drop the term containing $O(1/\sqrt{\mathrm{R}})$ as a multiplier, then

$$a_n = a_n^0 = \frac{\sigma b_n}{\omega_n \sinh \omega_n h}.$$

Hence

$$a_n = a_n^0 + O\left(\frac{1}{\sqrt{\mathrm{R}}}\right).$$

On this basis the first of Eqs. (8–65) can be rewritten in the form

$$a_n = \frac{\sigma b_n}{\omega_n \sinh \omega_n h} + \frac{\sigma}{\sqrt{\sigma \mathrm{R}} \cdot \omega_n \sinh \omega_n h} \sum_{m=1}^{N} \frac{b_m (\psi_m(0) + \psi_m(1))}{\omega_m \sinh \omega_m h}.$$

Using the expression thus obtained, we can eliminate a_n from the second equation of the system (8–65):

$$\frac{b_n}{\mathsf{F}}\left(\frac{\sigma^2}{\sigma_n^2}+1\right)+\frac{\sigma^2}{\sigma_n^2\mathsf{F}\sqrt{\sigma\mathsf{R}}}\sum_{m=1}^{N}\frac{b_m(\psi_m(0)+\psi_m(1))}{\omega_m\sinh\omega_m h}=\sigma^2 A\alpha_n, \tag{8–67}$$

where we have used the notation

$$\sigma_n^2=\frac{\omega_n\tanh\omega_n h}{\mathsf{F}}.$$

We now use the above procedure and simplify Eq. (8–67). If we drop in it numbers containing $1/\sqrt{\mathsf{R}}$ as a multiplier, it is easy to express b_n in terms of A:

$$b_n=A\mathsf{F}\frac{\sigma^2\sigma_n^2\alpha_n}{\sigma^2+\sigma_n^2}. \tag{8–68}$$

Replacing the quantities b_m under the summation sign by quantities (8–68) in (8–67), we shall create an error of the order of $O(1/\sqrt{\mathsf{R}})$. As a result, we have

$$b_n=A\mathsf{F}\left\{\frac{\sigma^2\sigma_n^2\alpha_n}{\sigma^2+\sigma_n^2}-\frac{\sigma^4}{(\sigma_n^2+\sigma^2)\sqrt{\sigma\mathsf{R}}}\sum_{m=1}^{N}\frac{\sigma_m^2\alpha_m(\psi_m(0)+\psi_m(1))}{(\sigma^2+\sigma_m^2)\,\omega_m\sinh\omega_m h}\right\}.$$

Now substituting this expression into (8–66), and dividing through by A, we will obtain the frequency equation

$$\sigma_m^2+k^2+\mathsf{F}\sum_{n=1}^{N}\frac{\sigma^4\sigma_n^2\alpha_n^2}{\sigma^2+\sigma_n^2}-\frac{\sigma^6\mathsf{F}}{\sqrt{\sigma\mathsf{R}}}\sum_{n=1}^{N}\sum_{m=1}^{N}\frac{\alpha_n\alpha_m(\psi_m(0)+\psi_m(1))}{(\sigma^2+\sigma_n^2)(\sigma^2+\sigma_m^2)\,\omega_m\sinh\omega_m h}=0. \tag{8–69}$$

Thus the problem again reduces to determining the zeros of some meromorphic function. Its poles are situated at points of the imaginary axis in the plane of the complex variable σ, *i.e.*, $\sigma=\pm i\sigma_n$. If $\mathsf{R}=\infty$, we have a problem in which the fluid is regarded as ideal, and all the zeros also lie on the imaginary axis. If $\mathsf{R}\neq 0$, the zeros are no longer on the imaginary axis but will be displaced from it into the left half-plane.

The roots of Eq. (8–67) for large Reynolds numbers can be calculated by the method of perturbation theory. This equation can be written as

$$\Psi(\sigma)=\varepsilon\Phi(\sigma), \tag{8–70}$$

where

$$\Psi(\sigma) = \sigma_m^2 + k^2 + \mathsf{F} \sum_{n=1}^{N} \frac{\sigma^4 \sigma_n^2 \alpha_n^2}{\sigma^2 + \sigma_n^2},$$

$$\Phi(\sigma) = \sigma^{11/2} \mathsf{F} \sum_{n=1}^{N} \sum_{m=1}^{N} \frac{\alpha_n \alpha_m (\psi_m(0) + \psi_m(1))}{(\sigma_m^2 + \sigma^2)\, \omega_m \sinh \omega_m h (\sigma^2 + \sigma_n^2)}, \quad \varepsilon = \frac{1}{\sqrt{\mathsf{R}}}.$$

We shall seek the solution of Eq. (8–70) in the form

$$\sigma = \sigma^* + \varepsilon \sigma' .$$

For σ^* we have

$$\Psi(\sigma^*) = 0. \tag{8–71}$$

The distribution of its roots is easiest studied graphically. Similar graphical constructions were used in Chapter 5. The roots of Eq. (8–71) lie on the intersection of the straight line

$$\Psi_1(\sigma^2) = k^2 + \sigma^2 m$$

and the function $\Psi_2(\sigma^2)$:

$$\Psi_2 = - \mathsf{F} \sum_{n=1}^{N} \frac{\sigma^4 \sigma_n^2 \alpha_n^2}{\sigma^2 + \sigma_n^2}.$$

For σ' we have the explicit expression

$$\sigma' = \frac{\Phi(\sigma^*)}{\left(\dfrac{\partial \Psi}{\partial \sigma}\right)_{\sigma = \sigma^*}}. \tag{8–72}$$

[8-3] THREE-DIMENSIONAL PROBLEMS OF THE THEORY OF VIBRATIONS OF A VISCOUS FLUID

[8-3.1] The Stream Function

Up to now we have considered only two-dimensional motion of a viscous fluid. We have succeeded in reducing the fluid flow problem for large Reynolds numbers to finding one harmonic function and the stream function. These functions satisfied various equations and were related only by means of boundary conditions. The assumption of large Reynolds numbers made it possible to construct a boundary-layer type asymptotic relationship for the stream function. Because of this, the final formulation of the problem contained only the boundary-value problem for harmonic functions. This procedure can also be implemented in the case of an arbitrary three-dimensional problem, except that we shall have to deal not with one stream function but with two.

Consider vibrations of a fluid subjected to the action of mass forces that have a potential U. We denote the complex frequency by σ and then write the linearized Navier-Stokes equation

$$\sigma \boldsymbol{v} = \nabla(U - p) + \nu \Delta \boldsymbol{v} . \tag{8–73}$$

As before, we set

$$\boldsymbol{v} = \nabla \varphi + \boldsymbol{u} . \tag{8–74}$$

We introduce the two stream functions ψ and χ, using the expressions

$$u_x = \frac{\partial \psi}{\partial z} - \frac{\partial \chi}{\partial y}, \quad u_y = \frac{\partial \chi}{\partial z} + \frac{\partial \chi}{\partial x}, \quad u_z = -\frac{\partial \psi}{\partial x} - \frac{\partial \chi}{\partial y} . \tag{8–75}$$

It is easy to check that the vector $\boldsymbol{u}$ that is introduced by means of the above expressions is solenoidal. If the function φ is harmonic, the vector field $\boldsymbol{v}$ defined by Eqs. (8–74) – (8–75) satisfies the equation

$$\operatorname{div} \boldsymbol{v} = 0 .$$

Further, we subject the function p to the condition

$$p = U - \sigma \varphi . \tag{8–76}$$

In order for the vector field (8–74) to satisfy Eq. (8–73), it is sufficient to require that the stream functions ψ and χ satisfy Helmholtz's equations

$$\sigma \psi = \nu \Delta \psi , \tag{8–77}$$

$$\sigma \chi = \nu \Delta \chi . \tag{8–78}$$

Thus, for three-dimensional linear vibrations of a viscous fluid, the problem can be reduced to determining three functions, of which one is harmonic and the two others satisfy the Helmholtz equation. These functions are interrelated solely by the boundary conditions.

The functions ψ and χ describe the vortex component of the velocity field. For low viscosity, these functions are boundary-layer type functions. In fact, let $\nu = 0$; then, according to Eqs. (8–77) and (8–78),

$$\psi \equiv 0, \quad \chi \equiv 0 .$$

Consequently, on this assumption of the boundary conditions, for example, the conditions for adhesion on the stationary boundary

$$v_x = \frac{\partial \varphi}{\partial x} + \frac{\partial \psi}{\partial z} - \frac{\partial \chi}{\partial y} = 0,$$
$$v_y = \frac{\partial \varphi}{\partial y} + \frac{\partial \chi}{\partial z} + \frac{\partial \chi}{\partial x} = 0, \tag{8–79}$$
$$v_z = \frac{\partial \varphi}{\partial z} - \frac{\partial \psi}{\partial x} - \frac{\partial \chi}{\partial y} = 0,$$

cannot be satisfied. Hence we arrive at the following problem of the asymptotic theory: for small ν, to construct functions ψ and χ, that will compensate for the mismatch in boundary conditions (8–79) and will decrease along the normal to the boundary of the region.

[8-3.2] Problems of Three-Dimensional Standing Waves

As our example, we consider the generalization of Lamb's problem, *i.e.*, the problem of damping of three-dimensional standing waves on the surface of a heavy unbounded fluid of infinite depth. The velocity field should satisfy the linearized Navier-Stokes equation and the continuity equation, and should become zero as $z \to -\infty$ (the condition of rest). On the free surface, in addition to the kinematic conditions, the requirements of absence of stresses should also be satisfied.

We present the formulation of these conditions using the previously introduced stream functions. In addition, we shall utilize dimensionless variables as introduced in Sec. 1 of this chapter. Since the surface of the fluid is free of stresses, the vector p_n, characterizing the stress on an area segment with normal $\boldsymbol{n}$, is zero. Since we are considering linear vibrations, $\boldsymbol{n} = \boldsymbol{z}^0$. In scalar notation, the requirement that $\boldsymbol{p}_z = 0$ is equivalent to the following three conditions:

$$p - \frac{2}{\mathsf{R}} \frac{\partial v_z}{\partial z} = 0, \quad \frac{\partial v_z}{\partial x} + \frac{\partial v_x}{\partial z} = 0, \quad \frac{\partial v_z}{\partial y} + \frac{\partial v_y}{\partial z} = 0. \tag{8–80}$$

We use the expressions for v_x, v_y, and v_z and the formula for pressure in which $U = -f/\mathsf{F}$, where f is the elevation of the free boundary, is the potential of the external forces. Then conditions (8–80) can be reduced to the following form, by analogy with (8–13) and (8–14):

$$\sigma \mathsf{F} \varphi + f = \frac{2\mathsf{F}}{\mathsf{R}} \frac{\partial}{\partial z} \left(\frac{\partial \psi}{\partial x} + \frac{\partial \chi}{\partial y} - \frac{\partial \varphi}{\partial z} \right), \tag{8–81}$$

$$2 \frac{\partial^2 \varphi}{\partial z \partial x} + \frac{\partial^2 \psi}{\partial z^2} - \frac{\partial^2 \psi}{\partial x^2} - \frac{\partial}{\partial y} \left(\frac{\partial \chi}{\partial x} - \frac{\partial \chi}{\partial z} \right) = 0, \tag{8–82}$$

$$2\frac{\partial^2\varphi}{\partial z\partial x}+\frac{\partial^2\chi}{\partial z^2}-\frac{\partial^2\chi}{\partial y^2}-\frac{\partial}{\partial x}\left(\frac{\partial\psi}{\partial y}-\frac{\partial\chi}{\partial y}\right)=0, \tag{8–83}$$

We now transform the last two equations. Since the function ψ satisfies the equation

$$\sigma\psi=\frac{1}{\mathsf{R}}\left(\frac{\partial^2\psi}{\partial x^2}+\frac{\partial^2\psi}{\partial y^2}+\frac{\partial^2\psi}{\partial z^2}\right),$$

after eliminating $\partial^2\psi/\partial z^2$ from (8–82), we shall reduce it to the form

$$\sigma\psi=\frac{2}{\mathsf{R}}\frac{\partial}{\partial x}\left(\frac{\partial\psi}{\partial x}-\frac{\partial\varphi}{\partial z}\right)+\frac{1}{\mathsf{R}}\frac{\partial}{\partial y}\left(\frac{\partial\chi}{\partial x}+\frac{\partial\psi}{\partial y}-\frac{\partial\chi}{\partial z}\right). \tag{8–84}$$

Similarly, transforming Eq. (8–83),

$$\sigma\chi=\frac{2}{\mathsf{R}}\frac{\partial}{\partial y}\left(\frac{\partial\chi}{\partial y}-\frac{\partial\varphi}{\partial z}\right)+\frac{1}{\mathsf{R}}\frac{\partial}{\partial x}\left(\frac{\partial\chi}{\partial x}+\frac{\partial\psi}{\partial y}-\frac{\partial\chi}{\partial y}\right). \tag{8–85}$$

We now write the kinematic condition

$$\sigma f=v_z.$$

Using the representations introduced, we obtain

$$\sigma f=\frac{\partial\varphi}{\partial z}-\frac{\partial\psi}{\partial x}-\frac{\partial\chi}{\partial y}. \tag{8–86}$$

With the aid of this expression we can eliminate from (8–81) the function f describing the free surface:

$$\sigma^2\mathsf{F}\varphi+\frac{\partial\varphi}{\partial z}=\frac{\partial\psi}{\partial x}+\frac{\partial\chi}{\partial y}+\frac{2\mathsf{F}\sigma}{\mathsf{R}}\frac{\partial}{\partial z}\left(\frac{\partial\psi}{\partial x}+\frac{\partial\chi}{\partial y}-\frac{\partial\varphi}{\partial z}\right). \tag{8–87}$$

The function φ is harmonic in the variables x, y, and z and is regular as $z\to-\infty$. Hence it is natural to take it in the form

$$\Phi=A\cos ax\cos bxe^{\omega z}, \tag{8–88}$$

where A is an arbitrary number, and b, a, and ω are positive numbers; here

$$\omega=\sqrt{a^2+b^2}.$$

The functions ψ and χ are boundary-layer type functions. Using the methods presented in the preceding section of this chapter, we obtain for them representations of the type of (8–24):

$$\psi=Ce^{\sqrt{\sigma\mathsf{R}}z},$$
$$\chi=De^{\sqrt{\sigma\mathsf{R}}z}.$$

Here C and D are some functions of x and y. To determine them we use Eqs. (8–84) and (8–85):

$$\begin{aligned}\sigma \mathsf{R} C - \frac{\partial^2 C}{\partial x^2} - \frac{\partial^2 D}{\partial x \partial y} + \sqrt{\sigma \mathsf{R}}\,\frac{\partial D}{\partial y} &= -2\,\frac{\partial^2 \varphi}{\partial z \partial x} \equiv A\omega a \sin ax \cos bx\,.\\ \sigma \mathsf{R} D - \frac{\partial^2 D}{\partial y^2} - \frac{\partial^2 C}{\partial x \partial y} + \frac{\partial^2 D}{\partial x \partial y} &= -2\,\frac{\partial^2 \varphi}{\partial z \partial y} \equiv A\omega b \cos ax \sin bx\,.\end{aligned} \tag{8–89}$$

The problem consists in determining bounded solutions of this system for $x, y \in (-\infty, +\infty)$.

The system of equations (8–89) contains a large parameter. Hence to construct the solution we can use asymptotic methods. Seeking C and D in the form of series in negative powers of R, and restricting ourselves to the first term of these series, we obtain

$$C = 2A\,\frac{\omega a}{\sigma \mathsf{R}} \sin ax \cos bx\,, \quad D = 2A\,\frac{\omega b}{\sigma \mathsf{R}} \cos ax \sin bx\,.$$

Thus

$$\begin{aligned}\psi &= 2A\,\frac{\omega a}{\sigma \mathsf{R}} \sin ax \cos bx e^{\sqrt{\sigma \mathsf{R}} z}\,,\\ \chi &= 2A\,\frac{\omega b}{\sigma \mathsf{R}} \cos ax \sin bx e^{\sqrt{\sigma \mathsf{R}} z}\,.\end{aligned} \tag{8–90}$$

By selecting the functions ψ and χ in the above form we have satisfied the requirement that tangential stresses must be absent, to a precision within quantities of the order of $1/\mathsf{R}$. We also have at our disposal the quantity σ, which will be treated so as to satisfy the requirement that normal stresses must be absent, as stated by Eq. (8–87).

Let us now consider this equation. We substitute in it expressions for φ, ψ, and χ and retain terms the order of which do not exceed $1/\mathsf{R}$. Dividing through by $A \cos ax \cos bx$, we have

$$\sigma^2 \mathsf{F} + \omega = \frac{2\omega a^2}{\sigma \mathsf{R}} + \frac{2\omega b^2}{\sigma \mathsf{R}} - \frac{2\mathsf{F}\sigma}{\mathsf{R}}\,\omega^2$$

or

$$\sigma^2 \mathsf{F} = \frac{2\omega^2}{\mathsf{R}}\left(\frac{\omega}{\sigma} - \mathsf{F}\sigma\right);$$

i.e., we have again arrived at Eq. (8–27). Thus Eq. (8–29) remains valid:

$$|\alpha| = \frac{2\omega^2}{\mathsf{R}}\,. \tag{8–27}$$

We introduce the dimensionless wavelengths

$$\lambda_1 = \frac{2\pi}{a}, \quad \lambda_2 = \frac{2\pi}{b}.$$

Then

$$\omega^2 = a^2 + b^2 = 4\pi^2 \left(\frac{1}{\lambda_1^2} + \frac{1}{\lambda_2^2}\right).$$

Consequently, the analog of Eq. (8–30) will be

$$|\alpha| = \frac{8\pi^2}{\mathsf{R}} \left(\frac{1}{\lambda_1^2} + \frac{1}{\lambda_2^2}\right). \tag{8–91}$$

The dimensional analog of Eq. (8–31) will be

$$|\alpha^*| = 8\pi\nu \left(\frac{1}{\lambda_1^{*2}} + \frac{1}{\lambda_2^{*2}}\right).$$

This equation shows that three-dimensional waves damp out more rapidly than two-dimensional ones.

In this chapter we have restricted ourselves to the examination of exponential solutions. This statement of the problem is justified in cases when purely vibrational processes are considered. However, the methods presented can be carried over in full to the Cauchy problem since, upon applying the Laplace transform, we will again obtain an equation of motion in the form

$$\sigma \boldsymbol{v} = -\nabla p + \nu \Delta \boldsymbol{v}.$$

This approach has made it possible to solve a large number of interesting problems of dynamics of a body containing a viscous fluid [75,87]. The same method was used to study Cauchy-Poisson problems for waves on the surface of a heavy viscous fluid [95].

In addition to studies of dynamics of a body containing a fluid at large Reynolds numbers, methods have been developing recently that make it possible to study the motion of a body containing a fluid for low Reynolds numbers [88,89].

This two-pronged approach to the problem describes a large range of engineering problems for which it is possible to suggest effective numerical and analytic methods.

EDITOR'S SUPPLEMENT TO REFERENCES

The Soviet journal *Prikladnaia Matematika i Mekhanika* (PMM) has been translated into English on a cover-to-cover basis, by Pergamon Press, N.Y. and Oxford, beginning with Volume 22, 1958; therefore, all of the references to articles in PMM which are now available in English have been designated by *, and in some instances corrections in bibliographical data have been made and page numbers added. Two very important English language papers dealing with the problem of stability of bodies containing fluids are:

K. Stewartson, On the Stability of a Spinning Top Containing Liquid, *J. Fluid Mechs.*, Vol. *5*, part 4, 1959.

E. H. Wedemeyer, The Unsteady Flow Within a Spinning Cylinder, *J. Fluid Mechs.*, Vol. *20*, part 3, 1964.

A recently published monograph covering many of the practical aspects of liquid behavior in moving containers is:

H. N. Abramson (editor), *The Dynamic Behavior of Liquids in Moving Containers*, SP-106, NASA, Washington, D.C., 1966.

REFERENCES

1. Lagrange, J. L., Analytical Mechanics, Vol. 2 (French Transl.), Gostekhizdat, 1950.
2. Sommerfeld, A., Mechanics of Deformable Bodies, Vol. 3, Academic Press, New York, 1950.
3. Lur'ye, A. I., Analiticheskaya mekhanika [Analytical Mechanics], Fizmatgiz, 1961.
4. Kochin, N. E., Vector and Tensor Analysis (in press), Chelsea Publishing Co., New York.
5. Rumyantsev, V. V., Uravneniya dvizheniya tverdogo tela, imeyushchego polosti, ne polnost'yu napolnennyye zhidkost'yu [Equations of Motion of a Rigid Body with Cavities not Completely Filled with a Fluid], PMM, Vol. 18, No. 6, 1954.
6. Kochin, N. E., I. A. Kibel and N. V. Rose, Theoretical Hydromechanics, Interscience Publishers, New York, 1964.
7. Gauss, K. F., Principia generalia theoriae figurae fluidorum in statu aequilibrii, Ges. Werke, Bd. 5, 1830.
8. Kirchhoff, G., Mechanics (German Transl.), Acad. Sci. SSSR, 1962.
9. Appel, P., Theoretical Mechanics (French Transl.), Vol. 2, Fizmatgiz, 1960.
10. Goryachev, D. N., Nekotoryye obshchiye integraly v zadache o dvizhenii tverdogo tela [Some General Solutions to the Problem of Motion of a Rigid Body], Warsaw, 1910.
11. Chetayev, N. G., Ob ustoychivosti vrashchatel'nykh dvizheniy tverdogo tela, polost' kotorogo napolnena ideal'noy zhidskot'yu [On the Stability of Rotational Motions of a Rigid Body with a Cavity Filled with an Ideal Fluid], PMM, Vol. 21, No. 2, 1957.

12. Beletskiy, V. V., Nekotoryye voprosy dvizheniya tverdogo tela v n'yutonovskom pole sil [Some Problems of the Motion of a Rigid Body in a Newtonian Force Field], PMM, Vol. 21, No. 6, 1957.
13. Poincaré, H., Sur une forme nouvelle des équations de la mecanique, Compt. rend. Acad. Sci., Paris, Vol. 132, 1901.
14. Rumyantsev, V. V., Ob uravneniyakh dvizheniya tverdogo tela s polost'yu, napolnennoy zhidskot'yu [Concerning the Equation of Motion of a Rigid Body with a Fluid-Filled Cavity], PMM, Vol. 19, No. 1, 1955.
15. Zhukovskiy, N. Ye., O dvizhenii tverdogo tela imeyushchego polosti, napolnennyye odnorodnoy kapel'noy zhidskost'yu [On the Motion of a Rigid Body with Cavities Filled with a Homogeneous Liquid Drop], Collected Works, Vol. 2, Gostekhizdat, 1948.
16. Landau, L. D. and Ye. M. Lifshitz, Mekhanika sploshnykh sred [Mechanics of Continuous Media], Gostekhizdat, 1953.
17. Lamb, Sir Horace, Hydrodynamics, sixth edition, Dover Publications, New York, 1945.
18. Keldysh, M. V. and M. A. Lavrent'yev, O yedinstvennosti zadachi Neymana [On the Uniqueness of Neumann's Problem], Doklady Akad. Nauk SSSR, Vol. 16, No. 3, 1937.
19. Stokes, G., Mathematical and Physical Papers, Vol. 1, Cambridge, 1880.
20. Neumann, C., Hydrodynamische Untersuchungen, Leipzig, 1883.
21. Muskhelishvili, N. I., Singulyarnyye integral'nyye uravneniya [Singular Integral Equations], Stechert-Hafner, New York, 1961.
22. Rumyantsev, V. V., Ustoychivost' vrashcheniya tverdogo tela s ellipsoidal'noy polost'yu, napolnennoy zhidskost'yu [The Stability of Rotation of a Rigid Body with an Ellipsoidal Cavity Filled with a Fluid], PMM, Vol. 21, No. 6, 1957.
23. Sludskiy, F. A., De la rotation de la terre supposée fluide à son intérieur, Bulletin de la Société des naturalistes de Moscou, Vol. 9, 1895.
24. Hough, The Oscillations of a Rotating Ellipsoidal Shell Containing Fluid, Phil. Transactions (A), Vol. 186, 1, 1895.
25. Poincaré, H., Sur la précéssion des corps déformables, Bulletin astronomique, Vol. 27, 1910.
26. Dirichlet, Untersuchungen über ein Problem der Hydrodynamik, Crelle, Vol. 58, 181, 1860.
27. Zhak, S. V., O vozmozhnosti kvazitverdogo vrashcheniya zhidkosti [On the Possibility of Rotations of a Fluid as a Semirigid Body], PMM, Vol. 21, No. 4, 1957.
28. Lyapunov, A. M., Obshchaya zadacha ob ustoychivosti dvizheniya [The General Problem of Stability of Motion], Collected Works, Vol. 2, Acad. Sci. USSR Press, 1956.
29. Chetayev, N. G., Stability of Motion, Pergamon Press, New York, 1961.
30. Skimel', V. N., K zadacham ustoychivosti dvizheniya tyazhelogo tverdogo tela vokrug nepodvizhnoy tochki [On the Problems of Stability of Motion of a Solid Rigid Body About a Stationary Point], PMM, Vol. 20, No. 1, 1956.

31. Chetayev, N. G., Ob odnom svoystve uravneniy Puankare [On One Property of Poincaré Equations], PMM, Vol. 19, No. 5, 1955.
32. Rumyantsev, V. V., Metody Lyapunova v issledovanii ustoychivosti dvizheniy tverdykh tel s polostyami, napolnennymi zhidkost'yu [Lyapunov's Methods in the Study of Stability of Motions of Rigid Bodies with Fluid-Filled Cavities], Izvestia Akad. Nauk SSSR, Mekhanika i Mashinostroyeniye, No. 6, 1953; see also Advances in Applied Mathematics, Vol. 8, Academic Press, N.Y.-London, 1964.
*33. Rumyantsev, V. V., Ob ustoychivosti dvizheniya girostatov [On the Stability of Motion of Gyrostats], PMM, Vol. 25, No. 1, pp. 9–19, 1961.
34. Volterra, V., Sur la théorie des variations des latitudes, Acta math., Vol. 22, 1899.
35. Kelvin, Lord, Mathematical and Physical Papers, Vol. 4, Cambridge, 1882.
36. Greenhill, A. G., On the general motion of a liquid ellipsoid, Proc. Camb. Phil. Soc., Vol. 4, 1880.
37. Basset, A. B., On the steady motion and stability of liquid in an ellipsoidal vessel, The Quarterly Journal of Math., Vol. 45, 1914.
38. Sobolev, S. L., O dvizhenii simmetrichnogo volchka s polost'yu napolnennoy zhidkost'yu [On the Motion of a Symmetrical Top with a Fluid-Filled Cavity], Prikl. Mekhan. i Tekh. Fiz., No. 3, 1960.
39. Ishlinskiy, A. Yu. and M. Ye. Temchenko, O malykh kolebaniyakh vertikal'noy osi volchka, tselikom napolnennuyu ideal'noy neszhimayemoy zhidskost'yu [On the Small Vibrations of the Vertical Axis of a Top with a Cavity Completely Filled with an Ideal Incompressible Fluid], Prikl. Mekhan. i Tekh. Fiz., No. 3, 1960.
*40. Zhak, S. V., Ob ustoychivosti nekotorykh chastnykh sluchayev dvizheniya simmetricheskogo giroskopa, soderzhashchego zhidkiye massy [On the Stability of Certain Particular Cases of Motion of a Symmetrical Gyroscope Containing Fluid Masses], PMM, Vol. 22, No. 2, pp. 330–337, 1958.
*41. Rumyantsev, V. V., Ob ustoychivosti vrashchatel'nykh dvizheniy tverdogo tela s zhidkim napolneniyem [On the Stability of Rotational Motions of a Rigid Body Filled with a Fluid], Coll.: Voprosy prikladnoy matematiki i mekhaniki [Problems of Applied Mathematics and Mechanics], No. 2, 1956; see also PMM, Vol. 23, No. 6, pp. 1512–1524, 1959.
42. Rumyantsev, V. V., Ob ustoychivosti dvizheniya po otnosheniyu k chasti peremennykh [On the Stability of Motion with Respect to a Part of the Variables], Vestnik Moscow State Univ., No. 4, 1957.
43. Gyunter, N. M., O dvizhenii zhidkosti, zaklyuchennoy v dannom peremeshchayushchemsya sosude [On the Motion of a Fluid Contained in a Given Moving Vessel], Izvestia Akad. Nauk SSSR, Nos. 13–14, 15–17, 1926; Nos. 7–8, 9–11, 18, 1927; No. 1, 1928.
44. Ladyzhenskaya, O. A., Mathematical Theory of Viscous Incompressible Flow, Gordon and Breach, New York, 1963.
*45. Movchan, A. A., O pryamom metode Lyapunova v zadachakh ustoychivosti uprugikh sistem [On the Direct Lyapunov Method in Problems of Stability of Elastic Systems], PMM, Vol. 23, No. 3, pp. 686–700, 1959.

*46. Rumyantsev, V. V., Odna teoriya ob ustoychivosti [One Theory on Stability], PMM, Vol. 24, No. 1, pp. 59–69, 1960.
*47. Rumyantsev, V. V., Ob ustoychivosti dvizheniya volchka s polost'yu, napolnennoy vyazkoy zhidskost'yu [On the Stability of Motion of a Top with a Cavity Filled with a Viscous Fluid], PMM, Vol. 24, No. 4, pp. 903–912, 1960.
48. Lyapunov, A. M., Ob ustoychivosti ellipsoidal'nykh form ravnovesiya vrashchayushcheysya zhidkosti [On the Stability of Ellipsoidal Equilibrium Configurations of a Rotating Fluid], Collected Works, Vol. 3, Acad. Sci. USSR Press, 1959.
49. Beletskiy, V. V., O libratsii sputnika [On the Libration of an Artificial Earth Satellite], in Iskusstvennyye sputniki zemli [Artificial Earth Satellites], No. 3, Acad. Sci. USSR Press, 1959.
*50. Kolesnikov, N. N., Ob ustoychivosti svobodnogo tverdogo tela z polost'yu, zapolnennoy neszhimayemoy vyazkoy zhidkost'yu [On the Stability of an Unconstrained Rigid Body with a Cavity Filled with an Incompressible Viscous Fluid], PMM, Vol. 26, No. 4, pp. 914–923, 1962.
51. Moiseyev, N. N., O dvukh mayatnikah, napolnennykh zhidkost'yu [On Two Fluid-filled Pendulums], PMM, Vol. 26, No. 6, 1952.
52. Narimanov, G. S., Odvizhenii tverdogo tela, polost' kotorogo chastichno zapolnena zhidskost'yu [On the Motion of a Rigid Body Whose Cavity is Partially Filled with a Viscous Fluid], PMM, Vol. 20, No. 1, 1956.
53. Rumyantsev, V. V., Ob ustoychivosti ravnomernykh vrashcheniy mekhanicheskikh sistem [On the Stability of Uniform Rotations of Mechanical Systems], Izvestia Akad. Nauk SSSR, Mekhanika i Mashinostroyeniye, No. 6, 1962.
*54. Pozharitskiy, G. K., Zadacha minimuma v zadache ob ustoychivosti ravnovesiya tverdogo tela s chastichnym zhidkim napolneniyem [The Problem of a Minimum in the Problem of Stability of a Rigid Body Partially Filled with a Fluid], PMM, Vol. 26, No. 4, pp. 895–913, 1962.
55. Lyapunov, A. M., Zadacha minimuma v odnom voprose ob ustoychivosti figur ravnovesiya vrashchayushcheysya zhidkosti [The Problem of a Minimum in One Problem of Stability of Equilibrium Configurations of a Rotating Fluid], Collected Works, Vol. 3, Acad. Sci. USSR Press, 1959.
56. Rumyantsev, V. V., Ob ustoychivosti ravnovesiya tverdogo tela, imeyushchego polosti, napolnennyye zhidskot'yu [On the Stability of Equilibrium of a Rigid Body Having Fluid-Filled Cavities], Doklady Akad. Nauk SSSR, Vol. 124, No. 2, 1959.
*57. Rumyantsev, V. V., Ob ustoychivosti ustanovivshikhsya dvizheniy tverdykh tel s polostyami, napolnennymi zhidkost'yu [On the Stability of Steady Motions of Rigid Bodies with Fluid-Filled Cavities], PMM, Vol. 26, No. 6, 1962.
*58. Pozharitskiy, G. K. and V. V. Rumyantsev, Zadacha minimuma v voprose ob ustoychivosti dvizheniya tverdogo tela s polost'yu, zapolnennoy zhidkos'tyu [The Problem of the Minimum in the Problem of Stability of Motion of a Rigid Body with a Fluid-Filled Cavity], PMM, Vol. 27, No. 1, pp. 18–32, 1963.

59. Appel, P., Equilibrium Configurations of a Rotating Homogeneous Fluid, ONTI, 1936.
60. Smirnov, V. I., Course of Higher Mathematics, Vol. 2, Addison-Wesley Publishing Co., Inc., Reading, Mass., 1964.
61. Moisseyev, N. N., Sur certains problémes mathématiques du mouvement relatif des satellites. Dynamique des satellites, Symposium, Paris, 1962.
62. Moiseyev, N. N., Zadacha o dvizhenii tverdogo tela, soderzhayushchego zhidkiye massy, imeyushcheye svobodnuyu poverkhnost' [Problem of the Motion of a Rigid Body Containing Fluid Masses with a Free Surface], Matematicheskiy Sbornik, Vol. 32, No. 1, 1953.
63. Krasovskiy, N. N., Nekotoryye zadachi teorii ustoychivosti dvizheniya [Certain Problems of the Theory of Stability of Motion], Fizmatgiz, 1959.
*64. Pozharitskiy, G. K., O vliyanii vyazkosti na ustoychivost' ravnovesiya i statsionarnykh vrashcheniy tverdogo tela s polost'yu, chastichno zapolnennoy vyazkoy zhidkost'yu [On the Effect of Viscosity on the Stability of Equilibrium and of Steady-State Rotations of a Rigid Body with a Cavity Partially Filled with a Viscous Fluid], PMM, Vol. 28, No. 1, pp. 67–76, 1964.
65. Mikhlin, S. G., Variational Methods in Mathematical Physics, second edition, Pergamon Press, New York, 1964.
66. Gunter, N. M., Potential Theory and Its Applications to Basic Problems of Mathematical Physics (in press), Frederick Ungar Publishing Co., New York.
67. Petrov, A. A., Priblizhennyy metod rascheta sobstvennykh kolebaniy zhidkosti v sosudakh proizvol'noy formy i potentsialov Zhukovskogo dlya etikh sosudov [Approximate Method for Calculating the Natural Vibrations of a Fluid in Vessels of Arbitrary Shape and of Zhukovskiy's Potentials for These Vessels], Zh. Vychisl. Matem. i Matem. Fiz., No. 5, 1963.
68. Moiseyev, N. N. and A. A. Petrov, Chislennyye metody rascheta sobstvennykh chastot kolebaniy ogranichennogo ob''yema zhidkosti [Numerical Methods of Calculating the Natural Frequencies of Vibrations of a Bounded Volume of Fluid], Acad. Sci. USSR Press, Moscow, 1965.
69. Moiseyev, N. N., Variatsionnyye zadachi teorii kolebaniy zhidkosti i tela s zhidkost'yu [Variational Problems of the Theory of Vibration of a Fluid and of a Body with a Fluid], in the collection Variatsionnyye metody v zadachakh o kolebanii zhidkosti i tela s zhidkost'yu [Variational Methods in Problems of Vibration of a Fluid and of a Body with a Fluid], Acad. Sci. USSR Press, Moscow, 1962.
70. Moiseyev, N. N., O kolebaniyakh tyazheloy ideal'noy i neszhimayemoy zhidkosti v sosude [On the Vibrations of a Heavy, Ideal and Incompressible Fluid in a Vessel], Doklady Akad. Nauk SSSR, Vol. 85, No. 5, 1952.
71. Moiseev, N. N., Introduction to the Theory of Oscillations of Liquid-Containing Bodies. Advances in applied mechanics, Vol. 8, Academic Press, New York and London, 1964.
72. Stoker, J., Nonlinear Vibrations in Mechanical and Electrical Systems, Interscience Publishers, New York, 1950.

73. Moiseyev, N. N., Dinamika korablya, imeyushchego zhidkiye gruzy [Dynamics of a Ship with Fluid Cargo], Izvestia Akad. Nauk SSSR, Otd. Tekh. Nauk, No. 7, 1952.
74. Bolotin, V. V., O dvizhenii zhidkosti v kolebluyushchemsya sosude [On the Motion of a Fluid in an Oscillating Vessel], PMM, Vol. 20, No. 2, 1956.
75. Krushinskaya, S. I., Kolebaniya tyazheloy vyazkoy zhidkosti v podvizhnom sosude [Vibrations of a Heavy Viscous Fluid in a Moving Vessel], Zh. Vychisl. Matem. i Matem. Fiz., No. 3, 1965.
76. Stokes, On some cases of Fluid Motion, Camb. Trans. *8*, 1843.
77. Okhotsimskiy, D. Ye., K teorii dvizheniya tela s polostyami, chastichno zapolnennymi zhidskost'yu [On the Theory of Motion of a Body with Cavities Partially Filled with a Fluid], PMM, Vol. 20, No. 1, 1956; available in English as NASA TT-F-33 (May 1960).
78. Kreyn, S. G. and N. N. Moiseyev, O kolebaniyakh tverdogo tela, soderzhashchego zhidkost' so svobodnoy poverkhnost'yu [On the Vibrations of a Rigid Body Containing a Fluid with a Free Surface], PMM, Vol. 21, No. 2, 1957.
79. Pontryagin, L. S., Ermitovy operatory v prostranstvakh s indefinitivnoy metrikoy [Hermitian Operators in Indefinite Metric Spaces], Izvestia Akad. Nauk SSSR, Ser. Matemat., No. 8, 1944.
80. Myshkis, A. D., M. A. Belyayeva and A. D. Tyuptsov, Gidrostatika v slabykh gravitatsionnykh polyakh [Hydrostatics in Weak Gravitational Fields], Izvestia Akad. Nauk SSSR, Mekhanika, No. 5, 1964.
81. Hobson, E. W., Spherical and Ellipsoidal Harmonics, Chelsea Publishing Co., New York, 1955.
82. Moiseyev, N. N., Asimptoticheskiye metody tipa uzkikh polos [Asymptotic methods of the Narrow-Strip Type], in the collection Nekotoryye problemy matematiki i mekhaniki [Some Problems of Mathematics and Mechanics], Siberian Division Acad. Sci. USSR Press, Novosibirsk, 1961.
83. Moiseyev, N. N., O krayevykh zadachakh dlya linearizovannykh uravneniy Nav'ye-Stoksa v sluchaye, kogda vyazkost' malaya [On Boundary-Value Problems for Linearized Navier-Stokes Equations for the Case of Low Viscosity], Zh. Vychisl. Matem. i Matem. Fiz., Vol. 1, No. 3, 1961.
84. Bagayeva, N. Ya. and N. N. Moiseyev, Tri zadachi o kolebanii vyazkoy zhidkosti [Three Problems of the Vibrations of a Viscous Fluid], Zh. Vychisl. Matem. i Matem. Fiz., Vol. 4, No. 2, 1964.
85. Moiseyev, N. N., O matematicheskikh metodakh issledovaniya nelineynykh kolebaniy [On Mathematical Methods in the Study of Nonlinear Vibrations]. Proceedings of the International Symposium on Nonlinear Vibrations, Vol. 2, Kiev, 1963.
*86. Krasnoshchekov, P. S., O kolebanii fizicheskogo mayatnika, imeyushchego polosti, zapolnennyye vyazkoy zhidkost'yu [On the Oscillations of a Physical Pendulum Containing Cavities Filled with a Viscous Fluid], PMM, Vol. 27, No. 2, pp. 289–302, 1963.
87. Krasnoshchekov, P. S., O vliyanii vyazkoy zhidkosti, nakhodyashcheysya

vnutri sputnika, na yego dinamiku [On the Effect Exerted on the Dynamics of an Artificial Earth Satellite by a Viscous Fluid Contained in It], Doklady Akad. Nauk SSSR.

*88. Rumyantsev, B. N., O dvizhenii tverdogo tela, soderzhayushchego polosti, zapolnennyye vyazkoy zhidkost'yu [On the Motion of a Rigid Body Containing Cavities Filled with a Viscous Fluid], PMM, Vol. 28, No. 6, pp. 1353–1358, 1964.

*89. Chernous'ko, F. L., Dvizheniye tverdogo tela s polostyami, zapolnennymi vyazkoy zhidkost'yu pri malykh chislakh Renol'dsa [Motion of a Rigid Body with Cavities Filled with a Viscous Fluid for Low Reynolds Numbers], Zh. Vychisl. Matem. i Matem. Fiz., Vol. 5, No. 6, 1965.

90. Sretenskiy, L. N., O volnakh na poverkhnosti zhidkosti [On Waves on the Surface of a Fluid], Trudy TsAGI, No. 541, 1941.

91. Vishik, M. I. and L. A. Lyusternik, Regulyarnoye vyrozhdeniye i pogranichnyy sloy dlya differentsialnykh uravneniy s malym parametrom [Regular Degeneration and the Boundary Layer for Differential Equations with a Small Parameter], Uspekhi Matem. Nauk, Vol. 12, No. 5, 1957.

Index

MIX
Papier aus verantwortungsvollen Quellen
Paper from responsible sources
FSC® C105338

If you have any concerns about our products,
you can contact us on
ProductSafety@springernature.com

In case Publisher is established outside the EU,
the EU authorized representative is:
Springer Nature Customer Service Center GmbH
Europaplatz 3, 69115 Heidelberg, Germany

Printed by Libri Plureos GmbH
in Hamburg, Germany